Contributors

Sari Autio-Sarasmo is a Senior Researcher at the Aleksanteri Institute (the Finnish Centre for Russian and East European Studies) and Adjunct Professor at the University of Tampere. Her recent publications include Sari Autio-Sarasmo and Katalin Miklóssy (eds) *Reassessing Cold War Europe* (Routledge, 2010).

R.W. Davies is Emeritus Professor at the Centre for Russian and East European Studies of the University of Birmingham. He was the co-author with E.H. Carr of *Foundations of a Planned Economy, 1926–1929*, volume 1, and is writing a series on 'The Industrialisation of Soviet Russia', the fifth volume of which, co-authored with S.G. Wheatcroft, is *The Years of Hunger: Soviet Agriculture, 1931–1933* (2004, pbk edition 2010). He has also published studies of *Soviet History in the Gorbachev Revolution* (1998) and *Soviet History in the Yeltsin Era* (1997).

Robert Hornsby is an Honorary Research Fellow in the School of Government and Society at the University of Birmingham. He completed his PhD at the Centre for Russian and East European Studies, University of Birmingham, on 'Political Protest and Dissent under Nikita Khrushchev' and is currently preparing a monograph based on this thesis, entitled *Citizens against the State: Political Dissent and Repression in Khrushchev's USSR*.

Melanie Ilic is Reader in History at the University of Gloucestershire and an Honorary Research Fellow at the Centre for Russian and East European Studies, University of Birmingham. She is author of *Women Workers in the Soviet Interwar Economy* (1999), editor of *Women in the Stalin Era* (2001) and *Stalin's Terror Revisited* (2006), and co-editor (with S.E. Reid and L. Attwood) of *Women in the Khrushchev Era* (2003) and (with Jeremy Smith) of *Soviet State and Society under Nikita Khrushchev* (2009).

Oleg Khlevniuk, Doctor of Historical Sciences, is Senior Specialist (*glavnyi spetsialist*) at the State Archive of the Russian Federation (GARF) in Moscow. He is the author of *The History of the GULAG. From Collectivization to the Great Terror* (2004); and with Yoram Gorlizki, *Cold Peace: Stalin and the Soviet Ruling Circle, 1945–1953* (2004) and *Master of the House. Stalin and His Inner Circle* (2008).

Nataliya Kibita is affiliated with the Department of Central and East European Studies, University of Glasgow. She completed her PhD at the Faculté des Lettres, University of Geneva on 'The *Sovnarkhoz* Reform in Ukraine: Evolution of the Economic Administrative System (1957–1965)'. She is currently working on a monograph on *Economic Decision Making under Khrushchev*.

Katalin Miklóssy is Adjunct Professor/Senior Lecturer at the University of Helsinki and a researcher at the Aleksanteri Institute (Finnish Centre for Russian and Eastern European Studies). She is the leader of the international research group 'Competition in Socialist Society'. Her most recent publications include the volume co-edited with Sari Autio-Sarasmo, *Reassessing Cold War Europe* (2011).

Nikolai Mitrokhin has been an academic researcher at the Research Centre for East European Studies at the University of Bremen (Germany) since December 2008. In 2005–2008 he was a research fellow at the Alexander von Humboldt Foundation and Gerda Henkel Foundation. He is author of *The Russian Party: Movement of the Russian Nationalists in the USSR. 1953–1985* (2003) and *The Russian Orthodox Church: Contemporary Constitution and Actual Problems* (2004, 2006). He is currently taking part in an academic project that is conducting research on 'The Activities of the Central Committee of the Communist Party of the Soviet Union Apparatus from 1953–1985'.

Jeremy Smith is Senior Researcher in Russian History at the Karelian Institute, University of Eastern Finland. From 1999 to 2010 he was Lecturer/Senior Lecturer in Russian History at the Centre for Russian and East European Studies, University of Birmingham. His research focuses on the non-Russian nationalities of the former Soviet Union in the twentieth century, especially in the South Caucasus. His publications include *The Bolsheviks and the National Question, 1917–1923* (1999), *The Fall of Soviet Communism, 1986–1991* (2005) and a forthcoming monograph *The Red Nations: The Nationalities Experience in and after the USSR*. He co-edited with Melanie Ilic *Soviet State and Society under Nikita Khrushchev* (2009).

Ian D. Thatcher is Professor in History at the University of Ulster, Coleraine. His main research interest is in Russian and Soviet history in the period 1894–1991. He is the author of *Leon Trotsky and World War One* (2000) and *Trotsky* (2003), and editor of numerous volumes. Recent articles have appeared in *History* (2009), *Slavonic and East European Review* (2009) and *Contemporary European History* (2010). His study of Brezhnev as leader was published in E. Bacon and M. Sandle (eds), *Brezhnev Reconsidered* (2002).

Alexander Titov is Teaching Fellow in Post-Soviet Politics at UCL-SSEES, having previously worked at the University of Birmingham on the AHRC-funded project 'Policy and Government in the Soviet Union under Nikita Khrushchev'. His doctoral thesis, completed at UCL-SSEES, was on Lev

Gumilev's role in the Eurasian intellectual and political movement. Recent publications include a chapter on the Third Party Programme of the CPSU in Jeremy Smith and Melanie Ilic (eds), *Soviet State and Society under Nikita Khrushchev* (2009), and an article on Lev Gumilev's relations with classical Eurasianism (co-authored with V. Ermolaev) in *Revue des Études Slaves* (2005).

Valery Vasiliev is an Assistant Professor and Chief of the Branch of Historical-Encyclopedic Research at the Institute of History of Ukraine of the National Academy of Science of Ukraine. He is co-editor of *Political Leadership of Ukraine, 1938–1989* (2006) and *Life under Occupation. Vinnitsa region, 1941–1944* (2010). He has also contributed regularly to projects at the Centre for Russian and East European Studies, University of Birmingham.

John Westwood is an honorary research fellow at the Centre for Russian and East European Studies of the University of Birmingham, and previously taught in the field of Russian studies at McGill, Florida State, and Sydney universities. His early career was as an economist with Canadian National Railways. He is the author of *Endurance and Endeavour; Russian History 1812–2001* (2002), *A History of Russian Railways* (1964), *Soviet Locomotive Technology during Industrialization* (1982) and *Soviet Railways to Russian Railways* (2002).

Acknowledgements

This book is the companion volume to Melanie Ilic and Jeremy Smith (eds), *Soviet State and Society Under Nikita Khrushchev*, published by Routledge in 2009. Both volumes were the product of a project running from 2005–2008, 'Policy and Governance in the Soviet Union under Nikita Khrushchev'. This project involved extensive research on Russian archival documents by a team based at the University of Birmingham's Centre for Russian and East European Studies (CREES), as well as coordination of a broader international group of scholars working on the Khrushchev era. The project was funded by the UK Arts and Humanities Research Council (Award number RRB011307), whose continued support for historical research is gratefully acknowledged. Administrative support for this project was provided by Marea Arries, Patricia Carr and Veta Douglas at CREES. Our thanks go to them as well as to Joshua Andy and Alexander Titov, who helped with the coordination of the project as well as contributing their own research. We also greatly appreciate the efforts of Nigel Hardware and other librarians who have kept the Alexander Baykov Library collection as such a valuable and user-friendly resource for scholars of Russia and Eastern Europe.

Earlier drafts of the chapters in this volume were discussed at a number of conferences and seminars, most notably a conference on 'Khrushchev in the Kremlin: State and Society in the Soviet Union' held at the University of Birmingham in December 2007, and the 38th Convention of the American Association for the Advancement of Slavonic and East European Studies held in Washington, DC, in November 2006. A large number of individuals provided comments and suggestions, and we are especially grateful to John Barber, Don Filzer, Yoram Gorlizki, Hope Harrison, Mark Harrison, Timothy Naftali, Michaela Pohl, Arfon Rees, Christopher Read and Susan Reid for their input. Nikolai Mitrokhin thanks the Gerda Henkel Fund (Germany) for its support under the project 'Telefonnoe pravo: gruppy vlyaniya i neformal'nye praktiki v apparate TsK KPSS 1953–1985gg'. Oleg Khlevniuk acknowledges the support of the UK Economic and Social Research Council for the project 'Networks and Hierarchies in the Soviet Provinces: The Role and Function of Regional Party Secretaries from Stalin to Brezhnev (1945–1970)', directed by Yoram Gorlizki (grant 000230880).

Derek Watson was involved in the planning of this project and had been due to contribute a chapter to this volume. Sadly, Derek passed away on 16 March 2006.

He was a stable feature of CREES for many years and took part in a number of history projects. His knowledge of the workings of the Soviet government under Stalin was unrivalled, and he was responsible for creating an extensive database of Soviet government decrees. His biography of Vyacheslav Molotov (Palgrave, 2005) is a testament to the thoroughness of his scholarship and an invaluable resource for historians of the Soviet Union. Derek is greatly missed, and we would like to dedicate this volume to his memory.

Glossary of Russian terms
and abbreviations

apparat	Communist Party administrative structure
BR	British Rail
BTC	British Transport Commission
CC	Central Committee
CIA	Central Intelligence Agency (USA)
CM	Council of Ministers
CMEA/Comecon	Council for Mutual Economic Assistance
CoCom	Coordinating Committee for Multilateral Export Controls
CPSU	Communist Party of the Soviet Union
CPU	Communist Party of Ukraine
GDR	German Democratic Republic (East Germany)
GKNT	Soviet State Committee for Science and Technology
Glavenergoprom	Chief Administration for the Energy Industry
Glavpur	Main Political Administration of the Soviet Army and Navy
Glavsnabsbyty	Chief Administration of Supply and Disposal
Glavstroiprom	Chief Administration for the Building Industry
Gosekonomkomissiya	State Economic Commission of the Council of Ministers of the USSR
Gosplan	State Planning Commission
Gosstroi	State Committee for Construction
Gulag	Chief Administration of Labour Camps
KGB	Committee of State Security
kolkhozy	collective farms
Komsomol	Communist Youth organization
korenizatsiya	nativizing policies
KPK	Commission of Party Control
MGK	Moscow City Committee (of the CPSU)
Mossoviet	Moscow City Soviet (Council)
MPS	Soviet Railways Ministry
MVD	Ministry of Internal Affairs
Narkomtyazhprom	People's Commissariat for Heavy Industry

Narkomles	People's Commissariat of the Timber Industry
NATO	North Atlantic Treaty Organization
nomenklatura	list of approved names used for official appointments
NTS	National Labour Alliance
OECD	Organization for Economic Co-operation and Development
Orgburo	Organizational Bureau of the CC CPSU
Orgraspred	Organization and Assignments Department of the CC CPSU
perestroika	restructuring
RSFSR	Russian Soviet Federative Socialist Republic
sovkhozy	state farms
Sovim	Council of Ministers
sovnarkhoz	council of the national economy
Sovnarkom	Council of People's Commissars
SRs	Socialist Revolutionaries
STR	scientific-technical revolution
Tekhsekretariat	Technical Secretariat
TsSU	Central Statistical Administration
Ukrsovnarkhoz	*sovnarkhoz* of Ukraine
USSR	Union of Soviet Socialist Republics

1   Introduction

Jeremy Smith

Nikita Khrushchev did not stride on to the stage of world history so much as wander casually on to it. Even with the elimination of the man regarded as the most powerful in the Soviet Union after Stalin's death – Lavrenti Beria – several greater luminaries outshone the man from Ukraine: not just the veterans of the Revolution and long-time Stalin lieutenants Vyacheslav Molotov, Lazar Kaganovich and Anastas Mikoyan, but also the rising star Georgy Malenkov. And yet Khrushchev not only rose to dominate them, he was eventually able to impose his own style of governance and range of policies. The extent to which this represented a genuine departure from the Stalin period has been hotly debated by scholars. Did the Communist system developed by Stalin undergo any fundamental revision? Did Khrushchev have any vision beyond the need to retain power for the Communist Party of the Soviet Union and make the economy operate more smoothly? How far was he prepared to go in reaching accommodation with the capitalist West? For Khrushchev's contemporaries, these were far from purely academic questions. For those on the left like Isaac Deutscher, Khrushchev offered the opportunity to redeem the values of the Russian Revolution and escape from the distortions of Stalinism.[1] At the other extreme, conservative politicians maintained that, since Communism was fundamentally evil, it was beyond reform. Khrushchev himself, the 'Butcher of Budapest' as William Buckley called him, was no different from Stalin, someone who 'murders people without regard to race, color, or creed'. A consequence of this view – thankfully one that did not prevail – was that negotiation with the Soviet leader was impossible, even in the extreme circumstances of the Cuban missile crisis.[2] US Presidents Eisenhower and Kennedy at least viewed Khrushchev as somebody they could do business with (to paraphrase British Prime Minister Margaret Thatcher's later assessment of Mikhail Gorbachev). A mutual respect of sorts grew up between Eisenhower and Khrushchev, while Kennedy's more cynical view was tempered by a realist recognition that Khrushchev was, after all, a human being: 'We all breathe the same air. We all cherish our children's future. And we are all mortal.'[3] This attitude saw the superpowers through some of the biggest international crises of the Cold War. Later assessments present Khrushchev as possibly well intentioned, but someone who was at best trapped within the confines of the system that had promoted him, at worst a Stalinist who had no real intention of reforming the system beyond a mild relaxation of the methods of Terror that had been central to Stalin's rule.

For the most part, this debate has focused on the significance of the 1956 Secret Speech denouncing Stalin, and the most obvious political characteristics of the regime: continuing use of repression, though in a milder form; a flourishing but heavily circumscribed cultural scene; selective restoration of nationality rights; public consultation without any real democracy; and so on. While these questions remain central, the recent research presented in this book and its companion volume, *Soviet State and Society under Nikita Khrushchev*, makes it possible to address the question of Khrushchev's place in the history of the Soviet Union from a variety of different angles.

From the middle of the 1980s and the launch of *perestroika* in the Soviet Union, a substantial part of the literature on Khrushchev and his era has been devoted to comparisons between Khrushchev and Mikhail Gorbachev. In personality and background the two could scarcely have been more different, but they did address similar challenges and faced similar obstacles. In particular, the attempt to bypass bureaucracy and improve economic efficiency through decentralization, and the reforms of the Communist Party of the Soviet Union itself – both issues that are addressed in this volume – hold important parallels with Gorbachev's efforts. While frequent allusion is made throughout the volume to the Gorbachev comparison, we set ourselves a different task in our project. The evident fact that Khrushchev failed in many aspects of his reform programme and was eventually overthrown because of it, while Gorbachev, by pushing reform efforts that bit further, contributed to the collapse of the communist system, appears to provide ample evidence that the system was essentially non-reformable.[4] A different perspective on this question can be achieved, however, by comparing the Khrushchev period with the Stalin period. Going beyond the relaxation of state Terror as a method of rule, in examining the utopianism of the Khrushchev project, the concern for welfare, rights and standards of living, and the emergence of a pluralism of sorts, and by looking at the short- and long-term consequences of reform, we find that the Soviet Union did change in important ways after Stalin's death. Khrushchev cannot take all the credit for this. The Soviet Union when he came to power was a very different place from how it had been when Stalin rose to prominence. It was in the 1950s that Russia became, for the first time, a predominantly urban society. It was, moreover, a highly educated one. Stalin, by contrast, had begun his rule over a largely illiterate peasant population. Less dramatic but equally significant social transformations had taken place by Gorbachev's time, and if we can associate a frustrated quasi-middle class with the eventual fall of communism, we can also credit the educated city dwellers of the 1950s with sustaining Khrushchev's reform programme. On the other hand, our research found some support for the notion that, already in the Khrushchev era, long inculcated Stalinist values and attitudes persisted and were at odds with the reforming tenor of the times, presenting obstacles to progress. At the very top, this was most of all evident in the brutal suppression of the 1956 Hungarian rising, which Miklóssy (Chapter 10) shows was in general out of character with the regime's policy in Eastern Europe. A more frequent problem was the persistence of the Stalinist hangover at lower levels of the Party, as is most clearly illustrated in the chapters by Khlevniuk and Vasiliev (Chapter 11 and 8, respectively), and which provided serious barriers to other areas of reform.

Greater (though not unlimited) access to Soviet archives from the period has provided the raw material from which a reassessment of the Khrushchev period can be made, and this is what we set out to achieve in a project funded by the UK Arts and Humanities Research Council from 2005 to 2008. But of equal importance was the precise object of our study. The focus in this volume is on a number of key policy areas: policy towards dissent; nationalism; economic decentralization (the *sovnarkhoz* reform); policy towards the West in technology exchange, and towards Eastern Europe in political and economic reform; agriculture; railways; and construction. While this list of case studies of significant policy initiatives in the Khrushchev era is by no means exhaustive, it is sufficient to lead us to a number of important conclusions.

We were not working on a blank canvas. In addition to the numerous studies produced during Khrushchev's time and by historians since, we were not the first to work in Soviet archives for the period. Most recent studies fall into one of two categories: biographical works, many of which are summarized by Ian Thatcher in Chapter 2 of this book; or culture and cultural politics, which are well represented in the volume *The Dilemmas of Destalinisation*.[5] The juxtaposition of both sets of writings poses something of a paradox. The biographies, very much in line with earlier writings, overwhelmingly provide a view that Khrushchev was inadequate as a leader, both in his leadership style and in his intentions. Consequently he was unable (or unwilling) to break fully with Stalinism, and what changes he did make served only to weaken the regime and his own position until that became untenable in 1964. The picture we now have of the culture of the period, by contrast, is extremely dynamic. It was here that the 'thaw' was most immediately felt, and for a while the Soviet Union witnessed an explosion of cultural originality that had not been seen under Stalin. Admittedly the process was short-lived and limited, leading to a cynicism among the original beneficiaries of the the thaw, which contributed to the later stagnation – a process that is wonderfully brought to life in Olga Grushin's recent novel, *The Dream Life of Sukhanov*.[6] The dynamism of the cultural sphere and the simultaneous failure of politics can be reconciled by reference to relative regime weakness of the time. But this volume presents a two-fold challenge to this characterization of the Khrushchev era: first by reconsidering the nature of politics from 1953–1964, and second by looking more closely at specific policy areas and seeking explanations for both the achievements and the failures.

In Chapter 2, Ian Thatcher presents an assessment of Khrushchev's leadership, which differs from that offered by most biographies. By looking beyond characterizations of Khrushchev's 'crudeness' and treating his memoirs as a serious source, Thatcher finds a leader who 'was constantly intrigued by real-world solutions to real-world problems' and who had a clear idea of how these solutions could be implemented, primarily by promoting the role of specialists above that of party apparatchiks in policy formulation and implementation. That he was ultimately unable to achieve this may have been down to the obstacles put in his path by a party *nomenklatura*, which had become, according to Thatcher, 'frightened ... by the leader's determination to bring it to account'. Khrushchev could hardly be accused of political ineptness. Having outmanoeuvred first Beria and then Malenkov for the leadership of the Soviet Union, he survived a concerted attempt to remove him in 1957, when he faced a hostile majority in the highest political body in the land, the Presidium of the Central Committee of the Communist Party of the Soviet Union. As Nikolai Mitrokhin describes in detail in Chapter 3, Khrushchev's political advantage accrued from the fact that he alone of all the leading survivors of Stalinism had succeeded in cultivating a group of powerful individuals who owed him personal loyalty and support – his 'clan', which he had cultivated since 1931 in Russia, and since 1938 in Ukraine. Mitrokhin calculates that by the end of the 1950s members of Khrushchev's clan shared between them about half of the managerial posts in government departments. This achievement required a combination of political skill and personal charisma, whatever the perceptions of Khrushchev as vulgar and naive. Perhaps inevitably, Mitrokhin finds evidence that, having helped secure Khrushchev's political position, the clan members were anxious to pursue opportunities for obtaining material advantages. Denial of this possibility, together with a general neglect of his clan base after 1957, laid the political basis for Khrushchev's eventual fall.

The relationship between politics and policies is best illustrated by looking at the periodization of Khrushchev's period in office as First Secretary of the CPSU. Khrushchev acted within political constraints, and traditional treatments mark out three distinct periods between 1953 and 1964. The first begins with the death of Stalin and/or arrest of Beria in 1953, which was followed by a period of political intrigue with Khrushchev and Malenkov the main protagonists. The second period begins with Khrushchev's 'Secret Speech' denouncing Stalin in February 1956, followed by a period of Destalinization up until 1961, when the new Programme of the CPSU formalized the dominant role of the Party, but after which Khrushchev's increasing impatience with that Party led to his own downfall. There is much to be said for this periodization, but by examining important policy initiatives we find that other events were of equal or greater importance in shaping the direction of the Khrushchev era. While the Secret Speech undoubtedly engendered a new atmosphere in the Soviet Union and its satellites, it was one of the consequences of that new atmosphere – the Hungarian rising of October 1956 – that had a greater impact on policy. On the one hand, the Hungarian events led to a harsher crackdown on all forms of political opposition (described by Robert Hornsby in Chapter 5), and on the other hand greatly increased the determination of the Soviet leadership to address the living standards of the population through fear of a similar revolt erupting in the USSR. The next landmark event was the defeat of the anti-Party group in May 1957, which was followed in short order by a number of Khrushchev's most significant reform drives – in regionalization of the economy, in housing, in education, in the legal system and in other areas. This is not to say that there were no significant reforms before May 1957, but the welter of legislation that followed it both indicates the extent of Khrushchev's reformist inclinations and suggests the limits to reform that politics had placed on the leader before then.

The 1961 Party Programme is put into a broader context by Alexander Titov in Chapter 4.[7] Although Beria's downfall in 1953 was linked to his preference for the ministries and security forces over the CPSU, which he dismissed as a propaganda machine, Titov shows that the role of the CPSU after 1953 was far from that of a mere accomplice in Khrushchev's struggle for power. Rather than being an end in itself, the revitalization of the Party was subordinate to Khrushchev's broader aims of reducing bureaucracy and decentralizing decision-making. The reform of the Party was, therefore, intimately linked with the introduction of policies like the *Sovnarkhoz* reform, which, Khrushchev believed, would improve Soviet economic efficiency and the life of Soviet citizens. Many of the successive reorganizations of the CPSU were aimed at giving the Party a greater role in economic life. Ultimately the effectiveness of the Party was undermined by the very measures taken to reinvigorate it – constant reorganizations led to disorientation and instability, and the final Khrushchev plan of 'bifurcation' of the Party, underlining its role of supervising the economy, turned its members against the leader.

What Titov shows, among other things, is that Khrushchev's manipulating was not unguided bureaucratic tinkering. The 1961 Programme itself was not a break, but a formal underlining of Khrushchev's continuing commitment to address what he viewed as the chief tasks of socialism – to overcome the increasingly evident and embarrassing gap with the USA in both economic productivity and citizens' living standards, in addition to the arms race. The early successes of the Soviet space programme, the Virgin Lands Campaign, and later the housing programme, illustrated how the targeted mobilization of resources and people could produce successful outcomes. R.W. Davies and Melanie Ilic illustrate in detail in Chapter 13 how such a carefully orchestrated campaign could work in practice in the case of the construction industry. The vision on which this programme rested was, as this chapter demonstrates, one that Khrushchev already clung on to in the 1930s. The contrast in aims, style and approach with Stalin's housing policy is of especial significance. The limited utopian aim of creating better housing conditions for Soviet citizens could be achieved, for all the problems the programme encountered. Here we see most clearly the differences with the Stalin period. In Chapter 12 John Westwood draws an equally significant contrast. The Soviet and British governments embarked on railway modernization programmes at more or less the same time. While both were state-led projects conceived and executed by extensive bureaucracies, it will come as a surprise to many that the Soviet bureaucracy seems to have been better suited to such an effort than was the British one.

Relations between the superpowers during the Khrushchev years have been the object of continuous study and are not treated further in this book.[8] The issue of Soviet intervention in Hungary in 1956 is also not treated separately, although the shadow it cast over domestic policies as well as over the international credibility of the Khrushchev regime is a recurring theme of several chapters. Two chapters do, however, deal with less explored aspects of international engagements, especially in Europe. In Chapter 9, Sari Autio-Sarasmo puts Khrushchev's obsession with technological innovation into an international context. The effort to acquire technologies from abroad in spite of a US-led embargo, especially from smaller European countries, had unintended outcomes: the economic benefits that had no doubt been hoped for were not, for the most part, forthcoming. But the regular scientific engagement with small European countries opened up a 'grey area' between the superpowers, which was to play a significant role in easing international tensions. Katalin Miklóssy, in Chapter 10, finds even more far-reaching consequences in Khrushchev's international engagements, this time with Eastern Europe. It was with Khrushchev's endorsement that a number of the satellite states embarked on their own reform programmes more deeply and for a longer period than was the case with the USSR. The successes of adjustments to the centrally planned economy, advanced systems of welfare, the achievement of greater pluralism, and the emergence of a genuine social sphere can lead us to speculate as to what might have happened had Khrushchev pursued his reforms more consistently and for longer. More tangibly, Miklóssy argues that the mood of change initiated by Khrushchev eventually came full circle, in time inspiring Gorbachev's programme of *perestroika* back in the USSR.

This volume underlines the fact that Khrushchev was a committed reformer. The achievements of some of his programmes also suggest that the picture of the Soviet Union as fundamentally unreformable is inaccurate. Where there was a will, there was a way. Against these successes, however, need to be weighed Khrushchev's failures. In the medium term, the Virgin Lands programme proved disastrous, as did the Maize campaign, while the Education reform of 1959 never achieved its desired ends. In part, these failures were down to poor scientific understanding, knowledge that was either not available or that Khrushchev and his advisers chose to ignore. In the case of the Education reform, lack of enthusiasm on the part of teachers, parents and administrators had much to do with the weakness of implementation. But, most of all, the attitudes and actions of farm and industry managers who did not share Khrushchev's vision of the link between education and work but sought rather to exploit the new schemes to acquire cheap labour, or else resented the presence of school students at their enterprises altogether, undermined any possibility of what were in any case largely unrealistic aims. The education reform is not treated in this volume, but a similar type of bureaucratic resistance can be seen in Oleg Khlevniuk's study of data-inflation in Chapter 11. In fact, the problem went beyond one of resistance to reform, to one of exploitation of new emphases in the economy in the interests of advancing the careers and fortunes of local officials. Khlevniuk demonstrates that the infamous Ryazan Affair, far from being an isolated incident, was the tip of a very large iceberg. Data-inflation was endemic in the USSR, to an extent that seriously undermined any serious attempt at reforming the economy. The campaign to catch up with the USA in meat production also illustrated another flaw in Khrushchev's approach. This relied on the enthusiastic mobilization on a voluntary basis of significant portions of the population. When such mobilization failed, as in the Education campaign as well, the default of reverting to bureaucratic direction from above not only signalled the failure of a particular campaign, it was failures such as these that gave the impression that the Soviet Union was incapable of breaking with key features of Stalinism. As Khlevniuk puts it, 'These measures testify to the fact that Khrushchev, disappointed in the mass enthusiasm organized from above, had again moved the accent on to the tested levers of centralized administrative control.'

The failure of the most significant economic reform attempt in the Khrushchev era – the decentralization of economic decision-making to regional *sovnarkhozy* – was down to related but somewhat different causes. In Chapter 7, Nataliya Kibita shows that, in the case of the Ukrainian republic, serious economic dislocation occurred as a result of the prioritization of the economic needs of the republic over those of the USSR. While the pursuit of local interest on the part of Party and state officials is also evident in Khlevniuk's chapter, here it is reinforced by the coincidence of the local with the national. Valery Vasiliev approaches the *sovnarkhoz* reform in Ukraine from a different angle in Chapter 8, showing how the problems posed by competing agencies involved in the Soviet planning process, which the reform was in part set to address, were simply replicated at the republican level. The tensions between Party, state and economic agencies exacerbated the difficulties of centre–periphery relations and undermined any economic advantages the reform might have produced.

The pursuit of republican interests perhaps should have been, but apparently was not, anticipated by the leadership. In Chapter 6, Jeremy Smith puts this difficulty into the broader context of the emergence of a broad national agenda on the part of the leaders of certain Soviet republics in the 1950s. This tendency replicated that which had emerged in the republics in the 1920s, and in spite of Khrushchev's efforts to reverse it in 1959, it remained an important element in Soviet developments thereafter. Here the failure to articulate any clear policy or ideology after 1953 contributed to a problem that might have been contained instead of getting out of hand. In Chapter 5, Robert Hornsby addresses an issue that likewise might have been anticipated, but apparently was not. The relaxation of terror and the relative freedom implied in the February 1956 Secret Speech led to the beginnings of the Soviet dissident movement. The extent of dissidence and disobedience in the Khrushchev era not only sheds light on a phenomenon more commonly associated with later years, but also tells us something about the regime's new approach. Not only was Terror no longer a feature, but the majority of cases of dissidence were dealt with by administrative means, or by 'prophylactic measures' that sought to anticipate the degeneration of mildly un-Soviet behaviour into outright dissent. Hornsby also shows that, for much of the time, the Khrushchev regime was thinking on its feet in dealing with dissent, before settling on a more stable policy from about 1960.

Allusion has already been made in this introduction to the relatively brief length of time allowed Khrushchev for the fruits of his programme to mature. The notion that the stagnation of the Brezhnev period meant outright abandonment of Khrushchevite innovation can be misleading, however. We have found through the studies in this volume, and through discussions of other topics at conferences and seminars, that a number of initiatives of the Khrushchev era came to fruition under Brezhnev. Policy towards dissidents was one of these areas, but on the more positive side, housing, agriculture, transport, welfare and technological development all progressed in later years, in no small part due to measures introduced by Khrushchev. Even the apparent failures in the early 1960s of the Virgin Lands campaign and education reform led eventually, after some tinkering, to improvements: the amount of arable land was significantly increased, while a long overdue rethink of the principles of education allowed the Soviet Union to continue to boast its primacy in this field. The more difficult issues that Khrushchev had tried to address were, by contrast, simply given up on by his successor: corruption, nationalism and bureaucratization became enduring features of Soviet life and contributed heavily to the failure of the communist project. After Kosygin's brief flirtation with economic decentralization, in the tradition of Khrushchev, the notion of adjusting or abandoning central planning was forgotten for another 20 years.

Although many of its consequences were unintended, the Khrushchev era left a lasting imprint on the Soviet Union. Some achievements there undoubtedly were, but in seeking to move the Soviet Union forward, Khrushchev exposed its major weaknesses. Examples from Eastern Europe and, more recently, China, suggest that Khrushchevite reform might have been more successful than it was. Khrushchev did at least recognize the problems he faced and sought to address. Ultimately, like Gorbachev a generation later, the will and vision of a leader was not enough, and it may be that Khrushchev's greatest failure was the inability to carry enough people with him with sufficient enthusiasm.

Notes

1. I. Deutscher, 'The Soviet Union Enters the Second Decade after Stalin', in I. Deutscher, *Russia, China and the West,* Oxford: Oxford University Press, 1970, pp. 251–8.
2. J.P. Scoblic, *US versus Them: How a Half Century of Conservatism has Undermined America's Security*, New York: Viking, 2008, pp. 1–5, 31–4.
3. A. Fursenko and T. Naftali, *Khrushchev's Cold War: The Inside Story of an American Adversary*, New York: W.W. Norton, 2006, p. 524 and *passim*.
4. For this view see, for example, M. McCauley, *The Khrushchev Era, 1953–1964*, London: Longman, 1995, pp. 98–9.
5. P. Jones (ed.) *The Dilemmas of Destalinisation: A Social and Cultural History of Reform in the Khrushchev Era*, London: Routledge, 2005.
6. O. Grushin, *The Dream Life of Sukhanov*, New York: Viking Press, 2006.
7. Titov treats the 1961 Programme itself in more detail in the first volume produced by our project: A. Titov, 'The 1961 Party Programme and the Fate of Khrushchev's Reforms', in M. Ilic and J. Smith (eds), *Soviet State and Society under Nikita Khrushchev*, London: Routledge, 2009, pp. 8–25.
8. The best treatment is Fursenko and Naftali, *Khrushchev's Cold War*.

2   Khrushchev as leader

Ian D. Thatcher

Khrushchev occupies a very special place in the roll call of leaders of the USSR. He was the only leader to be removed from power. He was the only leader to write his memoirs with the USSR still in existence. He witnessed his disappearance from official discourse and, unlike some other similarly deposed 'liberal leaders' of the communist bloc, did not live long enough to return to the political scene as communism Soviet-style collapsed. Khrushchev also faced the unenviable task of leading the USSR post-Stalin. How would such a leader, any leader, deal with Stalin's legacy, from Stalin the person to his remarkable policies, including industrialization, collectivization, the Great Terror, and the spreading of the Soviet Empire during and after the Second World War? Reaching a historical recognition and balance sheet of the Stalin era while taking the USSR into a post-Stalin period of development was quite an agenda. It was furthermore an agenda that could not be approached in a scholarly, 'objective' manner, but one that was conditioned by current politics, both domestic and foreign. This chapter will examine how Khrushchev dealt with the complex problem of establishing a post-Stalin leadership, or, to borrow again from recent British political discourse, of creating a 'New Soviet Communism'. It will begin with an exposition of Khrushchev's views on leadership expressed when he was First Secretary. The case against Khrushchev's leadership advanced by colleagues in the upper echelons of the Soviet elite will then be outlined. A historiographical survey will investigate how analysts of East and West have judged Khrushchev's leadership. We will then return to Khrushchev through the complete version of his memoirs that became available only after communism's collapse. These offer a credible, if overlooked, account of Khrushchev as leader.

Khrushchev on leadership

Khrushchev reflected on correct communist leadership most explicitly when he addressed the issue of Stalin's leadership. The key speeches were the famous 'Secret Speech' of the XX Party Congress and the address to the XXII Congress, after which Stalin's body was removed from the mausoleum.

The Secret Speech has been criticized for the absence of a full and truthful analysis of Stalin and Stalinism.[1] This was never its purpose. Khrushchev began by admitting that the 'objective of the present report is not a thorough evaluation of Stalin's life and activity'.[2] As matters stood Khrushchev had his work cut out by focusing upon his intended topic: how a correct communist leadership established under Lenin became subverted under Stalin. Much of the speech is given over to identification of what leadership under Lenin consisted. First of all, it did not mean attributing each and every success to the actions of one man. No matter how great and wise Lenin was as leader, Lenin, as a true Marxist, advocated modesty and due recognition of the efforts of the socialist movement more broadly. Leninism's strength was based, for Khrushchev, upon the unity between the creative masses and the Communist Party. The achievements of the world's first attempt at socialism were the outcome of the efforts of ordinary citizens led by the Communist Party. Moreover, the Communist Party itself was governed by norms and regulations that guaranteed its healthy relationship to the people. Above all, the Leninist Communist Party was characterized by collective leadership. The Central Committee and party congresses and conferences would meet regularly. Important issues facing the party and state would be discussed and resolved openly. Lenin sought consistently to convince comrades. Lenin would not threaten sanctions. Lenin would make every effort to keep loyal comrades within party ranks, even when they had erred temporarily in their views. In Lenin's time there was a pattern of leadership noted for its democracy and legality. There was not a denial of the importance of leadership, for 'Marxism does not negate the role of leaders of the working class in directing the revolutionary liberation movement'.[3]

Lenin clearly understood the importance of leadership, a topic to which he devoted his Political Testament. Here Lenin highlighted Stalin's negative characteristics, in particular his rudeness that may impact on Stalin's ability to be a good, communist leader. Indeed, in a subsequent postscript Lenin urged Stalin's removal from the key post of General Secretary. For Khrushchev, Stalin did heed Lenin's advice in the 1920s and early 1930s, and here Stalin made a vital contribution to the construction of socialism and the defeat of various deviations, from the Trotskyite Left to the rightists. Indeed, Stalin became an authority among comrades for his iron logic and genuine leadership. Unfortunately this changed from 1934 onwards. In the aftermath of the Kirov Affair, a decree was issued that formed the basis of Stalin's action against the party. Henceforth Stalin began to divorce himself and his rule from the party and its bodies for collective leadership (Central Committee and Political Bureau plenums and party conferences and congresses). There began a period of the 'cult of the personality'.

As Khrushchev saw it, the 'cult of the personality' did not happen overnight. In its initial phase it was constructed and strengthened in the period 1935–38. It reached its height in the last years of Stalin's life, years that were 'unbearable' for members of the top elite. The 'cult of personality' meant the breaking of party and Soviet laws. The rule of one man, divorced from reality, became the decisive factor in party and state policy. A propaganda machine that praised the genius of the leader Stalin created an atmosphere in which Stalin could not be questioned, even when data were available to disprove Stalin's policy. Despite not visiting the countryside since 1928, for example, in the post-war years Stalin insisted on agricultural policies that could only produce more harm to a seriously underfunded sector of the economy. Genius evidently did not have to take facts into account for it was able to see further than any fact. The detrimental consequences, on Khrushchev's reading, were many. First was the attack on the party. The Great Purges of 1936–38 eliminated numerous loyal communists on trumped-up charges backed by confessions achieved through torture. Second, government bodies, from the economic to the military, were purged of imagined enemies, resulting in a serious deterioration of their effectiveness. Third, Stalin made numerous tactical errors in the Second World War, from denying intelligence reports of a looming Nazi offensive to ignorant meddling with strategy that produced far more war losses than was necessary. Fourth, Stalin instigated crimes against the Soviet nationalities, including his 'home republic' of Georgia. Fifth, Stalin's weak grasp of reality led to many mistakes in foreign policy – the split with Yugoslavia was a completely unnecessary consequence of Stalin's arrogance. Sixth, in the post-war years Stalin continued to fabricate false plots – the Leningrad Affair, the Doctors' Plot – with the intention of moving once again against the top elite. Finally, Stalin allowed criminals like Beria to exert unjustified influence on policy. And, during all of these crimes, Stalin was praised as a great leader by a state information service directed by Stalin himself.

At several points in the speech Khrushchev raised the uncomfortable issue of the top elite's role in the 'cult of personality'. How could all of this happen during their watch? First, Stalin did not keep his colleagues well informed. Decisions were presented as a fait accompli, accompanied by confessions that were not open to refutation. Second, members of the Central Committee and Politburo were kept atomized. There were no meetings of the collective in which comrades could mount an effective attack on Stalin's mismanagement. Third, in one-to-one encounters with Stalin, an individual was in a constant state of panic over whether he would leave the meeting a free man. In these circumstances effective opposition to Stalin was impossible. Indeed, the elite became inculcated with certain aspects of the 'cult of personality'. Leaders immediately below Stalin began to treat state property as their personal chiefdom, evident in the number of towns, radio stations and so on named after them.

The critique of the cult of personality and the identification of true leadership under Lenin pointed to what had to be done under Khrushchev's leadership. Khrushchev would have to be a leader in the Lenin mould. A leader who knew the people and country he was leading through actual contact, regularly visiting factories and farms. A modest leader who pursued an agenda agreed on in collaboration with a collective leadership via properly constituted meetings of the Central Committee and party conferences and congresses. The practice of appointment by the *vozhd* (leader) would be replaced by elections of office holders. A leader aware of the limits of leadership, one who recognized that the power of the people and of the party could never be overtaken by one man. Indeed, it was due only to the deep moral strength of the party and of the people that, according to Khrushchev, even the period of the cult of personality was not devoid of important victories such as that over fascism.

The Secret Speech was a genuine and profound reflection on what had gone wrong in the Stalin period. Henceforth one could not deny the Purges. Khrushchev insisted that correcting the mistakes of the 'cult of personality' would be a gradual process in which the party would not allow its enemies to take advantage of this admission of past errors, but this overlooked just how radical his speech was. Khrushchev had called for an in-depth re-evaluation of the party's whole history. The distortions introduced by the cult of personality would have to be erased from all of the areas that it had impacted and, as Khrushchev had pointed out, these were very wide and varying. The demands placed on Khrushchev's leadership were correspondingly high.

Five years later, at the XXII congress of October 1961, Khrushchev congratulated the party for abandoning the cult of personality as a leadership principle.[4] It was difficult in 1956 to face up to this issue and admit the gross violations of legality that had occurred. Had the party not done so, however, it would have abrogated its responsibility of self-criticism. The party would have weakened its ties to the people still further. There would have been less progress in the economy. The party had been right in 1956 to trust the people. The restoration of Leninist legality had increased the trust between the party and the people, made the Communist Party law-governed, more genuinely revolutionary and more attractive to the people. This was why the party was more united in 1961 than ever before and why the path to communism was being traversed more speedily than before. It had not been easy. There was the attempt of those most responsible for repressions under Stalin to form an anti-Party group in 1957 to turn policy back to the cult of personality. There was still the disease of 'chance people' entering the CPSU for careerist reasons. These problems were best tackled by the regime established by Khrushchev, a genuine Leninist leadership that recognized the party and people as the 'real maker of history'. The CPSU was once again a properly functioning political organization. Government structures were more open to the people, with 'millions of Soviet men and women ... playing an increasing part in the administration of government and public affairs'.[5] In the midst of this upbeat account of the party leadership, the party and the people over the past five years, Khrushchev also examined the dangerous topic of how leaders can become outdated and harmful, in need of replacement. This is an interesting section of the address for it is not linked to the cult of personality. It was rather a more general reflection on the rise and fall of leaders:

> There have been many cases in history of particular leaders proving their worth at a certain period in their lives and playing a notable role but later stopping short in their tracks, as it were, gradually fading out.
> The reasons for this phenomenon may vary: some people become exhausted; others lose touch with reality, become conceited and do not work properly; still others turn out to be unprincipled, spineless people who have adapted themselves to circumstances and who lack staunchness in the struggle for

their party's cause. Meanwhile, in the course of the struggle, new political leaders emerge; they oppose all that hampers the development of the new, and overcome the resistance of the old. It is something akin to the phenomenon astronomers call light from extinct stars. Certain stars, which are very far removed from earth, seem to shine on even though they have been extinct for a long time. The trouble with some people who find themselves in the position of stars on the social horizon is that they imagine that they continue to radiate light even though they have long since become nothing but smouldering embers.[6]

In 1961 Khrushchev was not directing these words at himself but at the anti-Party group. He had however identified the Central Committee as the body that would monitor party members and deal with demotions and expulsions. He had also defined of what good communist leadership consists: 'leaders are worthy of the name only when they express the vital interests of the working people and follow the right path ... they serve the people and must be subject to control by the people'.[7] So if a leader, whatever past services, ceased to represent the interests of the party and of the people, they would be removed by new leaders who would 'overcome the resistance of the old'. In this indirect way Khrushchev raised the prospect for Soviet communism to experience a change of leader/leadership by a process other than the death or incapacitation of the *vozhd*.

Khrushchev as leader: the ouster

If we begin at the end, with the indictment of Khrushchev by the majority of the Central Committee, then it appears that Khrushchev did not meet his definition of a Leninist leader.[8] The charges against Khrushchev included policy failure, domestic and foreign. At home, industry and agriculture were underperforming. Abroad, relations with China had soured. Most importantly these policy failings were linked to Khrushchev's misdemeanours as leader. Khrushchev, it was claimed, was bypassing the Presidium and the Central Committee. He had taken to issuing decrees in the name of the Central Committee that were in fact on his own initiative. Khrushchev had surrounded himself with sycophants and family members that formed his inner staff. Presidium colleagues could not reach Khrushchev directly but had to deal with this entourage. Khrushchev simply ignored the advice of the Presidium, assigning key duties to his private circle outside the control of the party elite. In this sense Khrushchev broke party norms and even engaged in corruption. The award of honours to his son and son-in-law was noted, as well as the use of state money to fund family excursions abroad on what was supposed to be official business. Such irregularities were occurring because Khrushchev had concentrated power in his own hands. Moreover, Khrushchev did not know how to use this power sensibly. He interfered in areas about which he knew nothing, but considering himself an expert in agriculture, diplomacy, science, art, was quick to meddle, with often devastating consequences. Khrushchev defended the quack geneticist Lysenko, for example, despite warnings from eminent scientists. Khrushchev was unable to control his thoughts and most importantly his mouth. He had upset prominent friends within the socialist camp, causing trouble in relations with China, Albania, Romania and Poland. In the USSR Khrushchev engaged in constant reorganizations of economic and party bodies that brought only additional confusion and threatened to split the party. This sad story of failure and illegality was accompanied by excessive praise of Khrushchev in the media. Ignored and often insulted by a man who had turned meetings of the Presidium into 'empty formality', Khrushchev's colleagues had to act. Khrushchev's 'petty tyranny' unlike Stalin's was not based on terror, but this did not excuse it. If anything, it was 'harder to struggle with a living cult than with a dead one. If Stalin destroyed people physically, Khrushchev destroyed them morally'.[9]

The indictment of Khrushchev was a clever use of Khrushchev's denunciation of Stalin and the 'cult of personality' against Khrushchev himself. Indeed, the criticism of Khrushchev's talent for 'hare-brained' scheming came first from Stalin![10] Khrushchev found himself portrayed as a leader out of touch with reality, making a mess of policy and flouting party rules, ignoring and belittling comrades, while surviving in an artificial bubble of excessive praise from official propaganda and an inner coterie of toadies. It is an analysis of Khrushchev's leadership that most biographers of Khrushchev share.

Khrushchev and the historians

In the absence of any serious scholarship on the recently deposed leader in Brezhnev's USSR, Western publications dominated early assessments of Khrushchev. Several studies interpreted Khrushchev's leadership as a transitional stage from the excesses and imbalances of Stalinism to a stable pattern of communist development. Khrushchev was brave enough to attack and tame Stalinist terror, but he was too imbued with the prejudices of the Stalinist period to provide an effective, calm and regulated leadership. On the contrary a rush of contradictory and futile administrative reorganizations, combined with an aggressive rhetoric aimed at party officials, alienated the party from Khrushchev so there was near unanimity on the need to remove him from power. Alec Nove, who witnessed Khrushchev's uneducated preference for folk dancing over serious art at a Kremlin reception, was also shocked by Khrushchev breaking with party decorum, criticizing subordinates in front of subordinates. Typical examples of Khrushchev's speeches that were reported at the time include the following from a meeting with local party officials:

> Some might say: 'what's this, has Khrushchev come to criticise us and tell us off?' What do you think, that I'd come to read you Pushkin's poems? You can read poetry without me. I have come to show up defects, to urge you to freshen up some organisations, to blow some wind of change at the directing cadres.[11]

Nove summarized the impact of Khrushchev as leader thus:

The man was wilful, crude, lacking in dignity, unpredictable, a muddler … He interfered with privilege by his abortive educational reforms. He was a muddler in economic policy, in agriculture especially. He reorganised and disorganised the party. Therefore he must go.[12]

Russian historians who could arrange a Western publication supported Nove's conclusion. Key figures were the Medvedev brothers Roy and Zhores. Their joint study of Khrushchev emphasized failure in agriculture due to Khrushchev's inability to understand that only free labour under a system of economic rewards (i.e. capitalism) could rectify Soviet agriculture from the damage of collectivization. Roy Medvedev's subsequent biography assessed the indictment against Khrushchev and found that on all substantive points no defence could be offered. Indeed, Khrushchev's defects as leader included:

that he had little understanding of people or of their motives. As a result, he was often influenced by the unscrupulous and the venal and, having dismissed one corrupt official, frequently replaced him with another who was even worse. Khrushchev did not escape the corrosive effects of absolute power and adulation. In the last years of his leadership his manner was increasingly that of a bully, and he became less and less self-critical, compounding his mistakes by refusing to acknowledge his failures.[13]

For the Medvedevs Khrushchev's leadership illustrates that complex modern societies cannot be governed by the whims of a single man, especially when, as in Khrushchev's case, he has an 'inadequate background' and 'narrowness' of outlook.[14]

As with other forgotten faces from the Soviet past, interest in Khrushchev revived inside the USSR during Gorbachev's rule. Although it became possible to express contradictory evaluations of Khrushchev,[15] historians' views of Khrushchev in the Gorbachev period reflected authors' stances on *perestroika*.[16] Only in post-Soviet Russia did the Khrushchev period become an established topic for a more dispassionate historical investigation. Numerous documentary collections appeared, including the complete Khrushchev memoirs and the stenographic reports of Presidium sessions.[17] Assessments of Khrushchev remain, however, at best mixed. Former Moscow party chief V.V. Grishin's memoirs pay tribute to Khrushchev's 'organisational talents, intelligence, and boundless energy'.[18] Grishin points out Khrushchev's pride in the USSR, the genuine efforts made to improve the general standard of living, and how a more relaxed and democratic style of leadership ensued. Yes Khrushchev could encourage, and loudly so, but this was never rude. In contrast to other commentators Grishin argues that Khrushchev's encounters with the intelligentsia were calm. It was policy failures, particularly foreign, that undermined Khrushchev rather than any deep individual flaws. Grishin's appreciation is a minority view. V.N. Shevelev's biography states that Khrushchev's tragedy was a lack of education and culture typical of the second generation of party leaders.[19] Credit is at least often given to

Khrushchev for preparing the way for a more radical democratization of the USSR under Gorbachev.[20] The memoirs of the last leader of the USSR are less kind to his reformer predecessor. Gorbachev lists the problems of Khrushchev's leadership style, chiefly a tendency to become 'vulgar' – 'spontaneity and folksiness occasionally turned into open boorishness, not to mention the foul language and heavy drinking'. More serious still was Khrushchev's adventurism, triumphalist tone and the cultivation of a cult of personality. For Gorbachev, Khrushchev's political vision was 'hampered by stereotyped thinking and his inability to or reluctance to reveal the underlying causes of the contentious phenomena he was facing'. This is why, according to Gorbachev, democratization under Khrushchev was 'nipped in the bud'.[21]

The most damning post-Soviet Russian evaluation of Khrushchev as Stalinist is, however, by the late Dmitrii Volkogonov. According to Volkogonov, it would be a mistake to take such seemingly obvious and consequential acts as the XX Party Congress as self-evident acts of de-Stalinization. The Secret Speech, for example, was 'in many respects mendacious and superficial, and did not touch the foundations of the Leninist system'.[22] Volkogonov argued that there is a link from Leninism to Stalinism; the attempt to resurrect a pure form of the former and by so doing eliminating the latter was therefore a delusion. Khrushchev would 'die without realizing that in defending Lenin he was preserving Stalin'. For Volkogonov, much of what the accusers said about Khrushchev 'was true'.[23] Khrushchev's single historical service was to render a return to the Terror impossible, but it is unlikely that the Brezhnev leadership desired this. Otherwise, Khrushchev preserved the continuity from Lenin through Stalin.[24]

Outside the former USSR two recent biographies of Khrushchev, by William Tompson and William Taubman, stand out. Tompson is perhaps one of the most sympathetic biographers, seeking to contextualize and understand a peculiar political dilemma that conditioned Khrushchev's leadership. Tompson concurs with critics that Khrushchev thought that administrative reorganizations, not more fundamental reform, would be sufficient to reach his goals. This was not however out of Stalinism. Rather Tompson's focus is on Khrushchev's own style, coupled with the sort of political compromises that were made prior to 1958. For Tompson, Khrushchev was no ideologue. He was driven by results and the desire to see Soviet citizens live under communism, by which Khrushchev understood access to goods rather than a non-alienated existence. Khrushchev's experience of achieving results was as the hands-on party activist that characterized his own career. It was not surprising that as leader Khrushchev sought to transfer decision-making from central ministries to local party officials. This was not however something that could happen at the stroke of a pen. The central ministries would seek to ignore or subvert Khrushchev's intentions. To overcome the resistance of the central ministries, Tompson argues that more power had been transferred to the local party officials than Khrushchev ever intended. The result was that the central system through which Khrushchev issued orders was weakened, but the absence of terror meant that there was no effective control on the local party leadership. Khrushchev thus discovered that 'exercising power from the top offices of party and government was in many ways a greater challenge than winning them'.[25]

While Tompson roots Khrushchev's frustrations and solutions to a particular political situation, much of his biography of Khrushchev nevertheless provides evidence that the charges against Khrushchev in 1964 were correct. He was a liability both domestically and on the foreign policy stage. There was a Khrushchev cult, he would meddle in affairs beyond his grasp, he was crude, foolish and would harangue comrades in public.

This was not always the case with Khrushchev as leader. For Tompson there is a periodization to Khrushchev's leadership pre- and post-1961. Before the XXII Congress it was possible to criticize the leader's proposals within the context of free and open debate. After the XXII Congress Khrushchev was increasingly sensitive to criticism and acted more and more like a bully. This change occurred for Tompson not out of the arrogance born of holding power for too long but as the 'product of frustration, disillusionment and failure'.[26] A large part of Khrushchev's frustration issued from the inability to have meaningful control over events on the ground. It was on the party officials that Khrushchev placed so many expectations, and right to the end he sought the correct arrangement to make them efficient and competent implementers of policy. Colleagues in the Presidium and local party secretaries found Khrushchev's policy solutions to be increasingly troublesome and threatening, and it was this that united them to secure Khrushchev's removal: 'his impulsive and authoritarian style aggravated his colleagues, who were "stuffed to the throats" with ill-considered reorganisations and wanted, above all else, stable leadership'.[27]

William Taubman's award-winning biography of Khrushchev grants its subject one moment of political bravery. The Secret Speech at the XX Congress was 'the bravest and most reckless thing he ever did. The Soviet regime never fully recovered, and neither did he'.[28] Otherwise, Taubman perceives his subject as ill-suited to the career of political leader. The divergence into politics denied Khrushchev the education he sought. The result was a man bereft of a mind able to analyse events soberly and all too often out of his intellectual depth, but with a chip on his shoulder that expressed itself in outlandish behaviour and empty boasting.[29]

Taubman's massive work in essence consists of a chronicle of Khrushchev's many failings and failures. The Khrushchev leadership style was established, for Taubman, well before he ascended to first position in the Presidium. He liked to dominate meetings, hectoring and berating, often before a speaker had begun a speech. His approach was hands on, with a close and keen interest in all decisions. At the same time, he did not like criticism and surrounded himself with 'experts' who would confirm what Khrushchev wanted to hear. The outcome, for Taubman, was ill-conceived policies that Khrushchev had devised in a rush and that often were pursued to extremes.

Once the 'anti-party' coup of 1957 was seen off, Khrushchev felt more secure and confident. He 'stopped listening' to high-placed comrades, taking personal initiatives and surrounding himself with 'yes men'. Khrushchev relied upon a small circle of four assistants (one each for foreign policy, agriculture, culture and ideology, general affairs) and a 'Press Centre' of five for help with speeches. This became, according to Taubman, the 'informal centre of power' that was however incapable of controlling the 'vast party-state system'. There was a turbulent period of 'helter-skelter' policies, including for example, the crusade for corn that became an 'irrational obsession',[30] and a leader, evident from relations with the intelligentsia, who was simply too unsophisticated.[31] Undoubtedly the worst of all times was the last two years of Khrushchev's reign. These were, according to Taubman, a 'time of not so quiet desperation' in which 'much of his energy and initiative were gone. Khrushchev had learned at last that bluff and bluster didn't pay, but they had been his main weapons, and without them, he was lost ... as his miseries multiplied, he withdrew into an inner circle of personal aides and advisers, avoiding his colleagues, acting without informing them, and berating them in public and in private for what they regarded as *his* sins'.[32] The successful ouster of 1964 was a reflection of the fact that Khrushchev's colleagues could 'no longer stand him'.[33] It was in a way a recognition that the unsuccessful plotters of 1957 were right. Similar accusations were levelled, the only difference being the outcome.

Towards an understanding of Khrushchev's leadership: the memoirs

There seems to be a broad agreement between ousters and biographers on Khrushchev's failings as leader, even if there are differences in outlook and interpretation. In scholarly assessments of Khrushchev's leadership little regard is placed upon Khrushchev's memoirs as a source. A recent edited collection of international expertise on Khrushchev states:

> Khrushchev's memoirs offered a rich lode of reminiscence, but they were mostly dictated alone, long after the events in question, without any access to documents or archives, and with one eye on the internal security police, the KGB. Although Khrushchev had a remarkable memory and a burning desire to set the record straight, his recollections could not help but contain numerous omissions and errors.[34]

Taubman's biography argues that Khrushchev's memoirs were motivated by a 'deep need to justify himself to future generations'.[35] This is clear above all for Taubman in the sections on the Terror in which Khrushchev sought to obscure his own role through a blend of deception and self-deception.[36]

It is the case that the memoirs were in some instances a response to the indictment that ended Khrushchev's tenure as leader. This is not done directly, for the memoirs avoid discussion of the ouster. There is reference only to the fact that the memoir's author had been 'retired'. When Khrushchev emphasizes, for example, that it was Mikoyan's idea backed by the Central Committee that Nina Khrushchev should accompany her husband on foreign trips, this reads as a refutation to the charges of misappropriations and nepotism.[37] The memoirs are however much more than mere self-defence and justification. Khrushchev mentions the intention to help future generations avoid the mistakes of the past.[38] In the absence of any chance to return to active politics, the memoirs were also an opportunity for Khrushchev to reflect upon his leadership and explain what he thought he was attempting to achieve. Of course explanation can be tainted and linked to a justification, but the former can be separated from the latter. In the case of Khrushchev the memoirs contain important insights into Khrushchev as leader.

First of all, Khrushchev is reluctant to refer to himself as leader. The usual term is 'when I was in the leadership'. Indeed at one point he stresses that 'I never unilaterally did anything, nor could I have done anything, without permission and the decision of the government and of the party's Central Committee.'[39] This does not mean that Khrushchev did not take the initiative over policy, for the memoirs abound with examples of cases of pride that this or that decision was 'down to me'. But there was a clear break with Stalin's leadership in that the party rules under Khrushchev reiterated the regularity of meetings from the lower party structures through to the Central Committee and the Presidium.[40] Comrades in the elite may not have been happy with Khrushchev's conduct, from the publication of minutes and interruption of speakers to the presence of non-members, but nevertheless Khrushchev acted as leader within the rules. To Khrushchev this must have compared very favourably to the period of humiliating and uncomfortable appearances at 'Stalin's court'. Little wonder that one of Khrushchev's outbursts of self-defence at the time of his removal was to shout 'you call this a cult?'[41]

There was probably a mixture of factors at work in explaining why Khrushchev could be more than 'first among equals' in his behaviour as leader. As Khrushchev himself recognized it was hard to break with habits developed in Stalin's time. Even though Khrushchev was not sending comrades off to the Gulag, there was an authority that came with being first secretary, head of the government and armed forces. This must have been especially so from Khrushchev's point of view given the low esteem in which he seemingly held his colleagues. There is not one kind word, and quite a few unkind, uttered about his closest comrades. Khrushchev had to step up to the podium because, variously, Bulganin, Mikoyan, Malenkov, Molotov and so on were not up to the job. From the point of view of colleagues there must have been some subservience to the tradition of not objecting too strongly to what the leader was saying. This must have been especially so after the failure of the 1957 attempted coup. So Khrushchev was a forceful leader partly because this is what the Soviet political system demanded, both from the leader and from the elite entourage. The notion of collectivity, even on Khrushchev's reading, was still within a relatively narrow circle – the Central Committee had a membership of fewer than 150 full members.

Second, Khrushchev distinguishes his leadership from Stalin's not only in the importance placed on the integrity of party rules, but in the fact that if Stalin was a Marxist in theory but not in practice, Khrushchev, who could not lay any great claim to be an original theoretician of Marxism, was a master practician.

Khrushchev's time as leader was a particular part of a career devoted to 'building socialism':

> building factories and organizing municipal services – these are the concrete expressions of the Leninist idea of building socialism. It is not a matter of studying theory ... I was busy forming party organizations and orienting them toward solving the tasks of socialist construction ... I became one of those who turned the resolutions adopted by those plenums into reality.[42]

And as leader Khrushchev had different policy priorities to Stalin. Most notably, there was an emphasis upon improving the people's standard of living. The utopianism of the Khrushchev era was not about a leap into non-alienated existence as outlined by the Marx of the *Economic and Philosophical Manuscripts*, but the promise that Soviet citizens would enjoy the life of an American consumer by the 1980s. This brand of utopianism may have been encouraged by a population more willing to make demands upon the leadership. It may also have been an outcome of how the war affected Khrushchev. After the sufferings and tremendous achievements, the people deserved to be rewarded. There should be greater material comforts for Soviet citizens as well as a transformed relationship between the state and its citizens. Instead of viewing the people as potential spies and enemies, the political elite should trust the people and be responsible to them. The state should be an 'all-people's state'.[43]

Third, Khrushchev as leader expanded the ambitions of the government programme quite considerably. The war years left Khrushchev with a deep commitment to Soviet security. The USSR would not be invaded on his watch. Khrushchev would confound Stalin's prediction that devoid of Stalin the USSR would fall victim to imperialism. The Cold War and the need to maintain a credible deterrent would of itself have placed tremendous strain upon the Soviet economy. Indeed, Khrushchev accused the West of trying to 'bleed the USSR dry' by intensifying the arms race. But as leader Khrushchev had both to meet defence requirements and improve the domestic economy to prove to Soviet people that it was worth the effort of living under socialism ('Who is going to follow our example if we can't even satisfy the most basic needs of our people?' he asked).[44]

Khrushchev attempted to succeed in various ways, chiefly by doing things on the cheap and quickly. He was aware that the heritage from the Stalin period was not good. The leadership had not been trained in the practice of good governance, the collective farms were depressed and starved of investment, and the economy was full of distortions and imbalances (hidden inflation, statistical inaccuracies, poor labour discipline, etc.). The trips abroad, especially to the West, only gave numerous examples of comparative Soviet inefficiency and lagging behind, whether it was washing machines in Austria, or levels of milk yield in Scandinavia, or the quality of concrete mix on French runways. It was therefore natural that Khrushchev was always on the lookout for 'new and progressive methods that would be more economical'.[45] He takes most pride in his achievements in housing construction and in the Virgin Lands programme. These are policies that are either criticized or ridiculed in modern secondary literature. Yet Khrushchev explains how, while aware of difficulties and flaws including poor-quality construction and investment shortfalls, they were the best possible response to the demands to provide people with housing and food. It is bordering on the nonsensical to say of Khrushchev, as Taubman does, that 'all too often Khrushchev hadn't taken the time to do things right. Instead of thinking things through, he could rarely sit still.'[46] Time is the one thing that Khrushchev had in very short supply. Here he was no different from the Tsarist-era premier Peter Stolypin who was not granted the 20 years of peace that he sought to transform communal agriculture into individual farmsteads, nor Gorbachev who famously once asked for 'time, time, time'. It is also fairly meaningless to criticize the Khrushchev leadership for seeking administrative solutions to the problems of post-Stalin economic development rather than more fundamental structural reform. After all, there was no model of transition that outlined how to undertake fundamental reform to ensure a painless transformation of the command-administrative system to a new form of socialist economy, let alone while simultaneously winning the Cold War. Had such a model existed then one could criticize Khrushchev for ignoring it. In its absence Khrushchev adopted policies that made most sense to him, given his knowledge of the Soviet system as was.[47]

As leader Khrushchev tried to boost Soviet economic performance by depending upon expertise. The memoirs are full of fond memories not of Khrushchev's political allies, but of specialists, particularly engineers, designers and scientists. Here Khrushchev was drawing upon his experience of working on the Moscow metro or on immediate post-war reconstruction when he would set about any task by forming a team of specialists. In the context of the building programme Khrushchev recalls the construction engineer Lagutenko: 'this remarkable innovator, who skillfully promoted mechanization in the production of prefabricated building components of reinforced concrete. I had always searched for people like him'.[48] Khrushchev describes his various reforms, from the decentralization of economic councils to the division of the party into sections for industry and for agriculture, as attempts to ensure that the correct level of knowledge was being applied at an appropriate level of command, with the appropriate safeguards for responsibility and accountability. Over the period 1956–1964 Khrushchev learned some harsh facts about the reality of being head of the USSR. One of the most important was the relative independence of the *nomenklatura* (party and state officials) that could subvert the leader's policies in a number of ways. It could put on a show during a visit to create a false impression; it could nod its agreement and then simply ignore any instructions; it could take an encouragement (e.g. to plant corn) to extremes and compromise a policy by applying it to areas for which it was not intended. Khrushchev admits that in his time in office he was 'unable to find the appropriate lever that would enable us to move things forward'.[49] Khrushchev admired the profit motive in capitalism and how it helped make his favourite US farmer, Garst, an incredibly good agriculturalist. Retaining a belief in the superiority of the Soviet economic system over capitalism, Khrushchev thought that one could produce a host of

Russian Garsts by 'selecting cadres and training them'.[50] If Khrushchev had remained in office the search for the correct administrative form would no doubt have continued. The memoirs suggest that this would have been 'an agency that can supervise and manage production and monitor the economy'.[51]

It is not clear that this agency would have been the Communist Party. Khrushchev seems to have invited specialists to attend meetings of the Central Committee and Presidium because he thought that there was more sense to be gained from listening to specialists than to party comrades. When the Presidium discussed the issue of missile technology Khrushchev insisted that the rocket specialist V.N. Chelomei give a report. Khrushchev comments: 'Presidium members had poor knowledge of questions having to do with weapons, and so no one expressed great enthusiasm, but there were also no objections.'[52] If specialists had been barred from the Presidium Khrushchev may have taken to Trotsky's habit of reading French novels rather than listen to his uninformed colleagues! Khrushchev was not much more impressed by how local party secretaries gave insufficient care and attention to rational and profitable production. Party chiefs 'on the ground' wanted an 'easy life', they were more concerned with filling in forms to make themselves look good on paper than having a serious attitude towards work. In the midst of a discussion of opposition to his proposal to follow good practice in the US and establish agricultural colleges in the countryside so that graduates could have theoretical and practical knowledge, for example, Khrushchev despairs at officials who agreed but were not willing to put in the effort to make the suggestion happen:

> Alas, one man can't do everything, even if great power and influence is allotted to him. The most dangerous form of resistance is when they 'yes you to death', nodding their heads and agreeing. This is a tactic that has been assimilated by many in Soviet society, and it is widely used ... in comparing the American system and our own in agricultural education ... their system is more progressive. Capitalists know how to approach matters from a rational point of view. The remorseless law of profit is in operation there. But in our country not every government official has a highly principled understanding of the cause; often he displays philistine indifference, looking out solely for his own comfort.[53]

It may not be the case that Khrushchev had a penchant for hare-brained schemes but that he was continually looking for the best way to ensure expertise and accountability. In their absence his policies could be harmful not out of a personal failing (in Taubman's estimation he 'made a bad situation worse'),[54] but because they became subject to the irrationalities of the Soviet system, both political and economic. Khrushchev's removal was a sign of how frightened the party *nomenklatura* had become by the leader's determination to bring it to account. Time limits on office and new responsibilities were to be avoided. In its self-interest it may have acted just in time. Giulietto Chiesa contrasts Khrushchev and Gorbachev's approach to the party:

Gorbachev ... initiated the destruction of the monopoly of power of the Communist Party ... [Khrushchev] never put into question the power of the Party and its role as the sole ruling force in society. He was aware that the Party was thoroughly pervaded by Stalinism, but his objective was to replace one type of cadres with a more democratic type, and that was all. He certainly never intended to threaten the political structures through which the party apparatus exercised its power ... What provoked the criminal degeneration of the leader and of an immense army of killers subject to his orders? What produced the deformations of the socialist society? Khrushchev did not answer.[55]

In the memoirs Khrushchev did address these themes. He reflected on the fact that a dictator had established personal control with very negative consequences not only in the history of the Russian Revolution but more recently in China and in Albania. Indeed, any Communist Party as then constituted could, he reckoned, fall hostage to a dictator. This was because the organizational system of a 'centralized, disciplined party welded together by a single aspiration ... [can] allow a single individual to use it for the sake of his own personal power'.[56] Khrushchev recounts Lenin's attempt to deal with corruption through bureaucratic agencies such as the Central Control Commission, but these in turn became corrupt. His conclusion is that 'more effective means of control from below over the leaders is necessary – that is, genuine democracy is needed'.[57] This sounds very close to Gorbachev's starting point and we all now know where that led.

Conclusion

When Martin McCauley considered the topic of Khrushchev as leader over two decades ago he concluded with three interpretations of Khrushchev's leadership: (1) as a transitory leader; (2) as a transitional leader; (3) as an original leader.[58] He was all three. Transitory in that the experiment was cut short and its most radical direction halted by those that it seemed to most threaten. Transitional in that there were continuities in the post-Khrushchev period, from the taming of the Terror and the establishment of a 'law-governed socialism' to the notion of a 'social contract' between regime and society. There should be a feeling that Soviet citizens have 'never had it so good'. Above all, though, Khrushchev was original. He was the first Soviet leader who was not concerned to make a mark on theory. He had a class-based view of the world, but conceptualized very simply. Much more important for Khrushchev was what could be achieved in the 'real world'. He was constantly intrigued by real-world solutions to real-world problems, and he toured the world like no Soviet leader before him. He was willing to look at what worked and to try to learn from it. If only specialists could have the right input to the Soviet system it could overcome its distortions to out-produce and defeat capitalism. In this quest Khrushchev was more and more willing and motivated to think outside of the Stalinist framework. He did not have to physically annihilate 'wreckers'; they could be sacked or demoted or, if it came to it, face the popular vote. If he had continued as leader Soviet socialism could have unravelled 30 years before its eventual demise. Writing his memoirs in retirement Khrushchev was thinking of economic, political and international relations in a way very similar to the Gorbachev period. Khrushchev had to grapple with Stalin and Stalinism in a personal and in a political sense, in power and out of power. He regretted the elements of Stalinist thought and behaviour patterns that remained with him as leader. As a pensioner he was called into party offices and encouraged to stop recording his memoirs. He shouted at the interrogators: 'I was also infected by Stalin. I have freed myself of Stalin, but you haven't done so.'[59] Perhaps by the end of his life Khrushchev had finally laid Stalin to rest.

Notes

1 Isaac Deutscher, for example, compares Khrushchev's preference for a 'moderate' dose of distortion with Trotsky's analysis of Stalinism, which had 'depth', 'sweep' and 'the vigour of his critical thought'. I. Deutscher, *The Prophet Unarmed*, Oxford: Oxford University Press, 1959, pp. vi–vii.
2 N.S. Khrushchev, 'Secret Report to the 20th Party Congress of the CPSU', in T. Ali (ed.), *The Stalinist Legacy*, Harmondsworth: Pelican, 1984, pp. 221–72.
3 The tension between making the people the driving force of history but wanting to retain a special place for communist leaders is typical of new works on Marxism produced in the Khrushchev era. See, for example, O. Kuusinen (ed.), *Fundamentals of Marxism–Leninism*, London: Lawrence and Wishart, 1961, Chapter 6.
4 *The Road to Communism. Documents of the 22nd Congress of the Communist Party of the Soviet Union*, Moscow: Foreign Languages Publishing House, 1961, pp. 124–39.
5 *The Road to Communism*, p. 128.
6 *The Road to Communism*, p. 131.
7 *The Road to Communism*, p. 135.
8 M. McCauley, *The Khrushchev Era 1953–1964*, London: Longman, 1995, pp. 121–4.
9 R.V. Daniels, *A Documentary History of Communism volume 1*, London: I.B. Tauris, 1987, p. 354.
10 W. Tompson, *Khrushchev. A Political Life*, Basingstoke: Palgrave, 1995, p. 102; W. Taubman, *Khrushchev: The Man and His Era*, London: Free Press, 2003, p. 230.
11 A. Nove, *Stalinism and After*, London: HarperCollins, 1975, p. 127.
12 Nove, *Stalinism*, p. 155. See also C.A. Linden, *Khrushchev and the Soviet Leadership*, Baltimore: Johns Hopkins University Press, 1990, pp. 208–21; E. Crankshaw, *Khrushchev*, London: Collins, 1966.
13 R. Medvedev, *Khrushchev*, Oxford: Oxford University Press, 1982, p. ix.
14 R.A. Medvedev and Z.A. Medvedev, *Khrushchev: The Years in Power*, London: W.W. Norton, 1978, pp. 184, 188–9.
15 See, for example, the collection Yu.V. Aksyutin (ed.), *Nikita Sergeevich Khrushchev. Materialy k biografii*, Moscow: Izdatel'stvo politicheskoi literatury, 1989.
16 D. Norlander, 'Khrushchev's Image in the Light of Glasnost and Perestroika', *The Russian Review*, 52, 1993, pp. 248–64.
17 A.A. Fursenko (ed.), *Prezidium TsK KPSS 1954–1964 vol. 1*, Moscow: Rosspen, 2003.
18 V.V. Grishin, *Ot Khrushcheva do Gorbcheva*, Moscow: ASPOL, 1996, p. 8.
19 V.N. Shevelev, *N.S. Khrushchev*, Rostov-on-Don: Feniks, 1999, pp. 3–12.
20 See, for example, R. Medvedev, *Nikita Khrushchev. Otets ili otchim sovetskoi 'ottepeli'?*, Moscow: EKSMO, 2006, p. 424.

21 Mikhail Gorbachev, *Memoirs,* London: Doubleday, 1996, pp. 67–72.
22 D. Volkogonov, *The Rise and Fall of the Soviet Empire,* London: HarperCollins, 1998, p. 274.
23 Volkogonov, *The Rise and Fall*, pp. 212–54.
24 Volkogonov, *The Rise and Fall*, pp. 215–16.
25 Tompson, *Khrushchev*, p. 189.
26 Tompson, *Khrushchev*, p. 268.
27 Tompson, *Khrushchev*, p. 270.
28 Tompson, *Khrushchev*, p. 274.
29 Taubman, *Khrushchev*, p. 43.
30 Taubman, *Khrushchev*, p. 373.
31 Taubman, *Khrushchev*, p. 383.
32 Taubman, *Khrushchev*, pp. 581–2.
33 Taubman, *Khrushchev*, p. 615.
34 W. Taubman, S. Khrushchev and A. Gleason (eds), *Nikita Khrushchev,* New Haven: Yale University Press, 2000, p. 2.
35 Taubman, *Khrushchev*, p. xiv.
36 Taubman, *Khrushchev*, pp. 100–103.
37 S. Khrushchev (ed.), *Memoirs of Nikita Khrushchev. Volume 3. Statesman [1953–1964],* University Park, Pennsylvania: Pennsylvania State University Press, 2007, p. 95.
38 Sergei Khrushchev (ed.), *Memoirs of Nikita Khrushchev. Volume 2. Reformer [1945–1964],* University Park, Pennsylvania: Pennsylvania State University Press, 2006, p. 204.
39 Khrushchev (ed.), *Memoirs* 3, p. 849.
40 See, for example, G. Hodnett (ed.), *Resolutions and Decisions of the Communist Party of the Soviet Union. Volume 4. The Khrushchev Years 1953–1964*, Toronto: University of Toronto Press, 1974, pp. 51–2.
41 Cited in Taubman, *Khrushchev*, p. 15.
42 Khrushchev (ed.), *Memoirs 2*, pp. 247–50.
43 G.A. Brinkley, 'Khrushchev Remembered: On the Theory of Soviet Statehood', *Soviet Studies*, 24(3), 1973, pp. 387–401.
44 Khrushchev (ed.), *Memoirs 2*, p. 360.
45 Khrushchev (ed.), *Memoirs 2*, p. 261.
46 Taubman, *Khrushchev*, p. 5.
47 For some very sensible remarks on this point see Nove, *Stalinism and After*, p. 117.
48 Khrushchev (ed.), *Memoirs 2*, p. 270.
49 Khrushchev (ed.), *Memoirs 2*, p. 380.
50 Khrushchev (ed.), *Memoirs 2*, p. 414.
51 Khrushchev (ed.), *Memoirs 2*, p. 402.
52 Khrushchev (ed.), *Memoirs 2*, pp. 466–7.
53 Khrushchev (ed.), *Memoirs 3*, pp. 155–6.
54 Taubman, *Khrushchev*, p. xix.
55 G. Chiesa, 'Perestroika: a revival of Khrushchevian reform or a new idea of socialist society?' in T. Taranovski (ed.), *Reform in Modern Russian History. Progress or Cycle?*, Cambridge: Cambridge University Press, 1995, p. 377.
56 Khrushchev (ed.), *Memoirs 3*, p. 493.
57 Khrushchev (ed.), *Memoirs 3*, p. 494.
58 M. McCauley, 'Khrushchev as Leader' in M. McCauley (ed.), *Khrushchev and Khrushchevism*, Basingstoke: Macmillan, 1987, pp. 9–29.
59 S. Khrushchev (ed.), *Memoirs of Nikita Khrushchev. Volume 1. Commissar [1918–1945],* University Park, Pennsylvania: Pennsylvania State University Press, 2004, p. 836.

3 The rise of political clans in the era of Nikita Khrushchev

The first phase, 1953–1959[1]

Nikolai Mitrokhin

After the transfer of power from Iosif Stalin to Nikita Khrushchev, political life in the USSR was characterized by a change in the model of government. Instead of the country's leader personally guiding an important group of top managers, by the late 1950s there would take place a distribution of political power among several big clans.

Before discussing this further we should establish how the expression 'clan' is understood here in terms of the internal political life of the USSR. According to the analysis presented here, political clans in the USSR formed themselves according to regional or branch, *but not family*, label, to exert influence on the redistribution of power and resources, and were inseparable from the real-life situation of their memberships. As a rule, members of a clan had a certain common view of how to solve tasks facing their sphere of activity (at the regional or all-union level).

The expression of mutual respect between chiefs (leaders) and their junior colleagues is the most important practical manifestation in the formation and maintenance of a clan's existence. This expression can take the most diverse forms, from the rigidly ritualized to one hardly noticeable except to those participating in the performance. It can bear a quite hypocritical or a formal nature and it can be followed by things that are entirely unwelcome for one side. However, when this is the case it serves to confirm that a clan member remains as such. Rejection of such a situation (as a rule publicly or demonstratively) indicates in its turn exclusion from the clan, ostracization, of the person involved.

The question of the existence of such political clans in the first decades of Soviet power lies outside this survey, although to the author's mind they were there. However, as to the period immediately preceding that under consideration (that is, 1945–53) I can state that Iosif Stalin struggled energetically against the clannish inclinations of workers in the Central Committee (CC) apparatus and high political leadership. His administrative strategy consisted of personal interaction with key managers (party, state and sometimes regional); the regular reallocation of them from one position to another for the prevention of 'splicing' (that is, the formation of clans); and also the use of special representatives from his personal pool for the solution of especially important tasks and for checking the situation in individual branches of the economy or groups of regions.[2] In fact, after 1937 he had at hand a pack of about 300 more or less young and energetic managers, whom he shuffled as necessary. Most of them were not known to him before 1936, and were selected with the help of the CC apparatus (and partly by Politburo colleagues) as replacements for repressed cadres.[3]

In this context the 'Leningrad Affair' is a striking example of the struggle with clans. As we know, it involved hundreds of bureaucrats, including former bureaucrats, of the Leningrad party organization and administrative organs of the 1940s. The shooting of the leadership of the 'Leningrad clan', including the Voznesenskii brothers and sister (a Politburo member, an education minister and a secretary of one of the Leningrad districts), again emphasized to the Soviet political elite that 'familyism' was impermissible in the matter of selecting and allocating cadres for the political and state apparatus.[4]

An unspoken rule was introduced in the struggle against nepotism: children of members of the elite and especially of the top elite were in effect excluded from politics, especially as they often married among themselves and grew up in the same circle. They were denied the right to occupy political or administrative posts, although they could have a claim on quite high posts in technical, artistic or international spheres. The one exception, the appointment of Yurii, son of Andrei Zhdanov, to manage a section of one of the CC departments in the late 1940s, ended up with him exiled for decades as rector of Rostov State University. And, in the later years, the party leadership, with Mikhail Suslov to the fore, determined the permissible level of 'familyism'.

There are at least three testimonies about how Mikhail Suslov (who with Aleksei Kirilenko was responsible in 1960–70 for cadre policy and in particular completely controlled the ideological sphere) personally determined the permissible level of family connections. Thus Viktor Sukhanov in his memoirs tells the story of how Suslov in 1964 opposed, at a session of the CC Presidium, the appointment of Aleksei Adzhubei as CC secretary precisely because of his family connections with Khrushchev. This allegedly almost cost him his job.[5] The chairman of the USSR State Publishing Committee, Boris Stukhalin, tells of his namesake who had attained high positions in the Foreign Ministry:

> At the session of the CPSU CC secretariat at which Victor Fedorovich was confirmed as deputy minister, M.A. Suslov, presiding, turned to me and asked 'Is this a relative of yours?' This was said quite sternly. If the reply had been affirmative, Mikhail Andreevich would hardly have been pleased (at the time I was ... an all-union minister and he could have regarded the appointment of my 'relative' as some kind of 'familyism').[6]

A worker in the information department of the Communist Party of the Soviet Union (CPSU) CC, Lev Voznesenskii, son of the RSFSR (the Russian Soviet Republic) education minister Aleksandr Voznesenskii, shot in 1950 in the 'Leningrad Affair', and himself a camp inmate for three and a half years in connection with this, writes of his appointment in the 1960s: 'Suslov had said, when my candidacy was suggested to him, "Well, we must definitely take the son of Aleksandr Alekseevich into the CC apparatus."'[7]

It is revealing that the vast majority of Politburo members in the political crisis years of 1953–57 did not have clans backing them up and they were forced out of the struggle or became nominal figures subservient to the new leader. There were three exceptions to this: Lavrentii Beria, Anastas Mikoyan and Nikita Khrushchev.

Lavrentii Beria's clan, formed not so much on regional as on professional identification (MVD – Interior Ministry – administrators were first among its entrants and Mingrelians made up only a small part of it), was not very powerful.[8] Some of its members were shot (notable among them were the Azerbaijan leader Mirdzhafar Bagirov and his associates)[9] or condemned to imprisonment. Remaining members were quite quickly purged from the 'organs', and only a few of them managed to continue their careers in unimportant functions within the ministry of medium machine building – the one-time domain of Beria.[10]

Anastas Mikoyan's clan was even less important and included, besides one or two surviving friends of his youth (holding posts of third- or fourth-rate importance in the Soviet party-state elite and in science), just one person deserving attention: the first secretary of the Armenian Communist Party, Grigorii Arutinov.[11] As is known, Mikoyan took the side of Nikita Khrushchev almost immediately, and this allowed the clan to continue to exist. Only Grigorii Arutinov, against whom Khrushchev may have had a grudge,[12] lost his job, but unlike Mirdzhafar Bagirov he died in his own bed.

However, it was Khrushchev himself, without in March 1953 having any visible group of supporters behind him, who made a start on the restoration and later the victory of the clan system. Khrushchev's clan, hurriedly formed in 1953–57 and gaining its victory over opponents in the CC Presidium without many independents realizing what was going on, consisted of two sub-clans, Ukraine and Moscow. Later on, having survived 'overloading' and rid themselves of their creator, these clans carried on an independent struggle among themselves.

This chapter is devoted to the first stage in the formation of the clan system in the USSR, from the 1950s to the 1960s, and specifically the positions occupied by people personally devoted to Khrushchev in the all-union organs and above all in the CPSU CC apparatus.

The political career of Nikita Khrushchev and the evolution of his supporters' group

In his memoirs Khrushchev himself quite clearly links his emergence as a politician to help from the clan system. As is known, participation in the Industrial Academy in the late 1920s made him an avid devotee of Stalin, demonstrating his convictions to the General Secretary's daughter who was also studying there. However, to struggle successfully with rightists in the Academy and in the Bauman district committee to which he was subordinated, he needed support, and he got it among people from his own area:

> We came from the South. We had a quite big fellowship (people from the Donbas, Dnepropetrovsk, Lugansk, Artemovsk and Kharkov). We upheld the

positions of the Central Committee. A conflict broke out and I too was drawn quite actively into this struggle.[13]

However, judging from the fact that Khrushchev makes no substantial mention of any of these people in his memoirs and does not drag them in as he does with other people, he did not have any real authority in this milieu. In any case, despite being chosen as secretary of the organization, he was not regarded as a leader. It is worth noting that, according to Khrushchev's memoirs, there did exist at the Industrial Academy a regional clan, the Nizhnii Novgorodians, who aligned themselves with his enemies.

Khrushchev began the formation of his clan when he was elected secretary of the Bauman district party committee in January 1931. Here he acquired several valuable supporters, including not only the subsequently well-known Nikolai Bulganin (together with whom he soon began to lead Moscow and appear at Stalin's lunches), but also the vastly less known Demyan Korotchenko[14] and Semeon Zadionchenko.[15]

Khrushchev's stay in Moscow was, as is well known, successful and lengthy, lasting up to 1938. Leaving his post as first secretary of the Moscow party organization to become first secretary of the Ukrainian Communist Party (CPU) he took with him a whole group of Moscow Ukrainians. They had to take over leading posts in the party organization, which Lazar Kaganovich had laid waste. Even this, to all appearances the last manifestation of 'clannishness' before the final victory of the Stalinist administrative model, was arranged as an initiative of the General Secretary and his chief 'cadre man' of that period, Georgii Malenkov.

> My time of departure was fixed. I asked Malenkov to find me several Ukrainians from the Moscow party organization (there were lots of them) or from the Central Committee apparatus … The second secretary came up for selection. Malenkov named as second secretary [his deputy] Comrade Burmistenko[16]… I knew little of Burmistenko. We got acquainted. He produced on me a very good impression, and our characters matched up. I gave Burmistenko the task of selecting people whom we could take with us, 15–20 people. He chose, I think, about ten from CC departments and Moscow party organizations. From the latter we took Serdyuk[17] and someone else.[18]

Khrushchev's tactical cunning was that, in suggesting that cadres should be selected for him from Moscow or the central CC apparatus, he in fact was suggesting people who he more or less knew or people for whom he could get meaningful character references (and later reject if particularly unsuitable). This was not the same as blindly recruiting people he could have got from places like the Lower Volga, Siberia and the Far East, where there were likewise many people from Ukraine. In some cases, as when Semeon Zadionchenko was appointed first secretary of the Dnepropetrovsk regional committee, Khrushchev could himself consult Stalin.

Dnepropetrovsk region, which included the large industrial area of eastern Ukraine, then undergoing intensive development, was regarded as almost as important as Kharkov. Therefore Khrushchev's friends from the Bauman district committee, first Korotchenko and then Zadionchenko, in turn headed it and the first secretary himself was regularly there. He met Leonid Brezhnev, the future leader of the 'Dnepropetrovsk clan', as early as 1938.[19]

This was how the foundation of the 'Dnepropetrovsk clan' was laid even before the war, consisting of party and economic leaders of the region. Despite the separation that soon followed of the Zaporozhe region from the Dnepropetrovsk, the common industrial links (in particular the dependence of industrial enterprises on energy sources) made permanent the mutual working relationships between the leaders of these regions. Aside from Brezhnev, Andrei Kirilenko (from 1939 second secretary of the Zaporozhe regional committee)[20] and Nikolai Mironov[21] became key figures in the nascent clan. However, it was only in the 1960s that this clan began to play a substantial role.

Khrushchev's pre-war and post-war work in Ukraine and his years of activity as a member of the war councils of the South-Western, Stalingrad and Voronezh army groups naturally entailed active contacts with a large number of party and state officials and military personnel. Those who left a good impression with Khrushchev and survived in the peripheries of the war and the post-war repressions moved higher (or at least retained existing posts) after he came to power in 1953.

Akeksei Kirichenko became the best known of these people. He was formerly first among Khrushchev's permanent staff in Ukraine and became a candidate member of the CC Presidium as early as 6 June 1953 (and a full member from 1955). In the period separating the plenum of 5 March 1953, which shaped the new composition of the Presidium, and the next of July 1955, this was the sole instance of a newcomer being co-opted.

Khrushchev's post-war work in Moscow, and he was first secretary of the party's Moscow regional committee from 1949 to 1953, again turned his attention to the capital's cadres. And although after 1939 neither in Kiev nor Moscow was he still able to transfer people 'by order', he certainly had a definite idea of who 'his people' were, and he used this during the 1953–57 struggle for power in the CPSU CC Presidium.

An important feature of Khrushchev in this situation was his humanity; his readiness to maintain personal contacts with lower-ranking people, to 'befriend'. This very much distinguished him from practically all his colleagues of the CPSU CC Politburo/Presidium, who were marked, as a rule, by haughtiness and unsociability. According to memoirs, the champion boor and lout was Lazar Kaganovich, with the arrogant Molotov, 'the man in a spectacles case', in second place. Lavrentii Beria came third in unpopularity; his high competence was not enough to make up for his oppressive demeanour and the threat to send errant subordinates to Siberia.[22] Georgii Malenkov and Nikolai Bulganin were not blatant boors but rather introverted *apparatchiki*. Anastas Mikoyan had the reputation of a crafty fellow and, again, was an introvert, while his Caucasian origin did not make him popular amid the wholly nationalistic mood of the post-war elite. Kliment Voroshilov for more than a decade was regarded as a played-out symbolic figure and was not taken seriously by the politically involved.

Thus in the situation of political uncertainty, when for the first time in decades highly placed officials had received the freedom of political choice, the agreeable Khrushchev could rely on the support of the second echelon of the *nomenklatura* (party-approved appointees) who for 20 years had been disregarded by most members of the Presidium. The crucial episodes of 1950s political history, Beria's arrest carried out with the assistance of the high military and the summons to the plenum for the struggle with the 'anti-party group' in 1957, were achieved entirely by Khrushchev's friends. But for this to happen it had been necessary to behave in the right way in the 1940s. And violation of the 'humane' principle, the continual humiliation of members of his own clan which began after 1960, in the end played a key role in Khrushchev's retirement.

The first stage of clan formation, 1953–1959

So, by the time of the death of Stalin and the beginning of the share-out of his political heritage, Nikita Khrushchev had several groups of supporters. Migrants from the Ukraine party organization were one of them: officials from the Communist Party of Ukraine (CPU) apparatus, people from Kharkov and particularly the aforementioned 'Dnepropetrovskites', whose leader Leonid Brezhnev had already become a candidate member of the CC Presidium.

Even more important, although lacking an unambiguous leader, was the military group, with which Khrushchev had learned to negotiate during the war. This was not just Georgii Zhukov but included, for example, Marshal Rodion Malinovskii, who in 1953 commanded the Far East military district, with whom he had been particularly friendly in the war and with whom in 1957 he replaced Zhukov as defence minister. Also among these friends was Ivan Bagramyan,[23] commander of the Baltic district in 1953. Among the generals who came to arrest Beria in the Kremlin was not only Georgii Zhukov and Leonid Brezhnev,[24] but also the chief of the Moscow military district Kirill Moskalenko, an acquaintance of Khrushchev from as far back as July 1941, and later his subordinate in Stalingrad and the Battle of Kursk.[25] Moskalenko also supervised the imposition and carrying out of Beria's death sentence.

As is known, in the course of the struggle with the 'anti-party group' in 1957, Zhukov's support was again important when the defence ministry's aircraft brought the participants of the plenum to Moscow as a matter of urgency. At this time Zhukov's own party spirit was imitated by the former Stalingrad acquaintance of Khrushchev, Malinovskii's subordinate, the unassuming Lieutenant-Colonel Aleksei Zheltov who in April 1953 had been moved from his post as the main permanent official at the defence ministry to the position of head of *Glavpur* (the Main Political Administration of the Soviet Army and Navy). After taking an active role in the removal of Zhukov in 1957, he was able to achieve promotion to the very important post of head of the administrative department of the CPSU CC (1958–59), where he supervised the entire cadres policy in the armed forces and law-enforcement agencies. He was replaced in this post by another frontline acquaintance of Khrushchev (this time from the Voronezh army group) Vladimir Tishchenko, who was deputy manager of the administrative department from 1956 and then its head from 1959 to 1961.

In 1962 yet another old protégé of Khrushchev, Aleksei Yepishev, became head of *Glavpur*. With Khrushchev's help he had become first secretary of the Kharkov regional committee as early as 1940, and in the war was Khrushchev's subordinate as member of the war council of the 40th Army, which was part of the Voronezh and Ukraine army groups. In 1955–62 Yepishev (after getting his hands dirty as deputy minister in charge of cadres at the state security ministry in 1951–53) was the Soviet ambassador in Romania and Yugoslavia. Later Yepishev would attach himself to Brezhnev and would manage to hang on as head of *Glavpur* right up to his death in 1985.

Not so much a frontline friend as an acquaintance of Khrushchev in the Voronezh army group was the Moscow man Dmitrii Shepilov, whose successful career under Stalin continued under Khrushchev, with elevation to CC secretary in 1955. However, this was that rare instance when someone advanced by Khrushchev did not come up to expectations, and he took part in the anti-Khrushchev conspiracy of 1957.

However, Khrushchev had no supporters in the law-enforcement agencies, with the exception of the first deputy minister Ivan Serov in the MVD, who was known to Khrushchev from 'work' in the occupied territory of Western Ukraine,[26] and the USSR Procurator-General Roman Rudenko, whom he had himself appointed as Ukraine SSR procurator before the war and who was put in charge of the all-union procuracy immediately after the death of Stalin.[27]

Resources for renewing the leadership cadres of the law-enforcement agencies were found by Khrushchev in the party and state structure of Moscow, whose Khrushchev-sponsored representatives made up what was very probably the most powerful clan in the Soviet political leadership. Having taken control of the MVD after the arrest of Beria on 26 June 1953, Khrushchev was no longer content with confirming Kruglov and Serov as its leaders. At the head of the Ninth directorate of the MVD (charged with protecting the country's leaders) was placed K.F. Lunev,[28] the manager of the administrative organs department of the CPSU's Moscow committee who, moreover, was quickly promoted on 30 July 1953 to first deputy minister. He took an active part in the trial of Beria and personally shot his assistants, condemned during the second 'Beriavites' case. In 1954, after the creation of the KGB, he became second deputy to Ivan Serov and continued to work in this capacity up to 1959. Meanwhile in 1953, after his promotion, a Moscow man once more became head of the Ninth directorate (a crucial post for political leaders in this uncertain period); this was the former first secretary of Moscow's Proletarskyi district committee V.I. Ustinov.[29]

This was only the beginning. On 31 January 1956 Semeon Kruglov, loyal to Malenkov, was replaced as head of the MVD by the administrator of the CPSU CC construction department, Nikolai Dudorov, who from 1950 to 1954 had worked in an analogous post in the Moscow district committee and regional executive committee. After this almost all deputy ministers were replaced by people from Moscow.[30]

Equally serious movements took place in the purely political sphere, where Khrushchev made use of Moscow people to reinforce his own position; for example, in such a practical yet important business as the liquidation of his main enemy Beria. As we know, Beria was condemned by a special judicial convocation consisting of eight members. Apart from the aforementioned wartime friend of Khrushchev, General Moskalenko, at least three of these eight were members of the Moscow clan.[31]

But of course for Khrushchev the more important need was to secure support at the highest political level, among the CC Presidium and secretaries. As Rudolf Pikhoya has convincingly shown, the Presidium (Politburo) and the secretariat in fact were two parallel active power centres. Therefore in 1955–57 a whole series of new Khrushchev protégés appeared in both of them.

Immediately after Beria's arrest, Khrushchev introduced into Presidium membership two of his protégés, the above-mentioned Aleksei Kirilenko and the former instructor of the Industrial Academy (when Khrushchev was studying there) Mikhail Suslov (plenum of 12 July 1955). Doing this, he not unreasonably counted on the support of his two old Moscow acquaintances Nikolai Bulganin and Mikhail Pervukhin.[32] At the Twentieth Party Congress a quartet of close 'friends' of the first secretary became candidate members: Brezhnev, Zhukov, Shepilov and Ekaterina Furtseva, the former secretary of the Moscow city committee.

Khrushchev, Suslov, Suslov's minion Petr Pospelov, and Shepilov were CC secretaries from April 1953 to February 1956. In 1955 they were joined by Averkii Aristov, restored by Khrushchev (he had been removed in March 1953, possibly at the insistence of Beria).[33] After the Twentieth Party Congress the dependence of the secretariat on Khrushchev increased thanks to the inclusion in it of Ekaterina Furtseva. At the same time, in 1953–57 less notable, but important, administrative duties of the second order, first in the CPSU CC apparatus, were taken over by Khrushchev's people.[34]

Insofar as Khrushchev's basic organizational task in the struggle against Beria and later against Malenkov was a strengthening of the party apparatus in confrontation with the state's, it is not surprising that a large number of his supporters found a place precisely there; more precisely, in those departments concerned with cadre and organizational matters. Mention has already been made of his old Moscow acquaintance Zadionchenko and his military acquaintance Zheltov, who were given responsible work in the departments of party and administrative organs. The member of the war council of the 13th Army of the self-same South-West army group, Vladimir Malin, moved from being one of the regular instructors of the CC apparatus to managing the general department (CC document handling) and in this capacity lasted until 1965. Vladimir Tishchenko, member of the Voronezh army group's war council, was moved from his post as adviser to the Albanian workers' party (1954–56) to a post as deputy administrator of the CPSU CC's RSFSR department of party organs and then for almost two years (April 1959 to 1961) was administrator of this department. Valentin

Pivovarov, Khrushchev's personal bodyguard (from 1942), who obtained after the war a law diploma at Kiev University but had started out as a fitter in the same Donetsk workshop as his boss, in 1957 became deputy and from November 1960 leader, remarkably enough, of that most delicate of the CC's departments, its business administration (which controlled not only material supply for the party leadership but also the latter's finances). Notably, he tried personally to fix the wages of *apparat* workers.[35] Yet another Ukraine cadre was Viktor Churayev, who in 1940–48 was initially second and later first secretary of the Kharkov regional committee. He had already occupied the CC post of administrator of the mechanical engineering department from 1948, but in 1954 was found a more important duty as, initially, first deputy and then manager of the department of party organs for the RSFSR, where he stayed until 1959. Nor had Khrushchev left unemployed his assistant in the coal and heavy industry of Ukraine, Aleksandr Rudakov (also a former wartime subordinate), manager of the corresponding department of the CPU CC; in 1954–62 he was administrator of the CPSU CC's department for heavy industry and later became a CC secretary. The head from 1950 of the CPSU CC's agricultural department, Petr Doroshenko, known to Khrushchev as first secretary of the Kirovograd regional committee, headed in 1955 and up to 1959 the agricultural department for union republics. Yakob Kabkov, a member of the CPU CC apparatus from before the war, was moved in 1958 from the extremely insignificant post of minister for the Ukraine fish industry to become manager of the CPSU CC department of trade/finance and planning organs, and with breaks held on to this post right up to 1985.

Moscow people also received their share of power. Evgenii Gromov, deputy manager of the MGK (Moscow City Committee) organization department (1946–48), having continued to work in the same post but now in the CPSU CC, in April 1953 became full manager of the department of party organs. But evidently he was too close to Malenkov and in March 1957 found himself removed from his post and sent as ambassador to Hungary. Aleksandr Kidin, former deputy chairman of the executive committee of the Moscow regional soviet (1942–45), in 1955 became, first, manager of the CPSU CC's department of trade/finance and planning organs and, after its unification with the administrative department, was manager of the CPSU CC's department of administrative and trade/financial organs for the RSFSR (March 1956 to April 1959). Aleksandr Orlov, formerly first secretary of the Taganskii district committee (1938–41), in 1956 became head of the department of diplomatic and foreign trade cadres organ, enabling him to move on in 1959 to the position of deputy minister of foreign affairs for cadres. From this position, on the shattering of the 'Moscow' clan in 1968 he went off as an ambassador.[36] Ivan Serbin kept and strengthened his position. Having been first secretary of the Podol'sk district committee (Moscow region) in 1936–41 he was transferred in 1954 from his post as manager of the engineering industry department to become, first, deputy and, later, full manager of the defence department, already well known to him. This position he kept up until his death in 1981. Dmitrii Polikarpov, the Stalinist propagandist, demoted at the beginning of the 1950s to directorship of the Moscow City Pedagogical Institute, in 1955 became head of the newly formed CC department of culture and led it up to 1962. The secretary of the party committee of the Moscow Energy Institute (for 1954) Vladimir Kirillin, who had previously never held even a middle-level party or state post, was made manager of the CPSU CC's department of science and higher institutes in 1955. He kept this position until 1962. Nikolai Dudorov has already been mentioned, but before becoming minister of the interior he managed in the course of two years to have a spell as manager of the construction department. His place for a long time, up to 1962, was occupied by a Leningrader, whose first deputy, however, remained (to 1955) the former manager of the MGK construction department Vasilii Abyzov, who also replaced him for the 1962–64 period. In his place came the next representative of the Moscow construction complex, the secretary of the Moscow region party organization Anatolii Biryukov (1964–69).[37] Vladimir Mylarshchikov, secretary of the Moscow regional committee (from 1951) in 1954 became head of the CPSU CC's agricultural department for the RSFSR and kept his job until 1959.

At a very broad estimate, based on insufficient information (for example, we have almost no data about the deputies of department leaders, who also often carried solid political weight), it is possible to state the following: with the changes in the CPSU CC apparatus from summer 1953 to roughly 1959 (the retirement of the 'anti-party group' hardly touched the manifest loyalty to Khrushchev of the apparatus), the two groups linked with the first secretary shared among themselves at least half of the departmental managers' posts. One group was people from Ukraine, including his frontline friends (totalling seven) and there were also representatives of the Moscow party and state apparatus (nine). The distribution of posts in the first instance involved the departments supervising cadres and large financial flows (defence, construction, agriculture and trade). These groups were much less involved with 'ideological' departments, which were at the disposal of Mikhail Suslov's protégés.

'Housekeeping' or material interests of the Moscow clan were soon reflected in the attempt at a redistribution of resources. On 8 December 1954 Ekaterina Furtseva, then still MGK secretary, and Mikhail Yasnov,[38] chairman of Mossoviet's executive committee, sent to the CPSU CC an official request to 'widen the rights of the Moscow city executive committee and its subordinate administrations, departments and district soviets' executive committees'.[39] Really it was about the right to take decisions about construction, which, as is known, was and remains one of the most corrupted spheres of economic activity.

Included in this was the suggestion that estimates should be confirmed in a more streamlined manner and payment made for what was done *post factum*; that ownership by individual sub-divisions of the Moscow city executive committee should be alienated freely, and transferred to others; most important, that 'special funds be established … for personal bonuses to salaries, personal salaries for workers in the urban economy, including those raised to 4–5,000 roubles, irrespective of the size of the salaries';[40] that prices be set at enterprises working under the aegis of the Moscow city executive committee, and likewise with the mark-ups of supply organizations; that 100 per cent of the budgetary resources obtained from plan over-fulfilment should be used for city needs; and that the plans of the crafts cooperative be confirmed, with appropriation of up to 50 per cent of its 'excess resources'.[41]

Such proposed financial independence and almost unconcealed concession to corruption was not greeted with much sympathy in the USSR finance ministry,[42] and to all appearances the letter obtained no satisfaction.[43] However, it clearly shows what the clans were striving for once they had got hold of a bit of power; and it also shows that the movement 'upwards' was accompanied by assistance to those who so far were still 'below'.

That many of the 1953–57 appointments were made by Khrushchev in an endeavour to 'fill gaps' with his own people is another matter. Already after 1957, when he had become the all-powerful master of the Presidium, he began to part company with those who did not 'do the job', or who irritated him. The peculiarity of the situation was that while Moscow representatives could be regarded as a quite well-constituted clan, the greater part of the Ukraine promotees and frontline friends of Khrushchev (Dnepropetrovsk people excluded) did not present themselves as a clan as such. While the leader of the 'Moscow clan' of that period, Ekaterina Furtseva, was a quite well-respected politician (among *apparatchiki*), the Ukrainian Aleksei Kirichenko, raised rather higher by Khrushchev, was a typical boor and lout, and was incapable of uniting the promotees of his republic into a clan. His retirement in 1960 at the second stage of the Khrushchev cadres policy was foreordained.

All the same, just by analysing the first stage of Khrushchev's cadres policy it can be affirmed that he actively influenced the cadre arrangements, confidently advancing people he trusted on the basis of personal contacts in the 1930s to 1950s. Here, already in this first stage, there can be observed a sharing out of those spheres that were 'at the disposal' of those persons enjoying his trust. An indication of this, for example, is the activity of Mikhail Suslov in the ideological sphere[44] or of Nikolai Dudorov at the MVD.

Notes

1 This chapter has been prepared within the framework of the project 'Telefonnoe pravo: gruppy vlyaniya i neformal'nye praktiki v apparate TsK KPSS 1953–1985gg', supported in 2006–2008 by the Gerda Henkel Fund (Germany).

2 Good material for this can be found in interviews with ministers and deputy ministers of the 1940s. See G. Kumanev, *Govoryat stalinskie narkomy*, Smolensk: Rusich, 2005. Some of my views on the Stalinist cadres policy are expressed in my review of this book; see *Neprikosnovennyi zapas*, 2006, No. 50 (http//www.nz-online.ru/index.phtml?aid=80020168).

3 But an exception was made for one of the 1920s clans: the quite large group of former commanders of the First Cavalry Army, well known to Stalin from the Civil War (Voroshilov, Budennyi, Timoshenko, Gorodovikov, Kulik, Apanasenko, etc.) up to the war kept (and even consolidated) their positions in the army.

4 For the Voznesenskii family, see L. Voznesenskii, *Istiny radi...*, Moscow: Respublika, 2004.

5 V. Sukhanov, *Sovetskoe pokolenie i Gennadii Zyuganov. Vremiya reshitel'nykh.* Moscow: ITRK RSPP, 1999, p. 183.

6 B. Stukalin, *Gody, dorogi, litsa...*, Moscow: Fond imeni I.A. Sytina, 2002, p. 9.

7 L. Voznesenskii, *Istiny radi...*, p. 368.

8 In November 1951 a case was started against the Mingrelians, emigrants from the mountainous part of Georgia who formed a small ethnic group and had occupied high positions in Georgia, among them the organs of state security. They were accused of forming a nationalistic group. It is thought that it was directed against Beria, who was the highest-placed Mingrelian in the USSR. After Stalin's death the case was closed, although after the fall of Beria many of those involved were re-arrested and shot, this time accused of participating in mass repressions.

9 For details, see E. Ismailov, *Azerbaidzhan: 1953–1956gg. Pervye gody ottepeli.* Baku: Adil'ogly, 2006. Bagirov and Beria were stricken from membership of the CC Presidium at the plenum of 7 July 1953.

10 For a survey of a similar practice, see E. Zhirnov, 'Chisto chekisskaya chistka', *Kommersant-vlast'*, 2006, No. 36, pp. 64–9 (http://www.kommersant.ru/doc. aspx?DocsID=703829). In this connection the biography of one of the chief managers in the time of Lavrentii Beria, written by his son, is of some interest: Yu. Bogdanov, *Strogo sekretno. 30 let v OGPU-NKVD-MVD*, Moscow: Veche, 2002.

11 The beginnings of Mikoyan's clan (then quite numerous) are described in some detail, for example in the biography of one of its members, shot in 1938 (see O.I. Tregubenko, 'Sud'ba revolyutsionera', *Voprosy istorii*, No. 8, 2007, pp. 136–44). By the mid-1950s there remained from the rather large clan just four people, apart from Arutinov, in Mikoyan's close circle: his brother Artem (the well-known aircraft designer), Academician A.A. Arzumanyan (husband of his wife's sister, director of the party's Institute of World Economy and International Relations), General Gai Tumanyan (after the war, head of the political department of the Military Engineering Academy, in 1960 head of the political department of the Defence Ministry's cosmic resources office), Lev Shaumyan (son of one of the 'Baku commissars'). Quite detailed recollections of Anastas Mikoyan's eldest son and daughter-in-law about this circle have been preserved. Last in line was the niece and adopted daughter of Grigorii Arutinov. See S. Mikoyan, *Vospominaniya voennogo letchika-ispytatelya*, Moscow: Tekhnika molodezhi, 2002, p. 128, 213; N. Mikoyan, *Svoimi glazami*, Moscow: 2003, pp. 160–4.

12 For a view on this grudge, see N. Mikoyan, *Svoimi glazami...*, pp. 129–31.

13 N.S. Khrushchev, *Vremiya, Lyudi, Vlast'. Vospominaniya). Kniga 1*, Moscow: Moskovskie Novosti, 1999, p. 41.

14 Korotchenko, Dem'yan Semeonovich (1894–1969): peasant, later soldier, member of RCP(b) from 1918; from 1918 was in various party posts in Ukraine; in 1931–34 was chairman of the Bauman district executive committee in Moscow, then secretary of the Bauman and Pervomaiskii party district committees in Moscow; from 1936 secretary of the Moscow regional committee of the VKP(b); 1937–38 first secretary of the Zapadnii, later Dnepropetrovsk, regional party committee and chairman of the Ukraine Sovnarkom; 1939–47 secretary of the Ukraine CC; 1947–54 chairman of the Ukraine council of ministers, later chairman of the Ukraine Supreme Soviet Presidium (1954–69); member of the Ukraine CC from 1939; member of the Presidium (1952–53); and candidate and member of the CPSU CC (1957–61).

15 Zadionchenko, Semeon Borisovich (1898–1972): 1931–37 managed organization department, deputy and chairman of Bauman district executive committee, first secretary of the Bauman district committee; 1937–38 deputy chairman RSFSR council of ministers; 1938–41 first secretary of Dnepropetrovsk regional committee; 1942–46 first secretary of the Bashkir and Kemerovo regional committees; 1946–48, 1951–52, 1954–58 inspector of the CPSU CC; 1948–51 first deputy minister of procurements for USSR; 1952–54 section head of the CPSU CC department of party organs; from 1958 pensioner. 'When after the war Stalin let loose the struggle against cosmopolitanism Zadionchenko, like many others, suffered undeserved harassment.

He was removed from work, and the last years of his life were not the best. Only under Brezhnev was Semeon Borisovich accorded posthumous respect. All the central newspapers published an obituary signed by Politburo members, and his portrait was published.' Khanin, Il'ya, 'Sdelali vozmozhnoe i nevozmozhnoe', *Respublika Bashkortostan* (Ufa), 8 May 2007.

16 Burmistenko, Mikhail Alekseevich (1902–41): peasant background, member of RCP(b) from 1919; from 1932 second secretary of the Kalmytskii regional committee of the VKP(b); from 1941 one of the organizers of the partisan movement in Ukraine and member of the SW army group war council. Perished at the front.

17 Serdyuk, Zinovii Timofeevich (1903–82): in 1939–41, 1943–47 second secretary; in 1947–49 first secretary of the Kiev regional committee of the CP(b) of Ukraine; in 1941–43 political work in the Red Army; in 1949–51 secretary of the CC CP(b)U; in 1952–53 first secretary of the Lvov regional committee of the CP(b)U; in 1954–61 first secretary of the CC of the Moldavian communist party; later first deputy of the CPSU CC's KPK.

18 Khrushchev, *Vremiya Lyudi*, p. 146.

19 Moreover, nor did the Kharkov region escape his attention. As early as 1937 the second secretary of Khrushchev's native Donetsk region was the very experienced organizer Aleksandr Osipov (1899–after 1956), until then working in a most important Moscow job, personally supervised by Khrushchev – chairman of the 'Metrostroi' construction committee, and later as first secretary of the Stalinsk district committee in Moscow. In 1937 he also became first secretary of Kharkov region, but in the following year fell into the grinder of the Great Terror.

20 In a book devoted to one of the rank-and-file members of the clan, the second secretary of the Zaporozhe city committee Nikolai Moiseenko (based on his recollections and letters), it is asserted that even after the occupation of Zaporozhe by German troops in 1942–43, the families of regional and city committee workers lived together in Samarkand. There the wife of Kirilenko with her good rations and 'not-bad work' provided the less-favoured wives of party workers with food and got their children into educational establishments. See G. Moiseenko, *Pamyat'plamennykh let*, Moscow, 2006, p. 17.

21 Mironov, Nikolai Romanovich (1913–64): in 1937–41, 1946, student at Dnepropetrovsk University; fought in war; 1945–47 in Dnepropetrovsk regional committee of CP(b)U; 1947–49 first secretary of the Oktyabr'skii district committee of CP(b) U (Dnepropetrovsk); 1949–23 August 1951 secretary of the Kirovograd regional committee of VCP(b); 23 August 1951–15 March 1953 deputy minister of state security of USSR; 1956–June 1959 head of KGB administration for Leningrad region; June 1959–19 October 1964 manager of the administrative department of the CPSU CC.

22 In fact I do not know of any case when he carried out this threat.

23 He received the rank of marshal in 1954, and in 1955 the post of deputy defence minister.

24 To all appearances, he bore a serious grudge against Beria for the latter's initiation of a large reduction of Presidium members and the subsequent despatch of Brezhnev to a second-rate post as deputy head of the Main Political Administration of the Soviet army and navy.

25 Khrushchev wrote in his memoirs: 'I valued him for his unlimited devotion to the Motherland and his boundless courage.' Khrushchev, *Vremiya Lyudi*, p. 390.

26 Nikita Petrov also confirms this in his work, adding to the ranks of the Khrushchev–Malenkov bloc in 1953 the former subordinate of the latter, S. Kruglov. Nikita Petrov, *Pervyi predsedatel' KGB Ivan Serov*, Moscow: Materik, 2005, p. 141.

27 Of course this was helped by Rudenko's personal qualities: he had successfully performed his role of USSR prosecutor at the Nuremberg Trials.

28 The former pre-war secretary of the Pavlovo-Posadskii district committee, spending some time after the war as deputy administrator of the cadres department of the Moscow regional committee.

29 Yu.N. Bogdanov, *Strogo sekretno: 30 let v OGPU-NKVD-MVD*, Moscow: Veche, 2002, pp. 361–2.

30 The former deputy chairman of the Moscow regional council of trades unions, S.A. Vasil'ev, and the first secretary of Moscow's Oktyabrskii district committee of the CPSU, M.N. Kholodkov, became deputy ministers. Konstantin Chernyaev was appointed deputy minister for cadres and head of MVD cadres administration; he was a graduate of the Moscow Institute of Steel and Alloys, who in the early 1960s also occupied the more delicate post of the CPSU CC's business manager. The former head of material and technical supply of *Mossovet*, P.N. Bakin, was appointed head of *GULAG*. The former chairman of Moscow's Molotovskii district executive committee, I.I. Mityaev, headed the secretariat of MVD. The ministry's housekeeping administration was entrusted to the former head of *Mossovet*'s housing administration, I.I. Solodilov. See Bogdanov, *Strogo sekretno...*, pp. 383–4.

31 K.F. Lunev, first deputy minister of MVD; N.A. Mikhailov, secretary of the CPSU Moscow regional committee; A.A. Gromov, chairman of the Moscow city court. Apart from these, the members of the convocation were Marshal Ivan Koniev (chairman, likewise a good wartime acquaintance of Khrushchev, although not so close as Malinovskii or Moskalenko); VTsSPS (the central trades union organization) chairman N.M Shvernik (from 1956 chairman of the KPK along with Anastas Mikoyan, who was one of the few notable figures of Stalin's time remaining a member of the CC Presidium up to 1966); the first deputy chairman of the USSR supreme court and head of the military tribunal E.L. Zeidin; the chairman of the Georgian SSR council of trades unions M.I. Kuchava.

32 Promoted by Khrushchev in 1937 to the 'Moscow' level from being director of the Kashira hydro-electric station.

33 The political position and links of one more secretary, Nikolai Belyaev, are not known to us.

34 I will pass over the appointment by Khrushchev of a whole series of his pre-war and frontline friends to positions as first secretaries of the CCs of union republics, including Z. Serdyuk in 1954 in Moldavia and V. Mzhevanadze in Georgia.

35 And soon, on 19 June 1962, he paid with the loss of his post for excessive zeal displayed in this question, coming into conflict simultaneously with Frol Kozlov and Mikhail Suslov. See E. Zhirnov, 'Shizofraniya tsekovskogo tipa', *Vlast'*, No. 35 (538), 8 September 2003, http://www.kommersant.ru/doc.aspx?DocsID+409370.

36 At the same time, in 1968, Timofei Shevalyagin, who was regarded as Aleksandr Shelepin's man, lost his job as head of the CPSU CC's information department. (Interview of N. Mitrokhin with the former Politburo member Vadim Medvedev, Moscow 2007.)

37 For his recollections see A. Biryukov, *Ya dumal, chto rodilsya akrobatom...*, Moscow, 2004, and also the 2007 interview by N. Mitrokhin.

38 Mikhail Alekseevich Yasnov (1906–91): in 1936–38 was managing the Moscow waterfront construction trust; in 1938–56 he was deputy chairman of the Mossovet executive committee for construction; in 1956–66 he was first deputy chairman of the RSFSR council of ministers; and in 1966–85 chairman of the RSFSR Supreme Soviet's Presidium.

39 RGANI fond 556, CPSU CC bureau for RSFSR (1956–66). op. 14, d. 2, ll. 48–59.

40 RGANI f. 556, op. 14, d. 2, l. 53.

41 RGANI f. 556, op. 14, d. 2, ll. 57–8.

42 RGANI f. 556, op. 14, d. 2, pp. 61–79.

43 There is a second variant of the 16 April 1955 letter with an attached project for a council of ministers decree with a much more modest list of demands. It carries indecipherable comments by M. Suslov. See RGANI f. 556, op. 14, d. 2, pp. 80–7.

44 Naturally, Khrushchev himself often entered into ideological questions and there still existed at least one group having influence in this sphere (the so-called 'press group'). However, the majority of cadre decisions in this sphere, and everyday planning, remained with Suslov.

4 The Central Committee apparatus under Khrushchev

Alexander Titov

Khrushchev's rise to power in the post-Stalin period has traditionally been linked with the struggle for influence between the Communist Party of the Soviet Union (CPSU) apparatus and the state organs dominated by Khrushchev's rivals.[1] This chapter analyses the evolution of the Central Committee (CC) apparatus, one of the CPSU central decision-making organs, during Khrushchev's time in power.[2] I argue that the introduction of the territorial principle and attempts at simplification of the apparatus in the 1950s reflected broader aims of decentralization and fighting against bureaucracy. Together they constituted Khrushchev's project for a revival of the party's special role in society and government which was an important part of his political programme. However, tensions between alternative approaches meant that no single pattern of CC organization ever emerged. The economic difficulties of the early 1960s led to a radical reform of the party apparatus, which signalled the end of the revivalist project. Although the post-Khrushchev leadership returned to the pre-1956 organizational patterns, the party apparatus's retention of its dominant role over state institutions proved to be the lasting legacy of the Khrushchev period. However, its supremacy was that of a bureaucratic institution rather than a revived mobilizing force as Khrushchev had intended.

The Central Committee before 1948

An overview of the CC apparatus's history in the inter-war period is necessary to place Khrushchev's reforms of the apparatus in a wider ideological context of the evolution of the party's role and functions within the Soviet political system. From its establishment in 1918, the organization of the CC apparatus reflected the party's priority at this time of mobilization and allocation of party personnel, the supervision of local party organizations, and guidance of propaganda and agitation activities. The party's special task was to provide general guidance rather than involve itself in management of the economy or minute supervision of government bodies.[3] From the early 1920s the CC apparatus assumed responsibility for the appointment of party cadres to key posts in the Soviet party and government bodies, which henceforth was one of its key functions.[4] The key Organization and Assignments department (Orgraspred – *Organizatsionno-Raspredelitel'nyi*), headed by Kaganovich, played a major role in Stalin's use of the apparatus for strengthening his grip on the party. In the process, the Communist Party began its transformation from a mobilization machine into a government appointments machine.

From the early 1930s the role of the party and its apparatus was beginning to change. In 1930 Kaganovich split the Orgraspred into two departments, the Organization-Instruction and the Assignment departments, thereby decentralizing the cadre responsibilities within the Secretariat. This was the first sign that the party apparatus was moving towards greater direct involvement with economic matters, meeting the demand for more specialized appointments during the first Five-Year Plan.[5] At the XVII Party Congress in 1934 it was decided to reorganize the CC apparatus on a 'production-branch' basis. Every new branch of industry now had a parallel body of specialized party control and every sector of Soviet life was supervised by the corresponding party organ. The new departments were responsible for the supply of cadres, control over subordinate party organizations, checking of fulfilment of party orders and mass agitation.[6] The result was that the party assumed direct responsibility for economic management, which made it more akin to other government institutions.

This pattern was reversed five years later when at the XVIII Congress in 1939 the apparatus was reorganized on a functional basis – all personnel matters were now concentrated in a Cadres Directorate, and ideological matters in a Propaganda and Agitation Directorate, headed by Malenkov and Zhdanov respectively.[7] In a seminal paper on this topic, Jonathan Harris analysed the ideological background behind the 1939 reorganization.[8] Harris argues that two principal approaches to the apparatus organization reflected deeper differences in the party leadership about its role in the Soviet system, represented by G. Malenkov and A. Zhdanov respectively. Zhdanov championed the view of the party with a special role in leadership that would be undermined if it engaged in detailed management of the economy. He thought the party's role was to provide 'political leadership' to society as a whole rather than micromanage economic activity.

Malenkov, in contrast, stressed practice as the basis of party work, implying that the party officials had the right to intervene in the running of the economy by state organs. Malenkov maintained that a certain blurring of authority between party and state organs was essential for the maintenance of party supremacy and implementation of its decrees. Harris points out that the conflict between two different views of the party's role and relationship to state organs continued to play an important part in the decades after 1945. This thesis will serve as a starting point for the analysis of Khrushchev's reforms of the CC apparatus in this chapter.

Stalin periodically demanded that party officials shift emphasis between their economic and party-political role. When he wanted to give greater freedom to the Sovnarkoms (*Sovety Nakordnykh Kommisarov* – Councils of People's Commissars), he created functional departments; when he wanted to emphasize the Secretariat's economic work, he urged the creation of production branches.[9] However, these shifts of emphasis had profound implications for the role of the party in the Soviet system of government, which was never completely resolved under Stalin. The conflict between different functions of the party apparatus became one of the central themes in the history of the CPSU in the post-Stalin period.

The CC apparatus after 1945

A major reorganization of the CC apparatus took place in August 1948 when the functional organization of the apparatus was replaced by the production-branch one. The two Directorates, Propaganda and Cadres, were abolished and industry-based departments introduced. The abolition of the Directorates increased the role of the Orgburo (*Organizatsionnoe buro* – organizational bureau), which along with the Secretariat became the only two top party organs that met regularly. Malenkov, who by 1948 re-established his dominance of the Orgburo, regained his influence in the party apparatus.[10]

Altogether there were ten departments in the new CC apparatus, which employed 1,238 people.[11] The two most important departments were the department of propaganda and agitation and the department for party, trade union and Komsomol organs, which between them employed nearly half of all personnel of the CC apparatus. The economic block employed 383 people spread over six departments. The remaining two departments, external relations and administrative organs, employed 230 people. The allocation of personnel reflects the primary focus on propaganda issues understood in the broadest sense including supervision of science and education, and party organization including management of *nomenklatura* appointments.

Gorlizki and Khlevniuk have pointed to a parallelism with state institutions in the new organization of the CC. Four new departments had direct analogues in existing sectoral bureaus at the Sovmin (*Sovet Ministrov* – Council of Ministers) and a further two corresponded roughly with Sovmin bureaus.[12] This list could be extended to include principle ministries, which had parallel departments in the CC apparatus. The Foreign Ministry was matched by the department of external relations, while the security ministries and the defence ministry were shadowed by the administrative department.[13] The departments of Propaganda and Agitation and of party organs also had sectors that mirrored relevant ministries.[14]

The principle of parallelism was, therefore, the basis for the reorganization of 1948. On this principle, the CC departments mirrored the structure of government institutions, either with a department corresponding to a particular government ministry or bureau, as was the case with the four departments above, or a department responsible for a group of closely related ministries such as the administrative department. This principle remained at the centre of the CC organization until the early 1950s.

The XIX Party Congress and the CC apparatus

The XIX Party Congress that took place in October 1952 had a profound impact on the CC organization. Yoram Gorlizki has provided a careful analysis of the campaign on party democratization and its impact on the party apparatus.[15] The

Congress rhetoric on party democracy was particularly significant as it called for political as opposed to economic leadership from the party. This related to the renewed revivalist rhetoric reminiscent of Zhdanov's time before the war.

The post-Stalin struggle for power is often represented in historical literature as a conflict between the government presided over by Malenkov and the party headed by Khrushchev.[16] It is interesting to note that the balance of power between the state and party organs began to shift in the favour of the party around the XIX Party Congress. Gorlizki and Khlevniuk argue that, as the result of political reshuffles instigated by Stalin in 1952, the status of the Sovmin Bureau was lowered, while that of the newly enlarged Presidium and especially its Bureau was enhanced by transfer of political leaders from state to party bodies.[17]

In the wake of the XIX Party Congress, the structure and staff of the CC apparatus were also re-examined. The formation of the industrial-transport department in November 1952 by merging the departments of light industry, heavy industry, machine-building and transport was the most significant change. Gorlizki argues that this was the first attempt by the revivalists to reverse the trend to specialization and to emphasize instead the party's more general political role in government by forging together the partial perspectives of smaller departments. These changes in the CC structure after the XIX Party Congress went hand in hand with Khrushchev's revivalist emphasis on the importance of party power and with a focus on the party as the stimulant to other social and economic institutions; also it gave him an opportunity to place trusted personnel in key posts.[18]

The emerging conflict between Malenkov, who after March 1953 concentrated his authority in the Sovmin, and Khrushchev, who by then was the only Presidium member responsible for purely party affairs, centred on two conflicting visions of the role of the party. In Gorlizki's view, Khrushchev's success was due, first, to the continuity between pre-congress revivalism and Khrushchev's political rhetoric in the first months after Stalin's death. Second, it gave him a basis on which to reorganize the CC apparatus, enabling him to create new strategic vacancies to which he could appoint his allies.

In the period between the death of Stalin and the defeat of the 'anti-party' group in 1957, Khrushchev continued to use this strategy of the reorganization of the CC apparatus and a concurrent promotion of his men to positions of influence. Two main themes in the organization of the apparatus emerge in the post-1953 period. First, was the new regional focus in the organization of the CC apparatus, which became one of the most distinctive features of Khrushchev's time. Second, was the drive towards simplification of the apparatus's and individual departments' structures. This was linked to the related party revivalism and fight against the bureaucracy, which supports Gorlizki's argument about the importance of revivalist thinking in organization of the CC apparatus.

The first instance of the territorial approach in the organization of a central committee department goes back to March 1951, when 14 territorial sectors in the department of party organs were created. Reorganization on the territorial principle continued at full pace once Khrushchev was completely in charge of the CC. In January 1954, the agricultural department for the Russian Socialist

Federative Soviet Republic (RSFSR) was carved out from the department of agriculture thereby introducing the territorial principle at a departmental level. The new department had six territorial sectors according to the regional division of Russia and two general sectors, *sovkhozy* (state farms) and economy. On 28 May 1954 the Presidium decided to split the department for party organs into two departments of party organs for the union republics and for the RSFSR. The establishment of these two departments for the RSFSR in 1954 was a prelude to a major restructuring of the apparatus two years later.

The trend towards a regional focus and structural simplification can also be seen from the reorganizations of the department for propaganda and agitation. As a result of four reorganizations between 1950 and 1955, the number of sectors in the department was reduced from 14 in December 1950 to only six in September 1955. In addition to simplification of its structure, the emphasis on regional organization was also becoming more pronounced from May 1954 onwards when five territorial sectors were created.[19] The regional dimension was preserved in the next reorganization of the department in September 1955, when two regional sectors were created, for the union republics and the RSFSR. Finally, the trend towards increased regional focus had its logical conclusion with the creation of the Bureau for the RSFSR in April 1956. The department for propaganda and agitation was split into departments for the Union Republics and for the RSFSR.

There was, however, no single pattern in the organization of the CC apparatus at this period. The department of industry and transport had an extremely cumbersome structure, with ten sub-departments that in turn had 33 sectors employing 219 people. On 22 April 1953 its structure was simplified by the abolition of sub-departments and cutting down the number of sectors to 13. A year later the department was split into six departments, thus reversing the revivalist tendency of the earlier years by once again entrenching the production-branch principle within the CC apparatus.

One of the possible reasons for this break-up of the department could be the influence of the principle of parallelism. For example, on 19 April 1954 the ministry of transport and heavy machine building was broken up into four different ministries, thus providing reasons for the corresponding division of the relevant CC department of industry and transport.[20] Although this was the only example of splitting a major CC department, it is indicative of the complexities in the organization of the apparatus in the mid-1950s, when a combination of various approaches such as production-branch, functional and the new territorial principle co-existed at the same time.

Two new patterns emerged in this context. First, there was the increasing emphasis on territorial organization, starting within the individual departments, later extending to the whole CC apparatus. Second, there was a clear tendency towards simplification of the apparatus, seen in the reduction of the number of sectors in individual departments. This was linked with the continuing campaign against bureaucracy launched by Khrushchev. It is significant that this campaign affected the CC, although to a considerably lesser degree than state institutions, which were the main focus of the anti-bureaucracy drive.[21]

The Bureau for the RSFSR

The trend towards territorial organization of the CC apparatus was realized to the greatest degree by the creation of the Bureau for the RSFSR in 1956. The idea of establishing a Bureau for the RSFSR was not new. In the 1920s, there were proposals to give more prominence to the RSFSR bodies, including the establishment of a Russian communist party with its own CC. However, these proposals were never implemented.[22] The RSFSR Bureau at the CC briefly existed in the late 1930s but its role was marginal compared with the 1950s.[23]

Shortly after the war, in 1947, M. Rodionov, the Chairman of the RSFSR Council of Ministers, lobbied for the creation of a bureau for the RSFSR. The idea was to create a body for a preliminary consideration of issues about the RSFSR submitted to the CC and the Union Government, which could reduce the load of questions dealt with by the central organs. Such an organ would also facilitate consideration of economic and cultural matters in coordination with the RSFSR Council of Ministers. Finally, Rodionov argued that this would lead to an increased use of uncapped potential at the local level.[24] However, the proposal was not implemented and Rodionov himself perished in 1949 in the Leningrad affair, charged with attempts to create a regional patronage network centred on Leningrad. This showed the degree of Stalin's sensitivity about greater local responsibility, particularly at the RSFSR level.[25]

A theme of certain continuity with the Khrushchev era thus emerges. The unlocking of local potential was one of the principal reasons behind Khrushchev's drive towards decentralization in the 1950s through the creation of *sovnarkhozy* (*Sovety narodnogo khozyaistva* – national economic councils) and to a lesser extent in the reforms of the party apparatus. In this way, some of the key trends in policy making associated with Khrushchev were already present in the late Stalinist era, as Rodionov's proposal illustrates. However, these proposals were not implemented until after Stalin's death in 1953. In this respect, Khrushchev's era is important in that it saw the realization of policies that remained only proposals under Stalin.

In the post-Stalin period the trend towards decentralization of responsibilities was gathering pace and RSFSR matters were high on the agenda. The RSFSR was the only union republic without its own communist party. It therefore lacked a party organ that could take care of the less important issues that in other union republics were dealt with by their CC apparatuses. Instead, the central party organs had to deal with minor RSRFR issues rather than concentrate on more important and general problems.

There were many dangers in establishing a specific Russian centre in the party organization. From Lenin's time the existence of a Russian communist party was deemed a serious threat to the power and authority of the central union party. Khrushchev acknowledged this at the February Plenum in 1956, where the creation of the new Bureau for the RSFSR was debated. Khrushchev proposed to combine the post of the First Secretary of the CPSU and the Chairman of the Bureau as a safeguard against the Bureau's excessive independence.[26] The routine running of the Bureau was entrusted to the deputy chair, Belyaev.

The creation of the Bureau and the appointment of Khrushchev as its nominal head increased his weight in the apparatus and were used to promote some of his allies to positions of greater responsibility. For example, Churaev, an old ally from Ukraine and the head of the RSFSR department for party organs, and Mylarshchikov, a Khrushchev associate from Moscow, the newly appointed head of the RSFSR agricultural department, were both made members of the Bureau. In addition, some of the key figures in Khrushchev's leadership, such as Frol Kozlov and Andrei Kirilenko, made their first entry into the top governing bodies through the membership of the Bureau for the RSFSR in 1956.[27]

The Bureau consisted of six departments, which employed 316 people, with 271 responsible workers and 45 technical staff.[28] These new departments were carved out from the old CC departments, which were now renamed departments for union republics to distinguish them from their RSFSR counterparts.[29] The union departments had responsibilities for the union republics minus the RSFSR.

The history of the creation of the RSFSR departments also testifies to the importance of the anti-bureaucracy drive in reforms of the apparatus at that time. For example, the department for industry and transport was to consist of 11 sectors with 89 staff. However, the Bureau insisted on simplification of its structure and reduction in the number of staff. The number of employees was reduced by 38 per cent to 55 employees.[30] Similar staff cuts were made to other RSFSR departments. Taken within the context of the ongoing campaign against bureaucratization and centralism, the persistent concern with excessive staff numbers at the CC apparatus shows that Khrushchev did not exclude the party apparatus from the anti-bureaucracy campaign, although the main criticism was aimed at ministries.

The creation of the Bureau also testifies to the importance of the decentralization drive in the organization of the party apparatus. This trend also included some delegation of responsibilities to local party organizations.[31] While not as drastic or far reaching as the abolition of ministries, which is rightly seen as the culmination of the anti-bureaucracy campaign, the party and the CC apparatus were also affected by this campaign.[32]

The CC apparatus in 1957

By the time of Khrushchev's triumph over the 'anti-party' group in the summer of 1957, the CC apparatus assumed a stable structure that would remain unchanged in all important aspects until 1962, when Khrushchev responded to economic difficulties with a drastic reorganization of the party. Understanding the apparatus organization in 1957 is important for gaining a perspective on overall tendencies under the early Khrushchev leadership, as the CC apparatus was exposed to his influence from an earlier time compared with state institutions dominated by his rivals.

There were 2,204 employees at the CC apparatus on 1 August 1957; 1,114 were responsible employees with executive powers and there were 1,090 technical staff.[33] The principal decision makers in the apparatus consisted of 26 heads of departments, 16 first deputy heads and 52 deputy heads. Other responsible workers comprised 109 heads of sectors, 36 inspectors, 721 instructors, lecturers, deputy heads of sectors, assistants to heads of departments, heads of secretariats, and 118 referents and responsible consultants.

The majority of the CC staff dealt with domestic affairs, with 686 responsible and 181 technical staff employed in 19 CC departments dealing with the union republics and the RSFSR (see Tables 4.1. and 4.2). Some of the union republics had a wider remit as, for example, the union departments for party organs included the sectors for single party cards and the sectors of record keeping for leading cadres, which kept records of all *nomenklatura* appointments.

The other departments employed 406 responsible and 907 technical staff, or 35 per cent of the total responsible and 82 per cent of technical employees at the CC. Four departments dealt with foreign affairs and Soviet personnel deployed overseas. The three biggest departments were the Commission of Party Control with 104 responsible and 82 technical workers, the general department with 112 responsible and 229 technical staff, and the Administration of Affairs with 36 responsible and 457 technical employees. The Commission was primarily responsible for appellations about party membership expulsions and therefore did not have a significant role in the apparatus. The administration of affairs dealt with sensitive matters of allocation of material resources within the apparatus and also supervised their distribution at the local party level, giving its head substantial influence in party affairs.[34]

The importance of the general department did not go unnoticed by scholars, with several works devoted to analysing its history and role in the apparatus.[35] The most recent and important analysis of the general department's history was made by Yoram Gorlizki in his paper on party revivalism. Gorlizki traces the earliest reference to it within the protocols of the Secretariat from 12 November 1952.[36] This probably referred to the general department of the Secretariat, rather than the general department of the CC, which was formally set up on 13 March 1953.[37]

The general department was an important innovation after the abolition of the Orgburo in 1952. It was a direct successor to the *Tekhsekretariat* (*Tekhnicheskii sekretariat* – technical secretariat), the main chancellery of the dissolved Orgburo, but with much broader functions. By 1957 the general department had seven sectors and, with 341 staff, 112 of whom were responsible workers and 229 technical staff, was one of the largest departments in the CC apparatus.[38] It was responsible for the clerical work and record keeping of the apparatus as well as supervision of secrecy procedures in the party organs.[39] Another important function of this department was organization of the Secretariat's poll voting.[40] It also had the important function of dealing with matters that did not pertain to the responsibility of any specific department in the apparatus. In such cases the matters were settled directly between the general department and the relevant

Table 4.1 The Central Committee departments for the union republics, 1 August 1957

Name of department	Heads	First deputy	Deputies	Heads of sectors	Inspectors	Other staff*	Total for responsible employees	Total for technical staff	Total
Party organs	1	1	2	9	4	47	64	68	132
Propaganda and agitation	1		2	5		34	42	7	49
Heavy industry	1		1	3		20	25	3	28
Machine building	1	1	1	3		20	26	6	32
Defence industry	1	1	1	3	4	16	26	6	32
Construction	1	1	3			23	28	6	34
Transport and communication	1		1	3		14	19	5	24
Consumer goods and foodstuff	1		1	3		15	20	4	24
Agriculture	1	1	2	6		29	39	7	36
Administrative organs	1	1	1	6		48	57	11	66
Trade, finance and planning organs	1		1	3		15	20	4	24
Science, higher education establishments and schools	1	1	1	3		26	32	6	38
Culture	1		1	3		12	17	3	20
Total	13	7	18	50	8	319	415	136	551

Source: RGANI, f. 5, op. 31, d. 70, l. 82

* Instructors, lectors, deputy heads of sectors, assistants to head of department, head of secretariat

Table 4.2 The Central Committee departments for the RSFSR, 1 August 1957

Name of department	Heads	First deputy	Deputies	Heads of sectors	Inspectors	Other staff*	Total for responsible employees	Total for technical staff	Total
Party organs	1	1	2	8	8	65	85	15	100
Propaganda and agitation	1		2	3		27	33	5	38
Industry and transport	1	1	2	6		39	49	6	55
Agriculture	1		3	5		38	47	8	55
Administrative, trade and financial organs	1		2	5		23	31	7	38
Science, schools and culture	1		2	1		22	26	4	30
Total	6	2	13	28	8	214	271	45	316

Source: RGANI, f. 5, op. 31, d. 70, l. 82

* Instructors, lectors, deputy heads of sectors, assistants to heads, head of secretariat

Table 4.3 Other departments of the Central Committee, 1 August 1957

Name of department	Heads	First deputies	Deputies	Heads of sectors	Inspectors	Other staff*	Referents, consultants	Total for responsible employees	Total for technical staff	Total
General department	1		3	8		30	37	112	229	341
Administration of affairs	1	1	2	4		28		36	457	493
Department of diplomatic and foreign trade cadres	1	1	1	1		20		24	7	31
Department for relations with communist parties and workers parties of socialist countries	1	1	1	7		5	30	45	37	82
International department for relations with communist parties of capitalist countries	1	1	2	11		6	51	72	53	125
Committee of Party Control	1	2	11		20	70		104	82	186
Commission for travelling abroad	1	1	1			32		35	44	79
Total	7	7	21	31	20	191	118	428	909	1337

Source: RGANI, f. 5, op. 31, d. 70, l. 83

* Instructors, lectors, deputy heads of sectors, assistants to head of department, head of secretariat

CC Secretary. The head of the department and his first deputy also attended all meetings of the Presidium and the Secretariat.

The scope of the general department's workload can be seen from the following figures. In 1962, the general department received more than one million documents and dispatched over 850,000 documents. In comparison, the administration of affairs received over 500,000 documents and dispatched over 100,000 documents.[41] The general department both dealt with the Presidium paperwork and served as a chancellery for other departments and the Secretariat, thus performing the important and politically sensitive function of directing the paper flow in the CC. Crucially it was also responsible for maintaining secrecy procedures in the party organs and state departments, which gave it powers to interfere in any area of party work, making it a key organ in the CC apparatus.

Excluding departments for which no comparable data are available for the immediate post-Stalin period, the size of the apparatus was reduced by 14 per cent from 1,490 in April 1953 to 1,281 in August 1957.[42] The departments most affected by the cuts were the departments of party organs, reduced from 298 workers to 232, propaganda and agitation which lost 49 of its 136 staff, the department of administrative and trade and financial organs, which lost 28 employees, while the departments that dealt with culture and education lost almost half of their staff, having their numbers reduced from 161 to 88. It follows that the departments that dealt with traditional party matters (party organization and propaganda) were affected by the cuts most of all.

In contrast, the departments that dealt with economic matters increased in size. For example, the old industrial-transport department had 219 employees, while the six departments that replaced it on 9 June 1954 and the RSFSR department for industry and transport created in March 1956 had between them 229 employees. Similarly, the old agriculture department had 94 staff in 1953, while the two agriculture departments in 1957 had 101 workers.

Contrary to revivalist rhetoric employed by Khrushchev, the CC apparatus was incrementally shifting its focus to the economic sphere in the early post-Stalin period, with more specialization across a greater number of departments devoted to economic management. However, important features of this period, such as the fight against bureaucracy, maintained the momentum behind Khrushchev's revivalist project. The tension between conflicting views of the party, therefore, persisted in the Khrushchev era. The need to boost the CC apparatus's role in governing institutions led to a strengthening of the production-branch departments, while revivalist calls led to simplification of the existing departments.

The *Sovnarkhoz* reform and the CC apparatus

On 10 May 1957 the management of the Soviet economy was radically reorganized along the territorial principle through the creation of *sovnarkhozy* (national economic councils). Ten all-union ministries and 15 union-republican ministries were replaced by 70 *sovnarkhozy*. At the same time the role of the central planning bodies and of the union republics was strengthened, with their heads given seats at the Union Council of Ministers.

It would seem logical that the CC apparatus would be also reorganized to reflect the changes in the Soviet government. Indeed, a detailed proposal was jointly drafted by the two departments for party organs.[43] The gist of the proposal was in cutting down the number of union departments responsible for branches of industry (six at the time), while increasing the number of the RSFSR economic departments. Ideological matters were to be concentrated in larger departments of propaganda and agitation. This was a revivalist proposal par excellence, calling for concentration of responsibilities in fewer departments with wider responsibilities.

Overall, the proposals envisaged staff reductions of 49 responsible and 11 technical workers, less than 5 per cent of those employed at the time, which is not such an impressive figure compared to substantial cuts in the government apparatus happening at this time. However, this was to come on top of significant staff reductions in the previous years. Moskatov, the chair of the Central Revision Commission, claimed that overall staff numbers at the CC apparatus were reduced by 24.7 per cent between 1952 and 1956.[44]

These changes were not accepted by the leadership in 1957. It is not possible to establish detailed arguments for the preservation of the CC structure intact, but several suggestions can be offered for this. First, it could be argued that the CC apparatus was already reformed along regional lines with the creation of the Bureau for the RSFSR in 1956. In this way, the reforms in the CC apparatus preceded reforms of the central government.

However, the second possibility should also be considered. The abolition of central ministries was a severe blow to the power of Khrushchev's rivals in the Presidium, most of whom held ministerial posts. By keeping the structure of the CC intact, Khrushchev was in fact strengthening his own position as the First Secretary. With ministerial authority diffused across regional *sovnarkhozy*, the role of the CC apparatus as the central managing agency was significantly enhanced.

In the long run, the importance of the party over state apparatus was reasserted, echoing Khrushchev's rhetoric of party revivalism. The strengthening of party organs vis-à-vis the state was one of the principal pillars of Khrushchev's ideology, which was fully expressed at the peak of his career at the XXII Party Congress in October 1961. The imminent withering away of the state was announced and the importance of non-government agencies re-emphasized. Provisions for greater party democracy were envisaged, including the regular rotation of cadres. It was hoped that this would release the party's mobilizing potential and lead to the rapid advancement to communism. In this ideological matrix, the party was to assume the main governing function assisted by the soviet organs, the trades unions and various voluntary organizations.[45]

The bifurcation of the party apparatus

The renewed economic difficulties faced by the Soviet leadership in 1962 were particularly damaging against the backdrop of the new Party Programme adopted at the XXII Party Congress, which promised substantially increased standards of living and the imminent arrival of communism within 20 years. Khrushchev's response was to embark on the most radical reform of the party apparatus yet. However, this reorganization signalled a long-term defeat of Khrushchev's intention of reviving the party's special role in the economy and society, as it explicitly acknowledged that the main task of the party was management of the economy, with its other functions assuming a secondary role.

In fact, there were several reforms of the apparatus in 1962, which indicates that there was no clear plan for reforms. At the CC March Plenum it was decided to strengthen the party's control over agriculture by creating committees for agriculture at the local level headed by first secretaries of party organizations.[46] A major reorganization of the Department of Agriculture for the Union Republics followed in April 1962. More emphasis was placed on inspection and closer control of agriculture. In the summer of 1962 the sectors in the department of agriculture for RSFSR were replaced by inspector groups with a stronger territorial focus, with nine out of 13 inspector groups having regional specialization.

In June 1962 the reforms were extended to other economic departments when the department of industry and transport for the RSFSR was split into four separate departments.[47] This echoed some of the 1957 proposals for reorganization of the CC apparatus discussed above. Other departments were also reformed, such as the department of party organs for the RSFSR, which had the number of its regional sectors doubled. This shows that, faced with a crisis in economy, the leadership was abandoning its revivalist approach in the summer of 1962 by introducing more specialization in economic management, while at the same time encouraging a greater regional focus within the apparatus.

After the November Plenum in 1962, the CC apparatus was restructured on a new principle of the division between agriculture and industry, which affected nearly all departments, including non-economic ones. At the most fundamental level the bifurcation of party organs was intended to provide more concrete and detailed party management of industry, construction and agriculture. Khrushchev argued that

> the unification of communists according to the place of their economic activity gives party organizations the ability to concentrate their main attention on economic questions, subjecting all other forms of work – organizational, ideological, culture and educational – to solving the principal task.[48]

The principal task of the party organization, therefore, became economic management, while other areas of party activity were now deemed of secondary importance. Accordingly, the number of economic departments in the CC apparatus was nearly doubled.[49]

The leadership's need to strengthen its grip on ideological matters and to offset the excessive tilt towards economic specialization led to the establishment of a new ideology department through the merger of four union departments responsible for propaganda, culture and education.[50] The two corresponding RSFSR departments were replaced by the ideological departments for industry and agriculture. However, this was a minor concession to revivalist ideas, which were clearly undermined by the 1962 reform of the party apparatus. At the key event that was intended to put ideological matters at the forefront of party's work, the June Plenum on ideology in 1963, Khrushchev himself undermined the importance of ideology. After reading a long passage of ideology, a visibly tired Khrushchev sighed and said, 'now let's talk about real business' (*nu, a teper' pogovorim o dele*), leaving the impression that ideology was no longer 'real business' for him.[51]

The 1962 reorganization was the most radical reform of the Khrushchev era and, perhaps, in the history of the CC apparatus. Khrushchev rejected two traditional approaches to organization of the apparatus, the functional and the industry-branch. By splitting the party into agricultural and industrial branches Khrushchev unambiguously showed that he saw the party as primarily an agency responsible for boosting economic production.

The 1962 reorganization can be seen, therefore, as the final defeat of party revivalism. The overall increase in the size and number of departments by the end of Khrushchev's era reflected their greater involvement with micromanagement of the government business. This, together with a substantial increase in the bureaucratic functions of the CC apparatus, signalled the defeat of Khrushchev's anti-bureaucratic reforms and the revivalist drive that defined his early years in power.[52] In a twist of irony, Malenkov's old principles of party functions triumphed over Khrushchev and Zhdanov's hopes for the special role of the CPSU in Soviet society.

Although the bifurcation reform was reversed immediately after Khrushchev's fall, the party's primary focus on economic management remained. The new generation of party bosses, who came to the fore during Khrushchev's time in office and consolidated their position under Brezhnev, mainly had an economic or technical background. They relied more heavily on consultants and demanded from ideology practical solutions to concrete problems.[53] This preoccupation with economic functions was reflected in the structure of the CC apparatus. Out of 22 departments left after the reorganization of 1965, which drew a line under Khrushchev's reforms, almost half had an economic focus.[54]

The party was becoming increasingly indistinguishable from the state organs, whose organization and functions it mirrored and supervised. This proved to be one of the lasting legacies of the Khrushchev period for the CPSU and its role in the Soviet system. Specifically, the loss of the party's mobilizing force and increase in its bureaucratic function signalled the entrenchment of negative features associated with the Brezhnev era stagnation. The inertia and ossification of governing institutions that characterized the Soviet Union in the late 1960s and

1970s were the consequence of Khrushchev's failure to solve the conflict between the party's role as a guiding force in society and the party as a principal agent in the management of the economy and government institutions.

Conclusion

Three themes dominated reforms of the CC in the Khrushchev period: the decentralization seen in the introduction of the territorial principle, the anti-bureaucracy campaigns that led to continuous staff cuts and simplification of departmental structures, which together amounted to the third and more general trend of the revival of the party's importance within the Soviet system of government. The tension was always present in the CC apparatus between revivalist claims of the party's special role in society and the pragmatic need to increase party supervision of industry and agriculture. The former led to reductions in staff numbers and simplification of the apparatus's structure; the latter was manifested in a relative increase in the number of CC staff dealing with economic issues. This conflict was finally resolved after 1962, when the party's economic functions were prioritized over its other responsibilities.

The bifurcation of the party organs into industrial and agricultural branches signalled the defeat of revivalism as a factor in party organization. It also undermined Khrushchev's claim to leadership, until then built around revivalist rhetoric. His subsequent removal from power saw a return of organizational patterns familiar from the pre-1956 period centred on the industry-branch approach and parallelism between party and state organs. The legacy of Khrushchev's reforms was, however, the increased weight of the party apparatus vis-à-vis state organs, but on terms that made the party apparatus the most powerful bureaucratic organ in the Soviet system, an outcome Khrushchev hardly intended.

Notes

1 See for example R.G. Pikhoia, *Sovetskii Soiuz: Istoriia vlasti, 1945–1991*, Moscow: RAGS, 1998, pp. 101, 182; V.P. Naumov 'Bor'ba Khrushcheva za edinolichnuyu vlast'', *Novaya i noveishaya istoriya*, No. 2, 1996, pp. 10–31.
2 I wish to express my gratitude to Yoram Gorlizki for his comments on an earlier draft of this chapter, his advice on bibliography and for sharing his research notes from the Secretariat files, which are no longer accessible in Russian archives.
3 My account of the inter-war period is based on two classical works by M. Fainsod, *How Russia is Ruled,* Oxford: Oxford University Press, 1963, and L. Schapiro, *The Communist Party of the Soviet Union* 2nd edn, London: Methuen, 1970.
4 For the complex nature of the relationship between the Secretariat and local *nomenklatura* appointments in this period see J. Harris, 'Stalin as General Secretary: The Appointments Process and the Nature of Stalin's Power', in S. Davies and J. Harris (eds), *Stalin: A New History,* Cambridge: Cambridge University Press, 2005, pp. 63–82.
5 Fainsod, *How Russia is Ruled*, p. 192.
6 Fainsod, *How Russia is Ruled*, p. 194.
7 Schapiro, *The Communist Party*, p. 455.

8   J. Harris, 'The Origins of the Conflict between Malenkov and Zhdanov: 1939–1941', *Slavic Review*, 35, 2, 1976, pp. 287–303.
9   Jonathan Harris, 'The Origins of the Conflict Between Malenkov and Zhdanov, 1939–1941', *Slavic Review* 2 (1976), 287–303.
10 A.A. Danilov and V.A. Pyzhikov, *Rozhdenie sverkhderzhavy: SSSR v pervye poslevoennye gody*, Moscow: ROSSPEN, 2001, p. 233.
11 RGASPI, f. 17, op. 116, d. 365, ll. 9–43. The new departments were the department of propaganda and agitation, 300 employees; party, trades union and Komsomol organs, 300; external relations, 150; heavy industry, 80; light industry, 60; machine building, 70; transport, 50; agriculture, 80; administrative department, 80; planning, finance and trade, with 43 employees. These were provisional figures only as not all positions in the departments were necessarily filled. I am grateful to Yoram Gorlizki for drawing my attention to this fact. However, it is still reasonable to accept these figures as a reliable indicator of the size of the apparatus and its departments.
12 Yoram Gorlizki and Oleg Khlevniuk, *Cold Peace: Stalin and the Soviet Ruling Circle, 1945–1953*, Oxford: Oxford University Press, 2004, p. 194, n. 77.
13 The administrative department included the sector of the land forces, the sector of the air forces, the navy sector, the sector for the Ministry for State Security (MGB), the sector for the Interior Ministry (MVD), the sector of justice and state control and the sector of public health.
14 For example, the sectors of higher educational establishments and of cinematography and radio broadcasting in the propaganda and agitation department supervised respectively the ministries of Higher Education and Cinematography. Similarly, the party organs department had a sector responsible for the Komsomol and trades unions.
15 Yoram Gorlizki, 'Party Revivalism and the Death of Stalin', *Slavic Review*, 54, 1, 1995, pp. 1–22.
16 For an analysis of Khrushchev's use of anti-ministerial rhetoric against his rivals see Y. Gorlizki, 'Anti-ministerialism and the USSR Ministry of Justice, 1953–56: A Study in Organisational Decline', *Europe-Asia Studies*, 48, 8, 1996, pp. 1279–318
17 Gorlizki and Khlevniuk, *Cold Peace*, p. 153.
18 Gorlizki, 'Party Revivalism', pp. 15–16.
19 These were sectors of regions of the North and the Centre; of the Black earth belt and South; of Siberia and the Far East; Baltic republics, Belorussia and Karelo-Finnish SSR; Central Asian republics and Kazakhstan. The other sectors were the sector of propaganda, central newspapers and radio broadcasting, and the sector of publishing and journals. At each of these reorganizations the department's staff numbers were also reduced.
20 The six new departments were heavy industry, machine building, defence industry, construction, transport and communications, consumer goods and foodstuff.
21 For the impact of the anti-bureaucracy campaign and the role it played in the struggle for power between Khrushchev and his rivals see Gorlizki, 'Anti-ministerialism', pp. 1279–318.
22 T. Martin, *The Affirmative Action Empire: Nations and Nationalism in the Soviet Union, 1923–1939*, Ithaca and London: Cornell University Press, 2001, pp. 400–1.
23 G. Hosking, *Rulers and Victims. The Russians in the Soviet Union*, Cambridge, MA, and London: Belknap Press, 2006, p. 254.
24 RGASPI, f. 82, op. 2, d. 106, l. 2.
25 For recent accounts of the Leningrad affair see ch. 3 in Gorlizki and Khlevniuk, *Cold Peace*, pp. 69–96; and Hosking, Rulers and Victims, pp. 190–5.
26 The fact that Khrushchev was the driving force behind the creation of the RSFSR Bureau and was largely responsible for choosing its members becomes evident from his report at the February 1956 CC Plenum, which approved its creation. RGANI, f. 2, op. 1, d. 187, ll. 8–14.

27 Other members of the Bureau in 1956 were M.A. Yasnov, the chairman of the RSFSR Council of Ministers; I.V. Kapitonov, the first secretary of the Moscow obkom; F.R. Kozlov, the first secretary of the Leningrad obkom; Ignatov, the secretary of the Gorky obkom; A.M. Puzanov, Yasnov's first deputy; A.P. Kirilenko, then first secretary of the Sverdlovsk obkom; A.M. Aristov and P.N. Pospelov. Many of these newly promoted people were Khrushchev supporters and later would rise to great prominence; for example, Frol Kozlov would become the second figure in the party by the early 1960s, while Kirilenko and Ignatov would become members of the Presidium in 1957–61.

28 The six departments were: party organs; propaganda and agitation; industry and transport; agriculture; administrative, trade and financial organs; science, schools and culture.

29 A more detailed analysis of the routine work of the RSFSR Bureau is given in A.V. Pyzhikov, *Khruchshevskaya 'ottepel'', 1953–1964*, Moscow: Olma Press, 2002, pp. 103–14.

30 RGANI, f. 13, op. 1, d. 452, ll. 1–4, Materials for the Bureau proceeding No. 4 from 16 April 1956.

31 See Fainsod, *How Russia is Ruled*, pp. 201–3.

32 Yoram Gorlizki, 'Anti-ministerialism', pp. 1279–318.

33 Responsible workers (*otvetstvennye rabotniki*) were employees of the CC apparatus with executive powers, i.e. they had authority to make decisions within their sphere of responsibility that were necessary for implementation by their subordinates in the CC itself or in the bodies supervised by the CC. Technical workers (*tekhnicheskie rabotniki*) were support staff without any input into decision making. These included typists, chauffeurs, couriers, stenographers, archivists and secretaries. This is the reason for the large size of the general department and the administration of affairs, which by the nature of their work had large contingents of technical staff, e.g. the general department employed many clerical staff such as stenographers, typists and couriers, while the administration of affairs had among its staff a large number of service personnel such as chauffeurs, cooks and cleaners.

34 The Administration of Affairs was run by D.V. Krupin in 1938–59, followed by V.V. Pivovarov in 1960–62 and K.P. Chernyaev in 1963–65. See www.knowbysight.info for a full list of the heads of the CC departments.

35 See for example L. Schapiro, 'The General Department of the CC of the CPSU', *Survey*, 21, 3, 1975, pp. 53–65. There is also some discussion of the department's origins in Danilov, Pyzhikov, *Rozhdenie sverkhderzhavy*, p. 299. V.N. Malin made his party career in Belorussia and, after a brief sojourn in Leningrad, was promoted in 1952 to the post of a CC inspector before assuming the key role of the general department's head in 1954–65.

36 Gorlizki, 'Party Revivalism', p. 12.

37 Private correspondence with Yoram Gorlizki, March 2008.

38 The general department had the following structure: the first sector was responsible for the protocol and record keeping of the current Presidium's paperwork; the second sector dealt with the protocols and record keeping of the Secretariat's paperwork; the third sector was responsible for record keeping, supervision and coordination of paperwork flow in the apparatus; the fourth sector was responsible for cryptography of communications; the fifth sector handled the archive of the Comintern; the sixth sector managed the Presidium archive; the seventh sector was responsible for inspection of secret documents handling in local party organs and training of party staff in work with secret documents. In addition there was a sector of correspondence, which dealt with all general correspondence received by the CC, and the general office which carried out general clerical work such as copying of documents, typing and stenographic note taking. RGANI, f. 5, op. 31, d. 70, l. 79.

39 The importance of this side of the general department's work, albeit for a later period, is discussed in L. Onikov, *KPSS: anatomiya raspada. Vzglyad iznutri apparata TsK*, Moscow: Respublika, 1996, pp. 34–7, 77–80, 177–203.
40 The system of poll voting is discussed in M. Voslenskii, *Nomenklatura. Gospodstvuiushchii klass Sovetskogo Soiuza*, London: Overseas Publications Interchange Ltd, 1984, p. 374.
41 RGANI, f. 5, op. 26, d. 74, ll. 60–2.
42 The personnel numbers of the CC departments for April 1953 were published by Y. Gorlizki in 'Party Revivalism', p. 20. However, as acknowledged by the author, that list did not include all departments; specifically no data were available for the general department, the administration of affairs, the commission for travelling abroad, and the department for selection and assignment of cadres in party, public and state organs. The department for selection and assignment of cadres was abolished on 18 October 1955, its responsibilities transferred to the newly created sector of record-keeping of leading cadres in the department of party organs for the union republics.
43 RGANI, f. 5, op. 31, d. 70, ll. 60–80.
44 Report of the Central Revision Commission to the XX Congress of the CPSU, RGANI, f. 1, op. 2, d. 6, l. 11. The earlier figure of 14 per cent (p. 15, n. 42) refers to staff reductions in a limited number of CC departments for which detailed figures are available.
45 See A. Titov's chapter on the 1961 Party Programme in M. Ilic and J. Smith (eds), *Soviet State and Society under Nikita Khrushchev*, London: Routledge, 2009.
46 *KPSS v rezolutsiyakh i resheniyakh s'ezdov, koferentsii i plenumov TsK*, Moscow: Politizdat, 1986, 10, p. 225.
47 The new departments were the department of heavy industry, transport and communications for RSFSR; department of construction for RSFSR; department of machine building for RSFSR; department of light and food industry for RSFSR.
48 Khrushchev's report to the November Plenum, 1962. RGANI, f. 2, op. 1, d. 596, l. 33.
49 This was particularly the case for the RSFSR departments, whose number grew to 11:

heavy industry, communications and transport for the RSFSR
machine building for RSFSR
construction for RSFSR
light industry, food industry and trade for RSFSR
agriculture for RSFSR
processing industry of agricultural raw products and trade for RSFSR
ideological department for RSFSR industry
ideological department for RSFSR agriculture
party organs for RSFSR industry
party organs for RSFSR agriculture
administrative organs for RSFSR.

The union departments saw the merger of non-economic departments into the ideological department, while the number of economic departments was expanded, leaving the total at 13 departments:

heavy industry
chemical industry
machine building
defence industry
construction
transport and communications
light industry, food industry and trade
agriculture
agricultural department for scientific matters

processing industry of agricultural raw products
administrative organs
party organs
ideological department.

The rest of the apparatus remained the same, with the exception of a new department for economic cooperation with socialist countries:

international department
economic cooperation with socialist countries
diplomatic and foreign trade cadres
general department
administration of affairs.

The regional structures of the CC apparatus were also re-enforced by the creation of the Central Asian and Transcaucasian Bureaus, while two more Bureaus for the RSFSR were also added, for industry and agriculture respectively. RGANI, f. 2, op. 1, d. 668, ll. 5–21.

50 The abolished departments were propaganda and agitation for the Union Republics; science; higher education establishments and schools; and the department of culture.
51 G.L. Smirnov, 'Malen'kie sekrety bol'shogo doma. Vospominaniya o rabote v apparate TsK KPSS', in V.A. Kozlov (ed.), *Neizvestnaya Rossiia. XX vek*, vol. 3, Moscow: Istoricheskoe nasledie, 1993, pp. 361–82, 381.
52 Despite constant efforts at streamlining the CC apparatus, its staff numbers grew significantly by the end of Khrushchev period. For example, while in 1958 there were 767 clerical staff in the CC apparatus, by 1963 this number grew to 855, an 11 per cent rise over five years. And yet there were complaints that the CC apparatus could not handle the amount of paperwork it had, with its typing offices in the first five months of 1963 having to work 3,850 hours' overtime. RGANI, f. 5, op. 30, d. 408, l. 56.
53 Smirnov, 'Malen'kie sekrety bol'shogo doma,' pp. 381–2.
54 The economic block comprised nine departments including Heavy Industry, Machine Building, Chemical Industry, Light and Food Industry, Construction, Transport and Communications, Agriculture, Trade and Consumer Services, Planning and Financial Organs, which is very similar to the 1948 CC apparatus (see note 10 above).

5 The outer reaches of liberalization

Combating political dissent in the Khrushchev era

Robert Hornsby

Elena Papovyan was probably correct to suggest that for those who are not normally interested in the subject, the theme of political repression during the Khrushchev era can be quite surprising.[1] For example, despite its reputation as a period of liberalization and 'thaw', over twice as many people were jailed for 'anti-Soviet activity and propaganda' in the Khrushchev period than during the considerably longer Brezhnev era. Although never approaching the scale of state repression that took place during the Stalin years, the regime continued to meet its critics with a firm response.

This was a particularly dynamic field of policy in the Khrushchev era as the regime sought to respond to the challenges of dissenting behaviour without reliance upon the tried and tested method of mass terror. The early Khrushchev years witnessed what could be termed a 'firefighting' approach to policing dissent as the occasionally panicked and uncertain regime struggled to adapt to the post-Stalin environment. From the end of the 1950s, however, the situation began to settle down somewhat, and a more sophisticated and integrated series of measures were put in place to control protest and criticism. In fact, the measures that had been put in place to combat dissent by the end of the Khrushchev period went on to remain at the heart of the regime's efforts to tackle the problem throughout almost the entire post-Stalin era.

The most important point to proceed from is a clarification of the kinds of dissenting behaviour on which this study is based. Although many of the policies and attitudes raised here are also applicable to the way that the regime policed nationalist and religious dissent, these themes fall outside the main focus of the present work. Instead, the subject of this chapter is the way in which the authorities sought to combat 'political dissent'. Essentially, this meant acts of dissent that can broadly be defined as 'protest and criticism involving language and behaviours that either reflected or implied discontent at the policies, representatives and goals of the Soviet regime'. Political dissent involved a range of activities such as preparing and distributing anti-Soviet leaflets, calling for strikes and uprisings, publicly criticizing leaders of the Party or forming underground political groups. For a more substantial definition of what constituted political dissent, see my article 'Voicing Discontent: Political Dissent from the Secret Speech to Khrushchev's Ouster'.[2]

There were still some areas and ways in which it was possible to show disapproval legitimately. What the authorities did consider to be within the boundaries of acceptable criticism essentially consisted of 'businesslike' debate, through the correct channels, on proposed policies and complaints on specific abuses and shortcomings by individuals at lower levels of the political spectrum. Policies already in force and the political and social order were the most explicitly forbidden themes of criticism.[3] What these subjects implied to the regime was the presence of deep-seated hostility towards the most fundamental principles of Soviet rule: a not entirely unfounded, though undoubtedly extreme, viewpoint. Added to this trio of the most taboo domestic subjects was that of 'the West'. Since the authorities saw citizens' attitudes towards the West as a key indicator of their political loyalty to the Soviet regime, it naturally followed that any kind of dissenting behaviour involving the West, such as negative comparisons between Soviet and US living standards or attempts to communicate with Western organizations, instantly made any perceived transgression more grave.[4] The sphere of acceptable criticism had in fact expanded very little since the Stalin years; the main difference lay in the manner of response that the authorities deemed appropriate.

This intolerance towards criticism that presented the West in a positive light reflected the regime's long-standing belief that dissent inside the USSR was a result of foreign subversion. By effectively equating the activities of its critics with treason, the regime consistently overestimated the threat posed by dissent in general and failed to address the real causes of discontent. It should be pointed out, however, that this was not a completely irrational or paranoid assumption. For example, the CIA was covertly funding and supervising Radio Liberty, the White Émigré group NTS was sending hot-air balloons full of hostile propaganda on to Soviet territory and, by the 1960s, the Chinese regime too was sending masses of anti-Soviet propaganda into the USSR.

The aftermath of the secret speech

Surprisingly, no concrete plans were established before or immediately after the Secret Speech in regard to how to police critical responses to Khrushchev's report. In all likelihood this lack of pre-planning was a result of the short timescale that existed between the decision for Khrushchev to deliver Pospelov's report on Stalin and the actual event taking place. Nonetheless, it also demonstrated one of the central characteristics of policy against dissent in the early Khrushchev years: the authorities were rarely proactive in seeking to forestall outbursts of criticism and protest, and were therefore forced to respond to them afterwards.

With at least a few of its members already in a state of some trepidation over the potential consequences of the Secret Speech, it was entirely unsurprising that the Central Committee Presidium took an active interest in monitoring the lists of comments and questions that came back from meetings held to discuss the Secret Speech around the country.[5] It is interesting to note that when the centre received reports of dissenting remarks that were made at meetings held to discuss the Secret Speech, the response was, in Soviet terms, rather measured. This largely vindicates Papovyan's assertion that 1956 was characterized by 'an unusual liberalism in the punitive organs'.[6] This was true, yet political repression had already actually been dropping markedly since Stalin's death. This restraint was, therefore, reflective of a trend that had already been in place for some time and did not indicate any kind of new-found tolerance of political criticism.

At the lower levels of authority, various factors dictated these restrained responses to the spate of critical remarks that were heard at Party and *Komsomol* meetings in particular. First, according to Polly Jones, Party activists and officials were themselves usually too deeply embroiled in criticism and self-criticism to act decisively and promptly against dissenting remarks.[7] Second, the atmosphere of uncertainty that had facilitated many dissenting outbursts also meant that local Party organizations were often unsure as to what constituted acceptable and unacceptable criticism in the wake of Khrushchev's revelations. Furthermore, this was clearly not a time when one would want to be seen as an overzealous Stalinist.

Summaries of discussions that took place in all union republics were compiled by the Department of Party Organs and sent to the Central Committee for consideration.[8] According to Erik Kulavig, these reports from local Party branches quickly convinced the top leadership that even the limited liberalization that had already taken place had shaken the foundations of the system.[9] This seems like an exaggeration, however, as later events would show that when the authorities perceived a potential threat to the regime's stability they were far quicker to act decisively than was the case at this point in time. Nonetheless, it is already clear that the authorities were taking an active interest in monitoring the general public mood and were taking it into account.

With the majority of dissenting remarks being made in Party and *Komsomol* meetings, it was within those organizations that the first waves of what could be termed 'normalization' were to be carried out. One of the first attempts by the authorities to silence critics was a letter sent out to Party organizations on 5 April 1956 in response to critical speeches that were delivered at a meeting of the Party cell attached to Moscow's Thermo-Technical Institute, entitled 'On the harmful attacks at the meeting of the Thermo-Technical laboratory of the USSR Academy of Sciences Party organisation'. It stated that although the majority of discussion meetings had passed in the desired manner, 'the Central Committee has noted that there have been individual cases of harmful speeches by anti-Party elements that have tried to employ criticism and self-criticism for their own aims'.[10] That same day, the quartet – which included Yuri Orlov, later to be the founder of the Soviet branch of Amnesty International and of the Moscow Helsinki Group – were accused by *Izvestiya* of 'singing with the voices of Mensheviks and SRs' – parties that had been dead in the USSR for many years.[11]

On 30 June the Central Committee issued a further decree: 'On the overcoming of the Cult of Personality and its consequences'. According to Susanne Schattenberg this was intended as a further threat to those who persisted in their 'exaggerated' criticism.[12] As Gennadyi Kuzovkin has stated, the purpose of these documents was to demonstrate the new limits of acceptable criticism and discussion, to judge proper from improper in the new environment, and to establish that punishment would follow any transgression of these boundaries.[13]

The aim, therefore, was not to uncover those of differing opinions, as had been the case in the Stalin era, or to persuade them of the rightness of the Party line, but to enforce silence upon them. This was not quite the return to Leninism that Khrushchev had promised, but it was clearly a major divergence from Stalinism. The tactic was, to a large extent, successful. 'Unhealthy elements' were reprimanded or expelled from the Party and *Komsomol* and whole Party cells disbanded on occasion, as happened at the Thermo-Technical Institute. One ought not to underestimate the harmful impact that these measures could have on the life of a Soviet citizen, yet they would clearly have been preferable to a period in one of the remaining labour camps. Party members did continue to be involved in various forms of underground dissent but by the end of the year there were few traces of open criticism within the ranks of the CPSU.

Young people in particular became the subject of the regime's attention throughout 1956. Along with released prisoners and members of the artistic intelligentsia, the young generation was seen as being among the most likely sources of potential unrest. The authorities spoke of problems such as insufficient respect for the value of labour and disconnection from the Soviet revolutionary heritage, often looking to the children of the burgeoning middle classes as the root of the problem.[14] A December 1956 report from *Komsomol* secretary Shelepin to the CPSU Central Committee suggested that within the *Komsomol* at least there was some trace of Khrushchev's putative 'return to Leninism'. Emphasis was placed on educating those who were 'misguided', rather than taking punitive measures against them. Noting that a mood of pessimism and 'apoliticism' had emerged among young people since the XX Party Congress, he wrote that political-ideological work was being strengthened across all union republics. Agitators, teachers, workers and veterans of the revolution were being dispatched to workplaces, student dormitories and classrooms to explain the Party line and to provide a rebuff to bourgeois propaganda.[15]

The campaign against dissent

The letters sent out to Party organizations appear to have stemmed the tide of open dissent within the CPSU and *Komsomol*, yet a considerable volume of protest and criticism remained in wider society. Restoring discipline among the Communist Party and the *Komsomol* was clearly not going to be enough. By the end of 1956 the 'remarkable liberality' of that year, which Papovyan referred to, was replaced by a major clampdown on dissent.

The Secret Speech undoubtedly played a major role in bringing long-suppressed tensions to the fore, yet the immediate impetus for this change in direction was the uprising in Hungary. Mark Kramer has argued that, already well aware of ferment within the USSR, the regime feared events in Hungary could spill over into neighbouring countries and ultimately into the Soviet Union, potentially unravelling the entire socialist bloc.[16] The authorities' judgement in this instance was probably sound. Of course, one cannot say for sure that the region would have descended into chaos if the Soviet leadership had not taken matters in hand, yet the way that events in one East European satellite impacted on the next as the system began to collapse at the end of the 1980s gives some support to this assessment of events.

The Soviet Union had been able to maintain its Hungarian satellite regime by force of arms but no power would be able to prop up the Soviet regime if a similar situation were to arise inside the USSR: a fact of which all of the leadership must have been painfully aware. As Fursenko and Naftali have pointed out, 'The Hungarian effect could also be seen in the Kremlin's hardening attitude toward political dissent at home.'[17] The form that this 'hardening' took was a campaign of legal repression against dissent, initiated by a secret Central Committee letter that was sent out to all Party organizations on 19 December 1956, entitled 'On the strengthening of the political work of Party organizations in the masses and the suppression of attacks by anti-Soviet enemy elements'.[18] As the KGB's own internal history textbook, written in 1977, stated: 'the December letter began a merciless (*besposhadnyi*) campaign against anti-Soviet elements'.[19]

The letter asserted that the 'harmful atmosphere' in the USSR was a product of events taking place elsewhere, particularly in Hungary, where the imperialist powers had increased their efforts to undermine the socialist camp.[20] The letter went on to upbraid local Party organizations for allowing critical remarks to go without a decisive and timely response, and stated that bourgeois elements were attempting to 'hijack' the struggle with the Cult of Personality for their own ends. The real crux of the letter's message could be found in statements such as 'each and every communist must play their part in fighting for the Party line and defending its interests' and 'in the struggle against anti-Soviet elements we must be strong and unrelenting'.[21]

Although the Central Committee Presidium set the tone for the forthcoming clampdown in the December letter, responsibility for its implementation was left to lower-level officials. They correctly perceived that they had not only been shown the green light to take measures against critics but that there would be negative consequences for themselves if they did not. As Boris Firsov has written, 'the call was heard. All the links of the Party and state apparatus began to move and to reply, just like in the old days.'[22] The campaign that followed was not conducted on anything like the scale of the Great Terror of the late 1930s, yet neither did it resemble the kind of 'thaw' that has so often been used to characterize the Khrushchev years.

Table 5.1 shows the number of individuals sentenced for anti-Soviet activity during the Khrushchev period. The surge of convictions in 1957 and 1958 is particularly noticeable. The right-hand column has been added to give a sense of the chronological spread of sentences.

As Fedor Burlatsky wrote in his memoir of the era:

> Later I learned that under Khrushchev many hundreds of people had suffered for so-called political crimes, that is, for voicing disagreement with his

Table 5.1 Individuals sentenced for anti-Soviet activity, 1956–1964

Year	Total sentenced	Proportion of all political sentences during the Khrushchev period
1956	384	6.7%
1957	1,964	34.3%
1958	1,416	24.7%
1959	750	13.1%
1960	162	2.8%
1961	207	3.6%
1962	323	5.6%
1963	341	6.0%
1964	181	3.2%
Total	5,728	100.0%

Source: *Istochnik*, 1995, No. 6, p. 153

policies. Brezhnev developed this practice on a massive scale and with even greater deceit, but it must be acknowledged that it began under Khrushchev.[23]

In fact, the practice of imprisoning dissenters in this way not only began under Khrushchev but was actually more prevalent during his time as First Secretary – a fact that would probably come as a surprise to most people.

At no other time in the entire post-Secret Speech era were a comparable amount of citizens arrested and sentenced for dissenting activity. Although many times lower than the number sentenced for anti-Soviet activity and propaganda in the Stalin years, the total of 1,964 sentences in 1957 alone by far outstripped that of any subsequent year. However, one must exercise some caution in making sweeping comparisons between the two periods by reference to this measure alone because there was a general move away from such large-scale custodial sentencing in later years.

By providing only a rudimentary outline of what constituted anti-Soviet behaviour, the ensuing campaign saw a considerable degree of unpredictability and arbitrariness return to the Soviet repressive apparatus. The fact that provincial officials conducting the clampdown were generally poorly trained and eager to appear vigilant was a cause of considerable unpredictability and one of the reasons why the December Letter spawned a full-blown campaign.[24] The result was that *oblast'* procurators and KGB branches frequently erred on the side of caution and employed article 58-10 in a wide-ranging and often unsuitable fashion.

Most notable was the sheer volume of those sentenced as a result of apparently isolated, and often drunken, outbursts: over 50 per cent of all convictions. These tended to be instances of citizens drunkenly calling members of the militia 'fascists' or shouting slogans such as 'long live Eisenhower'. There were also a considerable number of cases where the KGB uncovered individuals who had been producing and distributing anti-Soviet leaflets or were involved in underground group activity yet there was practically no distinction made between these two kinds of dissenting behaviour. The fact that the authorities failed to distinguish between isolated drunken outbursts and more persistent, purposive acts of dissent shows that they were still looking at dissent in an overly simplistic way.

It is interesting to note that this campaign progressed throughout 1957 without any kind of noticeable interruptions. First, it suggests that this was not something that had simply been forced upon Khrushchev by hard-line Stalinists in the leadership, since the most senior among them were removed after the anti-Party affair in June 1957 while the campaign continued for another year afterwards. This may offer support for Carl Linden's hypothesis that Khrushchev's victory over the anti-Party group had been achieved after he agreed to shelve, or at least to slow down, the process of destalinization.[25]

The campaign initiated by the December letter lasted until the middle of 1958. By that time over 3,000 individuals had been jailed for anti-Soviet activity since the letter had been circulated. This was a total far lower than for any 18-month period under Stalin, yet also far higher than any comparable period under Brezhnev. This did not necessarily show that the Khrushchev era was more conservative or repressive than that of Brezhnev, but that social control was still at an earlier stage in its evolution away from Stalinism, when more crude forms of response such as labour camps and prisons had been staples of repressive activity. Moshe Lewin's assertion that 'historians ought to view the changes in repressive policy [under Khrushchev] not as a new stage but as a step toward another stage' clearly offers some support for this suggestion.[26]

By the middle of 1958 the main stimulus for the campaign against dissent – the threat of instability prompted by events in Hungary – had all but evaporated. The cause of the original crackdown may have disappeared but it still required some kind of catalyst for the campaign to be wound down. The Supreme Court had, in fact, been expressing reservations about the legal basis of the campaign for some months prior to the summer of 1958 but with little success. One particular document in which doubts were raised in respect of several cases, and calls were made for sentences to be reviewed, was a Supreme Court review entitled 'Information on the results of legal practice in cases of counter-revolutionary crimes'.[27] Compiled in early 1958 and drawing upon numerous cases from late 1956 and early 1957, the report essentially argued that too many of those who were being sentenced under article 58-10 should not have been branded 'anti-Soviet' but dealt with in some way that was more applicable.

This was then followed by a further review on the application of article 58-10 by the Procurator's office that was presumably sanctioned by the very highest political authorities. The Procurator review began by presenting detailed information on the numbers and social composition of those who had been sentenced. It went on to cite numerous individual cases of citizens arrested and jailed for anti-Soviet agitation and propaganda since Stalin's death, and particularly since the December letter. It pointed to the uprising in Hungary and the unmasking of Stalin as the two main catalysts for raised levels of dissenting activity, and stated that the increased number of convictions showed that the KGB and Procurator's office had been effective and vigilant in following the new guidelines set out in December 1956.[28]

However, after the 'sugar coating' that was traditional at the beginning of such reports, it then painted a more complete picture. In its concluding remarks the review stated that 'the [security] organs are essentially conducting the struggle well but are sometimes apprehending people who are not truly anti-Soviet', before proceeding to assert that 'complaints about individual shortages or problems are not anti-Soviet. This can entail gossip about leaders, jokes of a political character, complaints about agriculture – all of which can be without counter-revolutionary meaning.'[29] The closing lines of the review proved to be the most significant of all: 'Mistakes are being made in cases of counter-revolutionary crimes. The courts require an explanation from the Plenum of the Supreme Court in regard to what does and does not constitute anti-Soviet behaviour.'[30]

The Supreme Court resolution duly arrived on 13 June 1958; its message can be summed up by the following line: 'for an act to be considered anti-Soviet it has to be consciously aimed at harming the Soviet state'. It recommended that those who drunkenly cursed the authorities or protested out of material discontent should not *necessarily* (my italics) be charged under article 58-10, and that courts and investigators should look at individuals' biographies, including their work and war record, social status and age, in order to help distinguish between genuine anti-Soviet activity and a 'faulty attitude toward certain events or policies'.[31] This came to be an important feature of the way that the regime dealt with acts of protest and criticism throughout the Brezhnev era and beyond. At the end of the year these guidelines were included in the new 'Law on State Crimes' that set out general principles on how to deal with what were now termed 'crimes against the state' rather than 'counter-revolutionary crimes'.

This was a crucial step in the creation of a more sophisticated and effective corpus of policy against dissent. It marked the point where the regime's 'firefighting' approach to policing dissent began to be replaced with a more sophisticated and less outwardly repressive approach. It showed that by the end of the 1950s the authorities themselves had privately begun to distinguish between conscious acts of dissent such as those often carried out by members of the intelligentsia and the spontaneous expressions of frustration and anger that tended to feature more among workers, and to tailor their response accordingly.

A wider point to be flagged up in regard to the way that the campaign was brought to a close is the role played by the Soviet legal establishment. This was not the first or last time that Soviet jurists were able to have a restraining effect on Party policy. As Harold Berman pointed out in regard to the parasitism laws that Khrushchev later attempted to force through, Soviet jurists were able to exert a degree of pressure on the leadership preventing a return to the arbitrariness and mass illegality that Khrushchev's proposals had the potential to unleash – something that was arguably also the case in regard to the persecution of dissenters.[32] This is not to suggest that one should consider the legal establishment to have been somehow 'liberal' but instead one should probably see it as an attempt to become more 'professional' as well as perhaps displaying an instinct for self-preservation. However, it is probably true that the legal establishment was able to have an influence such as this only when the leadership allowed them to do so, meaning that any gains in this area could always be reversed at a stroke.

The 1960s

While the development of policy against dissent in the early Khrushchev period was largely characterized by a sense of trial and error, the first half of the 1960s saw the authorities employing a more consistent and confident approach. The regime began to develop a more sophisticated and varied corpus of policy whereby greater efforts were made to prevent dissenting behaviour among the wider population in the long term as well as continuing to employ punitive responses against those who were perceived as being 'genuinely anti-Soviet'. Ultimately this added up to a polarization of responses whereby less serious transgressions, such as drunken outbursts, were treated with greater lenience than before, while more serious offences continued to be met with a harsh punitive response.[33]

In many ways the system of responses to dissenting behaviour was becoming markedly less Stalinist. The ever-growing reliance on 'soft' methods of social control, such as prophylactic measures, succinctly demonstrated this fact. The authorities had come tacitly to acknowledge that society's obedience could not be taken for granted but would have to be earned and that material inducements, for example, were sometimes a more effective means of social control than naked coercion. Furthermore, they were also obliged to accept that the goals of society were not always the same as those professed by the regime and had to factor this in to the way in which policy was conducted. As Vladimir Kozlov wrote: 'The lessons of the early 1960s pushed the Party leadership into a search for means of compromise in the conflict between authority and the people.'[34] Although still heavily in the regime's favour, the balance of power in the relationship between state and society had moved away from the authorities a little.

For the most part, members of the top leadership were rarely involved in the minutiae of the struggle against dissent. They undoubtedly were kept abreast of the most important developments around the country, particularly during times of significant unrest such as summer 1962, yet there was no repeat of December 1956 when top leaders had sat together and worked out their plan of attack. According to the KGB, it was they rather than the Central Committee Presidium that began to take the more active role in drawing up measures to deal with dissent.[35]

The most notable instance where the leadership undoubtedly did play a direct role in responding to dissent was in regard to the rising at Novocherkassk. The fact that a delegation composed of high-ranking figures (including four Presidium members) was quickly dispatched to the town showed just how concerned the leadership were about the developing situation.[36] Its most senior figures were the hard-line conservative Frol Kozlov and the more moderate Anastas Mikoyan. These two men practically embodied the nature of the authorities' policy against dissent. As Samuel Baron concluded, 'While one [Mikoyan] could explore the chances for a peaceful solution, the other [Kozlov] could be relied upon to crack down should the situation warrant'.[37] Khrushchev wished to avoid violence but, as in Hungary, he eventually found the exhortations of conservatives the more convincing and sanctioned the use of force, which ultimately saw the protest end with an estimated 24 dead and over 100 wounded.[38]

What we can see, therefore, is that while Khrushchev was willing to take daring and even reckless risks in foreign policy, he was less inclined to do so at home, especially when the potential for instability arose. This not only demonstrates that domestic stability continued to be the regime's single overriding priority but also emphasizes the point that notions of Khrushchev as a 'liberal' and of the era as one of 'thaw' can be misleading. As had previously been shown in the campaign against dissent of 1957–58, the authorities were quick to revert to a more repressive approach when the possibility of serious unrest surfaced.

It seems that the demonstration at Novocherkassk was forcefully put down not just because it was a protest and therefore deserved to be punished but because of the authorities' fear that unrest could spread to the surrounding area and further. This was amply demonstrated by the fact that all of the town's road and telephone links with the outside world were immediately cut by the authorities and the town held in an effective state of quarantine for some months afterwards. As events died down and investigations against the demonstration's ringleaders concluded, the authorities broadcast the harsh sentences against those found guilty (which included seven who were subjected to the death penalty) on local radio in order to intimidate others into silence. However, they also made a special point of shipping additional food supplies to the town in order to placate the population there – showing that the authorities wished to restore calm among the masses just as much as they wanted to punish those who were guilty.

On 19 July 1962 a report was sent from the Administrative Organs Department to the Central Committee that effectively constituted a review of the regime's struggle against dissent. It stated that the KGB had penetrated deeper into workers' organizations and had improved its 'prophylactic work', but conceded that there was still much to be done. It linked the growth in anti-Soviet activity to an increase in imperialist intelligence work, and stated that there were only a few 'anti-social elements' who, under the influence of foreign propaganda, were 'continuing to try to use temporary hardships for their own ends'.[39] Cooperation between the KGB and militia was described as 'weak' and it acknowledged that the security organs were not 'mobilizationally prepared' for major disturbances such as Novocherkassk, conceding that they had struggled to influence events once the disturbance was already under way.[40]

The review then demanded that decisive measures were taken to strengthen the work of the KGB against anti-Soviet elements. Surveillance of 'suspicious elements' and released prisoners was to be stepped up, the recruitment of informers and KGB agents increased and specific training undertaken for future scenarios of mass disturbances in built-up areas. To combat weaknesses in the placing and work of undercover agents it called for an increase in the availability of technological services as well as improvements in the training and political education of agents.[41]

Using the media against dissent

Among the authorities' main propaganda targets in the late Khrushchev era were citizens who dared to voice positive views of the West. An example of this could be seen in a September 1963 article in *Kazakhstanskaya pravda* entitled 'From an Alien Voice', in which two miners from Leninogorsk were attacked for 'lavishing praise upon life in America at every opportune moment' and 'being unashamed to slander their homeland'.[42] Media excoriation was not restricted to those who praised the West, however. In Russia a notable drive against public criticism began in March 1964 when *Izvestiya* published a letter, purporting to be from a Magadan miner named Nikolai Kuritsyn, entitled 'This Must Be Fought!'. After allegedly overhearing two young people mocking that year's poor harvest while waiting in line at a bank, Kuritsyn wrote to *Izvestiya* describing such people as 'toadstools' and insisting that the Soviet way of life must be defended: 'we cannot act like our woodcutter acted, passing himself off as a gardener for 30 years. However, we must fight them, disgrace them, shame them, unmask them in front of honest people.'[43] The disavowal of Stalin ('the woodcutter') was central to the letter's message: we are not reverting to the 'bad old days' but criticism of the system will still not be tolerated.

Probably the best example of this general whipping up of ill feeling against dissenters could be seen in the February 1964 edition of *Trud*, where it was said of the 'anonymous calumniator' G.R. Levitin (who had apparently written a series of anonymous letters to *Trud*, though their content was never discussed), 'he poured dirt on Soviet reality and blackened the state which gave him an education, a well-built home and guaranteed him a pension' and concluded that 'the anonymous calumniator is not only abominable but he is also dangerous. Here is an evil which we must not tolerate.'[44]

The aim of such letters and articles was essentially to intimidate rather than to persuade. For the most part, no arguments were put forward or official positions explained other than the simple message that criticism would not be tolerated. However, it is also important not to overlook the point that under Stalin, and even at times under Khrushchev, instances of citizens 'lavishing praise' upon foreign powers would most probably have resulted in a lengthy spell of corrective labour instead of scathing press coverage.

One notable factor in the media attacks on dissenters was the change of tone in discourse on dissent. Now the newspapers took a more subtle approach to tackling the subject and abandoned their previous tone of ideological outrage, which had been employed in *Izvestiya*'s attack on Yuri Orlov and his colleagues in May 1956. Instead, the media employed language intended to invoke hostility against dissenters by way of patriotic, moral or material grounds. This undoubtedly showed that the authorities were aware that the ideological rhetoric of previous years no longer had the same resonance in society.

The theme of appealing to citizens' patriotism instead of ideological values is one that can be seen at work in regard to propaganda attacking foreign broadcasts being beamed into the USSR. For example, media attacks on Radio Liberty regularly saw its Soviet émigré staff branded 'fascist riff-raff' and 'Vlasovites' – a label that Gene Sosin conceded was not entirely inaccurate.[45] In regard to Chinese anti-Soviet broadcasts and leaflets, the authorities employed equally bellicose rhetoric, at one stage suggesting in *Izvestiya* that the Chinese regime was comparable to that of Hitler, Napoleon or Genghis Khan.[46] The intention of such remarks to stir up some kind of patriotic fervour as a means of combating dissent was quite clear. To what extent it would have resounded among the USSR's non-Russian population is less clear, however.

Pressure to conform did not come solely from above but was also increasingly embedded in the very fabric of life in the USSR. The Khrushchev regime had become even more invasive into citizens' everyday lives than Stalin's had been and nowhere was this aspect of social control more evident than in the form of the *kollektiv* (collective). This was the basic unit of Soviet society and all citizens were automatically members of several collectives, in workplaces, housing blocks, recreational societies and elsewhere.[47] Members of any given collective were increasingly expected to take an active interest in the ideological lives of their fellow members; as Elena Zubkova (1993) has argued, 'personal life was considered a public matter'.[48] By placing a degree of responsibility on the collective as a whole for the actions of its members, an unseen but powerful deterrent was added to help prevent undesirable behaviour.

Another powerful but largely unseen deterrent was introduced when Khrushchev announced at the XXI CPSU Congress that 'prophylaxis' would become the main theme of what he referred to as the state's 'educational work'. The semantic imagery that invoked dissent as some kind of unhealthy phenomenon from which society had to be protected is immediately apparent, but the precise meaning of the term 'prophylactic measures' was less clear. Vasily Mitrokhin's guide to KGB terminology defined *profilaktika* as: 'Activity carried out by Soviet state bodies and social organizations aimed at the prevention of crimes against the state, politically harmful misdemeanours and other acts which affect the interests of the state security of the USSR.'[49]

The central feature of prophylaxis as it touched upon the lives of individual Soviet citizens was the 'prophylactic chat' (*profilakticheskaya beseda*). What this usually involved was for individuals who were considered to be ideologically wayward or potentially troublesome to be summoned to local KGB offices for a 'chat'. In many cases this would involve people who had played a minor role in an underground group, for example, or whose names had simply come up in connection with an ongoing investigation or whom KGB agents and informants had picked out as being in some way politically unreliable.

During the course of their meeting with the KGB, the person in question was usually bullied and cajoled into admitting and then renouncing any kind of dissenting behaviour. It was made clear that they were being watched by the security organs and that a resumption of 'undesirable behaviour' could have serious consequences, such as the loss of one's job, refusal of a university place to one's children, or the threat of imprisonment for either the interviewee or friends and family members. While Stalin's *Gulag* and arbitrary terror had done untold damage to the fabric of the regime and society, this was a practice that came at little economic, demographic or political cost. Furthermore it avoided the kind of large-scale antagonism of society and international opprobrium that mass imprisonment would have incurred and which the authorities were desperate to avoid. Most importantly, these measures appear to have been highly effective at stifling criticism.[50]

Exactly how many of these 'prophylactic chats' were undertaken by the KGB is hard to say, because only a limited amount of data has been made available.[51] Nonetheless, there are sufficient materials to show conclusively that this became the most common form of official response to breaches of political norms from the early 1960s onwards. A KGB report to the Central Committee from 25 July 1962 stated that in the first half of that year 105 people had been sentenced for preparation and distribution of anti-Soviet documents, while a further 568 had been subjected to prophylactic measures.[52] An analogous report sent a little under two years later stated that out of the 385 authors of anti-Soviet documents uncovered in the first five months of 1964, 39 had been jailed while 225 had been subjected to prophylactic measures.[53] The above figures would, therefore, suggest that the ratio of prophylaxis to imprisonment was approximately 5:1.

Punishing dissent

As F.J. Feldbrugge pointed out in a 1963 article reviewing Soviet legal developments: 'The social straggler is invited to rejoin the ranks immediately, and if he cannot or will not do so, he is annihilated.'[54] Feldbrugge's assessment may have been a little hyperbolic but it was not misleading. Even though the authorities were showing a tendency to resist using custodial sentences against dissenters, the number of people jailed for anti-Soviet agitation and propaganda during the early 1960s was far from insignificant, as the figures provided by *Istochnik* in Table 5.1 demonstrate.

The 1958 'Law on State Crimes' was codified in the new criminal codes that began to emerge in the union republics around the turn of the decade. Moshe Lewin has suggested that this overhaul of the legal system can be seen as an attempt to create a 'proper justice system' that was the product of extensive professional discussion and subjected to rigorous drafting and redrafting.[55] However, the Soviet concept of justice continued to be centred upon the ideals and aims of the CPSU and, although the authorities had begun to make a more convincing pretence at operating within these new laws, they could always be bent to the Party's will when necessary. It is worth restating that the law-enforcement apparatus and judiciary were very much under the sway of the Party. A more accurate assessment of the post-Stalin legal system has been provided by Louise Shelley: 'commitment to the rule of law, intrinsic to democratic policing, was conspicuously absent from the Khrushchev reforms'.[56]

For those who were jailed, the situation in labour camps and prisons remained particularly harsh. One need only read Anatoly Marchenko's *My Testimony* (1969) or the relevant sections of *Gulag Archipelago* (1978) to see that any notion of liberalization in this sphere of life was strictly relative to what had gone before.[57] While the immediate period following Stalin's death has been described by Solzhenitsyn as 'the mildest three years in the history of the archipelago', around 1961–62 conditions deteriorated to the point where the same author could write that the difference between the camps of the Khrushchev period and those of the Stalin period lay in their composition rather than the regime.[58]

One of the less commonly acknowledged aspects of punitive policy against dissenters in the Khrushchev period was that of psychiatric detention. Generally associated with the Brezhnev period, during which time it came to be more widely employed, the practice of systematically confining dissenters in psychiatric wards actually had very distinct roots in the late 1950s and early 1960s. Several of the most notorious institutions of the 1960s and 1970s including the Serbsky Institute, Leningrad SPH (Special Psychiatric Hospital), Kazan SPH and Mordova SPH were already holding dissenters in the Khrushchev years. Practitioners such as Daniil Lunts, Georgii Morozov, Andrei Snezhnevsky and numerous other 'psychiatrist-executioners' were already becoming dominant in the field during the period.[59] Finally, prominent Brezhnev era dissidents diagnosed as mentally ill under Khrushchev included Vladimir Bukovsky, Aleksei Dobrovolskii, Natal'ya Gorbanevskaya and Petr Grigorenko.[60]

It is possible that members of the leadership may have been unaware of the use of punitive psychiatry in the early stages of the Khrushchev era, but it seems clear that they knew of it by the turn of the decade. The fact that the practice had gained a degree of approval at the highest level could be seen when a May 1959 *Pravda* article made explicit the supposed link between political non-conformity and mental illness: 'to those who start calling for opposition to communism … we can say that now, too, there are people who fight against communism … but clearly the mental state of such people is not normal'.[61] One could not expect that everything contained in the pages of *Pravda* or any other newspaper was literally dictated by the highest authorities yet neither were they in the habit of printing anything that might be considered objectionable by the leadership.

Although conclusive evidence on the matter is yet to surface, it seems that there may in fact be a reasonable case for ascribing the growth of this practice to Khrushchev himself. As a caveat to his numerous declarations that there were no longer any political prisoners in the USSR, Khrushchev was known to remark that anyone dissatisfied at life under the Soviet political system must by definition be mentally ill.[62] Whether this had been intended as an off-the-cuff quip by the General Secretary or was a genuine signal to those charged with policing dissent remains unclear. However, in a system where the utterances of a single leader carried so much authority it seems doubtful that such remarks would have been entirely inconsequential.

There were many advantages for the authorities in branding critics mentally unhealthy: it explicitly rejected the validity of dissenters' criticisms and bypassed all legal requirements such as the authorities' need to establish evidence of any kind of anti-Soviet activity before sentencing.[63] This in turn meant that potentially embarrassing political trials could be avoided – a not unimportant consideration while Khrushchev continued to insist that there were no political prisoners in the USSR. Furthermore, because those imprisoned in this way received a diagnosis rather than a sentence, it was possible to hold them indefinitely or, in many cases, until they recanted their former views and behaviour.

Conclusion

This was a period in which the Soviet regime faced the most widespread domestic unrest between the rise of Stalin and the commencement of *glasnost*. If one were to combine the totals of those who were sentenced for dissenting activity with those who were subjected to prophylactic measures the figure would most likely be well in excess of 10,000 Soviet citizens. The renunciation of mass terror meant that the authorities were forced to find new ways of keeping society in check. From the uncertain and ad hoc policing of dissent in the early Khrushchev years through to the more subtle and precise methods of the 1960s, one can see that the regime steadily became more sure-footed and sophisticated in suppressing acts of protest and criticism.

The general depiction of the period in question as one of 'thaw' is misleading in some respects. KGB figures showing that over five and a half thousand people were jailed for dissent under Khrushchev prove that although political repression had indeed dropped massively since Stalin's time, it still remained high enough to indicate a strongly authoritarian state. The policies and practices against dissent that came to be established under Khrushchev went on to remain at the centre of the authorities' efforts to tackle the problem throughout almost the entire post-Stalin era.

Perhaps the key change in the relationship between state and society came after the uncertain situation of the late 1950s began to stabilize. From the end of the decade onwards, the authorities no longer sought to control the thoughts of its citizens but instead tried to control their public behaviour.[64] Worker discontent was effectively 'bought off' by improving living standards, and intelligentsia dissent was reduced to small and embattled communities on the fringes of society.

The general periodization of the Khrushchev years into a more liberal early period and an increasingly conservative second half finds both support and contradiction in the evidence presented by the present study.[65] The early Khrushchev years in particular proved to be something of a 'mixed bag' in this respect. The evidence of growing conservatism in official policy against dissenters is not hard to find during the early 1960s. However, we can also see that a growing majority of those who were found to have been involved in dissenting activity were no longer arrested and sent to camps but were dealt with administratively. The many advantages that this new approach had in regard to issues such as the economy and the social structure, suggest that one can view it as having been a case of rationalization just as much as it was one of liberalization.

Notes

1 E. Papovyan, 'Primenenie stat'i 58-10 UK RSFSR v 1957–1958gg. Po materialam Verkhovnogo suda SSSR i Prokuratury SSSR v GARF', in L.S. Eremina and E.B. Zhemkova (eds), *Korni Travy: sbornik statei molodykh istorikov*, Moskva: Obshchestvo 'Memorial', 1996, p. 73.

2 In M. Ilic and J. Smith (eds), *Soviet State and Society in the Khrushchev Era*, London: Routledge, 2009.

3 V. Shlapentokh, *Soviet Intellectuals and Political Power: The Post-Stalin Era*, Princeton: Princeton University Press, 1990, p. 78.

4 V. Shlapentokh, *Public and Private Life of the Soviet People: Changing Values in Post-Stalin Russia*, New York: Oxford University Press, 1989, p. 139.

5 See E. Kulavig, *Dissent in the Years of Khrushchev: Nine Stories About Disobedient Russians*, Basingstoke: Palgrave Macmillan, 2002.

6 Papovyan, 'Primenenie stati 58-10', p. 73.

7 P. Jones, 'From the Secret Speech to the Burial of Stalin', in P. Jones (ed.), *The Dilemmas of De-Stalinization: Negotiating Social Change in the Khrushchev Era*, London: Routledge, 2006, p. 46.

8 See K. Aimermakher et al. (eds) *Doklad N.S. Khrushcheva o kul'te lichnosti Stalina na XX s"ezde KPSS: Dokumenty*, Moskva: Rosspen, 2002.

9 Kulavig, *Dissent in the Years of Khrushchev*, p. 16.

10 RGANI, f. 3, op. 14, d. 13, ll. 76–9.

11 *Pravda*, 5 April 1956, and Interview with Yuri Orlov, Ithaca, December 2006.

12 S. Schattenberg, '"Democracy" or "Despotism"? How the Secret Speech Was Translated into Everyday Life', in Jones (ed.), *The Dilemmas of De-Stalinization*, p. 66.

13 Kuzovkin, 'Partiino-Komsomol'skie presledovaniya po politicheskim, motivam v period rannei "ottepeli"' in L.S. Eremina and E.B. Zhemkova (eds), *Korni travy*, p. 90.

14 See J. Furst, 'The Arrival of Spring?' in Jones (ed.), *The Dilemmas of De-Stalinization*, pp. 135–53.

15 RGANI, f. 5, op. 30, d. 233, ll. 1–73.

16 M. Kramer, 'The Soviet Union and the 1956 Crises in Hungary and Poland: Reassessments and New Findings', *Journal of Contemporary History*, 33, 2, April 1998, p. 192.

17 A. Fursenko and T. Naftali, *Khrushchev's Cold War: The Story of an American Adversary*, New York: W.W. Norton, 2006, p. 141.

18 RGANI, f. 89, op. 6, d. 2, ll. 1–15.

19 V. Chebrikov et al., *Istoriya sovetskikh organov gosudarstvennoi bezopasnosti: Uchebnik*, Moskva: Vysshaya krasnoznamenskaya shkola komiteta gosudarstvennoi bezopasnosti pri sovete ministerov SSSR, 1977, p. 527.

20 RGANI, f. 89, op. 6, d. 2, ll. 1–5.

21 GARF, f. 8131, op. 32, d. 5080, ll. 13–43.

22 B. Firsov, *Raznomyslie v SSSR 1940–1960 gody: Istoriya, teoriya i praktika*, Sankt Peterburg: Izdatel'stvo Evropeiskogo universiteta v Sankt Peterburge, 2008, p. 261.

23 F. Burlatsky, *Khrushchev and the First Russian Spring: the Era of Khrushchev Through the Eyes of his Adviser*, London: Weidenfeld and Nicolson, 1991, p. 97.

24 Papovyan and Papovyan, 'Uchastie verkhovnogo suda SSSR v vyrabotke repressivnoi politiki, 1957–1958', in Eremina and Zhemkova (eds), *Korni Travy*, p. 86.

25 C. Linden, *Khrushchev and the Soviet Leadership*, Baltimore: Johns Hopkins University Press, 1990, p. 47.

26 M. Lewin, *The Soviet Century*, London: Verso, 2005, p. 119.

27 GARF, f. 9474, op. 16c, e.kh. 648, ll. 1–73.

28 GARF, f. 8131, op. 32, d. 5080, l. 17.

29 GARF, f. 8131, op. 32, d. 5080, l. 42.

30 GARF, f. 8131, op. 32, d. 5080, l. 43.
31 GARF, f. 8131, op. 32, d. 5080, l. 64.
32 See H. Berman, 'The Struggle of Soviet Jurists', *Slavic Review*, 22, 1, June 1963, pp. 314–20. Yoram Gorlizki has also shown that justice officials resisted Khrushchev's attempts to give Comrades' Courts the power to exile citizens for up to five years. See Y. Gorlizki, 'Delegalization in Russia: Soviet Comrades' Courts in Retrospect', *American Journal of Comparative Law*, 46, 3, Summer 1998, pp. 403–25.
33 F. Feldbrugge, 'Soviet Criminal Law. The Last Six Years', *Journal of Criminal Law, Criminology and Police Science*, 54, 3, September 1963, p. 263.
34 V. Kozlov, *Neizvestnyi SSSR: Protivostoyanie naroda i vlasti 1953–1985*, Moskva: Olma-Press, 2006, p. 408.
35 Chebrikov et al., *Istoriya sovetskikh organov gosudarstvennoi bezopasnosti*, (1977) p. 581.
36 The delegation included four of the eleven Central Committee Presidium members: Frol Kozlov, Anastas Mikoyan, Andrei Kirilenko and Dmitrii Polyanskii. Alongside this quartet were included former KGB chairman Aleksandr Shelepin and Leonid Il'ichev.
37 S. Baron, *Bloody Saturday in the Soviet Union: Novocherkassk, 1962*, Stanford: Stanford University Press, 2001, p. 47.
38 See Baron, *Bloody Saturday in the Soviet Union*.
39 RGANI, f. 89, op. 6, d. 20, ll. 1–9.
40 RGANI, f. 89, op. 6, d. 20, ll. 1–6.
41 RGANI, f. 89, op. 6, d. 20, ll. 6–9.
42 *Kazakhstanskaya Pravda*, 27 September 1963.
43 *Izvestiya*, 1 March 1964.
44 *Trud*, 25 February 1964.
45 G. Sosin, *Sparks of Liberty: an Insider's Memoirs of Radio Liberty*, Pennsylvania, Pennsylvania State University Press, 1999. General Andrei Vlasov had been at the head of the Committee for the Liberation of the People's of Russia – an army of Soviet prisoners of war that had changed sides to fight against the Soviet regime.
46 *New York Times*, 24 August 1963.
47 See O. Kharkhodin, *The Collective and the Individual in Russia: A Study of Practices*, London: University of California Press, 1999.
48 Zubkova, Elena, *Obshchestvo i reformy 1945-1964*, Moscow: Rossiia molodaia, 1993, p.121.
49 V. Mitrokhin, *KGB Lexicon: The Soviet Intelligence Officer's Handbook*, London: Frank Cass, 2002, p. 329.
50 See Chebrikov et al., *Istoriya sovetskikh organov gosudarstvennoi bezopasnosti*.
51 The majority of the relevant information is held in the archives of the KGB as well as the Presidential Archive of the Russian Federation. The latter was briefly opened up to a number of researchers during the early 1990s but both are now closed.
52 RGANI, f. 89, op. 51, d. 1, ll. 1–4.
53 RGANI, f. 5, op. 30, d. 454, l. 110. The remaining 121 cases were still in progress at the time the report was written.
54 Feldbrugge, 'Soviet Criminal Law. The Last Six Years', *Journal of Criminal Law, Criminology, and Police Science*, 54, 3, September 1963, p. 263.
55 Lewin, *The Soviet Century*, p. 161.
56 L. Shelley, *Policing Soviet Society: The Evolution of State Control*, London: Routledge, 1996 p. 45.
57 See Marchenko, *My Testimony*, and A. Solzhenitsyn, *The Gulag Archipelago*, 3, London: Collins/Fontana, 1978.
58 A. Solzhenitsyn, *The Gulag Archipelago*, 3, pp. 427, 493. The difference in composition that Solzhenitsyn referred to was that the majority of inmates during the Khrushchev years tended to be from Ukraine or the Baltic States rather than from

Russia. This was largely a result of the fact that most who had been arrested for nationalism were not included in the amnesties that drained the *Gulag* of much of its population after Stalin's death.

59 The label 'psychiatrist-executioner' was designated by a US Senate committee charged with investigating accusations of psychiatric abuse against dissidents in the Soviet Union during 1972.

60 See A. Koppers (ed.), *A Biographical Dictionary on the Political Abuse of Psychiatry in the USSR,* Amsterdam: International Association on the Political Use of Psychiatry, 1990.

61 *Pravda,* 24 May 1959.

62 See, for example, V. Bukovsky, *To Build a Castle: My Life as a Dissenter*, London: Andre Deutsch, 1978.

63 See A. Pyzhikov, *Khrushchevskaya 'ottepel'*, Moskva: Olma Press, 2002.

64 See, for example, Shlapentokh, *Public and Private Life of the Soviet People.*

65 One encounters this broad periodization in numerous secondary sources. See, for example, W. Taubman, *Khrushchev: the Man and His Era*, London: Free Press, 2003.

6 Leadership and nationalism in the Soviet Republics, 1951–1959

Jeremy Smith

The death of Stalin in March 1953 was the first of a series of events leading to a renegotiation of the relationship between the non-Russian republics of the USSR and the central authorities in Moscow. One of the most concrete outcomes of de-Stalinization was the partial or full rehabilitation of the Chechen, Ingush, Crimean Tatar, German and other peoples deported from their homelands on Stalin's orders before and during the Great Patriotic War. Before then, the 'gift' of the Crimean peninsula to the Ukrainian Soviet Socialist Republic at the expense of the Russian Soviet Federation of Socialist Republics (RSFSR) in 1954 signalled an end to the increasingly russo-centric political, cultural, demographic and ideological shifts of the 1930s and 1940s. On the other hand, the Virgin Lands campaign centred on settling Slavic farmers into northern Kazakhstan, with little effort made to refute the appearance of a return to colonialism. The decentralizing *sovnarkhoz* reform of 1957, as discussed in other chapters in this volume, further altered the balance of authority in favour of the republics and away from the centre, only to be partially reversed by the language provisions of the education reform the following year and the abandoning of the *sovnarkhoz* reform in the early 1960s.

This chapter does not seek to relate the experience of non-Russian inhabitants of the USSR in the 1950s, or to treat all of the major issues mentioned above. Its aim is rather to focus on one aspect that proved particularly troublesome to the nationalities agenda during the Khrushchev years, and that was to re-emerge with a vengeance in the 1980s. The implementation of Soviet policies was in the hands of regional communist party leaders. In the non-Russian Soviet Socialist Republics (SSRs), these individuals were not only representatives of Soviet power, but were leaders of national groups – each of these had a clearly defined territory governed by powerful local organs, which were dominated by the titular nationality, and which had the promotion of national language and culture as one of its core aims. The tension inherent in the dual roles of republican party leaders as implementers of central decisions and representatives of their nationality led to a number of cases where these leaders pushed against the limits of the freer hand that was permitted to them after Stalin's death. The relaxation of the more repressive aspects of Stalinism resulted, among other things, in a growth of expressions of grass-roots nationalism, especially in Georgia and the three

Baltic republics of Estonia, Latvia and Lithuania. By the end of the 1950s, it was clear that the convergence of nationalist expressions among the population and an increasingly independent line on the part of republican leaders was threatening a major crisis in centre–republic relations. Khrushchev sought to address this through the exemplary purges carried out in Azerbaijan and Latvia in 1959, with only partial success. Displays of local particularism did not generally go as far as the crises of national leadership of the 1920s, but they were in many ways analogous and were products of the same federal system.

Between 1946 and 1950 leadership in the non-Russian republics remained relatively stable, with only three changes of first secretary (two of which resulted from the promotions of Khrushchev and Ponomarenko to the Secretariat of the CC CPSU). By contrast, in 1950 the first secretaries in Estonia, Karelia, Kirghizia, Moldavia and Uzbekistan were replaced amid widespread purges. In 1951–52 a more extensive purge took place in Georgia. The so-called 'Mingrelian affair', which Khrushchev later described as a personal initiative of Stalin's, removed a number of top Georgian party and government officials. These officials were united not so much by their regional Mingrelian identity, as by their membership of the network established by the former boss of the south Caucasian party organizations Lavrenti Beria, now one of the most powerful individuals in Moscow. Consequently, historians have tended to depict the Mingrelian Affair as a purge engineered by Stalin specifically to undermine Beria, of whom he was increasingly distrustful.[1] Be that as it may, the purge also tackled head-on the most entrenched and nationalist leadership in the Soviet republics, and there is some evidence that signs of nationalism among Georgian youth, encouraged or at least tolerated by the republic's leaders, were of sufficient concern to motivate the purge: a subsequent reorganization of the Georgian Communist Party into two territorial branches further weakened the party structure there and was specifically targeted against the corruption seen to be endemic throughout Georgia and the anti-Soviet and nationalist activities and opinions of Georgian youth in particular.[2]

Against this background, the measures taken by Beria during his brief ascendancy to restore the position of the republics were particularly significant. While careful to lay the blame at the feet of local leaders in violating the 'Leninist–Stalinist' nationality policy of the Communist Party, Beria's analysis of developments in Western Ukraine led in effect to a wholesale revision of the Party's policy. Here senior posts in the Communist Party, the administration and Higher Education (which had effectively been russified) were dominated by Russians and Eastern Ukrainians. Between 1944 and 1952, according to a Presidium resolution of May 1953, up to half a million people had suffered repression in Western Ukraine.[3] The Presidium resolved to replace the Russian Mel'nikov with the Ukrainian Kirichenko as first secretary of Ukraine but went further in calling for a policy of training and promoting local cadres in Western Ukraine as well as Ukrainizing Higher Education.[4] A second resolution adopted on the same day regarding Lithuania affirmed the generalization of this policy. Here the top leadership escaped punishment, but it was ordered to undertake a wholesale revision of cadre policies. This resolution explicitly linked the development of local national cadres and the more widespread use of the Lithuanian language to undermine the propaganda of 'bourgeois-nationalist elements'.[5] Not just the policies, but the language employed in the resolution, were reminiscent of the policies of *korenizatsiya* or, in Terry Martin's phrase, affirmative action, in the 1920s.[6]

These resolutions amounted, in effect, to a manifesto of a new nationality policy most closely associated with Beria, elements of which remained in force for the remainder of the Soviet period. Most notably, the policy of appointing republican First Secretaries from the local nationality prevailed in general up until 1986. As Pikoya correctly points out, however, this policy was never absolute and, even before Beria's arrest, there were clear signs that not all members of the Presidium shared his new approach. A resolution on Belorussia, adopted on 12 June 1953, was similar to those on Ukraine and Lithuania, but its recommendation to replace the Russian first secretary, Patolichev, with the Belorussian Zimyanin, was never acted on. Not long afterwards the Ukrainian Ponomarenko was appointed to the top post in Kazakhstan, where he was later succeeded by Brezhnev.[7]

Although other leaders, including Khrushchev, had experience in the republics and some base of support there, Beria presented himself as the leading Politburo authority on nationality questions given his years of experience in the Caucasus. Thus his efforts to reinforce the authority of the titular nationalities at the local level dovetailed with moves to strengthen his own position at the centre. Beria had already succeeded in promoting both of these aims at the time of the reorganization of membership of the Politburo immediately after Stalin's death, on 5 March 1953. On the basis that the Politburo needed a current representative of the southern Caucasus and a representative of the Muslim republics, Beria was able to secure the election of his close ally, M.D. Bagirov, to the Politburo (now renamed the Presidium).[8]

Although his nationality policy was not a major reason for the arrest of Beria, it figured high on the list of charges against him and the implicit repudiation of the Presidium's resolutions in the months following Stalin's death left the new leadership without a clear nationality policy. Although, unlike Beria, Khrushchev never elaborated a clear nationality policy, he is generally credited with a pro-nationality line illustrated by the gift of Crimea to Ukraine in 1954, the rehabilitation of peoples deported by Stalin and Beria, and decentralization of economic decision-making to the republics. As a result of these policies and his personal connections, Khrushchev was able to call on crucial non-Russian (especially Ukrainian) votes in the Central Committee during his confrontation with the anti-Party group in 1957. But Khrushchev's rehabilitations and economic decentralization were policies subordinated to the broader aims of de-Stalinization and improving economic efficiency. He had little to say about the national question either at the time or in his later memoirs. The question of nationality was widely discussed in the context of the new Party programme of 1961, but it occupied little place in the Constitution itself and what there was remained essentially neutral in tone.[9] Life for non-Russians in the republics was governed by a series of actions and reactions with little coherence and no ideological underpinning. Although a crucial turning point was reached in the years 1958–59, Khrushchev's relative neglect of nationality affairs thereafter allowed the return to a watchful but fairly *laissez-faire* attitude to the activities of republican leaders which was only to deepen under the inertia of the Brezhnev years.

This is not to say that Khrushchev was unaware of nationality issues. He had, after all, himself served two periods as First Secretary of the Ukrainian Communist Party. His memoirs and other writings show that he was aware of the nationality of all the individuals he dealt with, suggesting a crude and stereotypical view of nationality. In addition to the decentralization of economic and financial ministries,[10] Khrushchev oversaw an extensive devolution of judicial powers and procedures to the republics in 1956–57.[11] While the republics played an important part in his administrative thinking, there is little to suggest that he had any coherent policy either way when it came to promoting or denying national identity. The absence of a clear policy on the one hand opened the space for republican leaders to pursue their own agendas, and on the other hand left a weakness at the centre that allowed for certain abuses. The personal prejudices and inclinations of central personnel, from Khrushchev downwards, could influence a whole range of day-to-day decisions, from administrative appointments and university admissions to the allocation of economic resources and the development of industrial and agricultural schemes. Differences in approach were already evident in 1953 when Beria introduced his measures and personnel changes. Differences at the top go some way to explaining the inertia over nationality policy, but it was still possible for the dispositions of individual leaders to influence serious policy stances, such as the practice of assigning only Slavs (Russians, Ukrainians and Belorussians) to military service on the frontier, a position that came under attack from Republican leaders in 1957. At the same time, little was done to prosecute or remove nationalists from official positions, in contrast to the continuing repression of anyone accused of other forms of 'anti-Soviet activity'.[12]

Within a month of Beria's arrest, his close ally and protégé, Bagirov, had been deposed from his position as First Secretary of the Communist Party of Azerbaijan. Although charged with a series of errors and anti-Party activities, no doubt was left that Bagirov's fate was down to his association with Beria.[13] Otherwise, the new leadership took care not to upset established republican leaders or to give any impression of reversing the concessions introduced by Beria. Although reports of growing signs of nationalism were reaching the centre, especially from the Baltic republics, it took a major crisis to push the leadership to address nationality questions in any detail. The most extensive discussion in the Presidium of non-Russian nationalism between the arrest of Beria and 1958 took place in the wake of demonstrations in Tbilisi and elsewhere in Georgia over the failure to officially observe the third anniversary of Stalin's death in March 1956. Beginning with the spontaneous laying of wreaths at Stalin's monument in Tbilisi, events spiralled under the opposition of the authorities to any commemoration and rumours spreading about the contents of the Secret Speech. By the evening of 9 March, demonstrators had effectively taken control of the centre of Tbilisi and the first shots were fired at the Communications Building as a delegation attempted to enter to send a telegram to Moscow. The disturbances spread to Gori, Sukhumi and Batumi, and were not finally suppressed until 11 March, a week after they had started.[14]

Only brief notes are available of the Presidium's discussions on 23 May 1956 of the report presented by the Central Committee of the Communist Party of Georgia. These notes indicate that although the Georgian leaders were given a thorough grilling by Khrushchev, Bulganin and Kaganovich, in general the report of the Georgians was accepted and they escaped with no more than a ticking off. Khrushchev, while critical of mistakes made by the current leadership, accepted that there was no organized nationalist counter-revolutionary centre in Georgia, placed most of the blame for rising Georgian nationalism on Beria and implicitly accepted the responsibility of the CC CPSU for allowing Georgia to slip under their radar.[15]

The Georgians, who had held the top positions in the republic for only a few years, were able to blame the disturbances on a limited section of the population, taken mainly from the intelligentsia and student youth, among whom the cult of Stalin had thrived and taken on a particularly extreme form as a result of Beria's earlier influence. They admitted to weaknesses in the Party's political work and a lack of attention to the political education of young people, but also pointed to the energetic measures they had recently been taking to implement a correct national policy in Abkhazia and elsewhere, which had inflamed the passions of nationalistic elements.[16] In this and a subsequent report, national tensions were admitted, but downplayed, and any antagonism between Russians and Georgians was denied altogether prior to and during the March events.[17]

In the view of at least one scholar, First Secretary of the Georgian Communist Party Vasily Mzhavanadze and his associates may have initially encouraged the March 1956 events in order to put pressure on the centre. If this was the case, they succeeded in disguising their role and were not held accountable.[18] The Georgians did admit to cases of national antagonisms after the March 1956 demonstrations, ascribing them mostly to criminal and hooligan elements and in any case downplaying their frequency.[19] But other sources indicate that these cases fitted into a general trend of growing local nationalism that, most alarmingly from the leadership's point of view, was beginning to carry the leaders of certain republics along with it. From 1956 onwards the Central Committee was receiving a number of reports detailing examples of local official and non-official nationalism. The most worrying indications came from the three Baltic republics, where Sovietization, mass deportations and colonization by Russians only heightened the hostility of those nationalists who had managed to remain and a new generation of youth. Latent hostility was made worse by the failure to implement the May 1953 resolutions on cadres policy: in 1956 in the 10 to 12 raions of Lithuania where Poles made up a majority of the population, not a single raikom First Secretary and only one Second Secretary was Polish, while in many of the *kolkhozy* and rural soviets with a Polish majority the chairman was a Russian or a Lithuanian. In the republic, where 80 per cent of the population was Lithuanian, only 46.6 per cent of CP secretaries were Lithuanian – an improvement on the 27 per cent in 1950 but not enough to satisfy the population that they were not being russified, while the under-representation of national minorities, especially Poles, added another dimension to the problem.[20] In Estonia the figure for Estonians was 46.6 per cent, in Latvia for Latvians as low as 32 per cent.[21]

Against this background, workers were reported as calling for the replacement of Russian cadres by local ones. Students and the intelligentsia were growing increasingly restless, uniting their national cause with the fate of Hungary and Poland. Thus two slogans appeared simultaneously at Vilnius University in October 1956: 'Long live the revolution in Hungary, let's follow their example!' and 'Lithuanians! Lithuania is for Lithuanians, Russians are occupiers, throw them out!' Elsewhere in Lithuania the following slogans were seen: 'Give freedom to Lithuania'; 'Let's follow Hungary's example'; 'Down with communists'; 'Russians, get out of Lithuania'. In Estonia protests against collectivization and military service were accompanied by similar graffiti: 'Death to the occupiers'; 'Down with the Russian governors. Death to Russian occupiers. Throw the Russians out of Estonia'. The secret police reported numerous anti-Russian comments in public places. Perhaps more worrying were signs of official collusion with these anti-Russian sentiments: according to the same report, the rector of a Lithuanian university was preventing non-Lithuanians entering the university by setting Lithuanian-only entrance tests, while the most disturbing signs of all emanated from Latvia, where Second Secretary Eduard Berklāvs was promoting an exclusivist Latvian language policy.[22]

The signs of nationalism in the Latvian party pre-dated the Secret Speech. In a report to the CC CPSU, E.I. Gromov described the XIV Congress of the Communist Party of Latvia (CPLat) in January 1956 as 'the most critical congress ever', although here the main bones of contention were agricultural policy and housing.[23] By the end of the year, however, Gromov could report that language was becoming the key issue around which nationalist tendencies in the CPLat leadership were rallying. Berklāvs was allied with the chair of the Latvian Supreme Soviet, Ozolin', and the Deputy Chair of the Latvian Sovmin, Krumin'sh, in condemning the influx of Russian cadres who made no effort to learn Latvian, meaning that 'Riga was becoming a Russian city and losing its national character.' Russian party secretaries were compared to the German barons of tsarist times. In response, cadres who had no knowledge of Latvian were already being asked to leave the republic, while Berklāvs and his colleagues proposed that Russian administrators be given two years to learn Latvian or else leave.[24]

Nationalism in the Baltic republics was no surprise. What was clear, however, was that its open expression was becoming more commonplace in the wake of the Secret Speech and was given a further boost by events in Hungary. What gave real cause for concern, however, was the evidently anti-Russian line being taken by certain leading Latvian communists. The Latvian leaders had mostly been in place since Sovietization, most had lived in independent Latvia and some of them had fought as partisans against Nazi occupation. They therefore had deep local roots and even a certain independent legitimacy, a combination that left them both susceptible to the mood of the local population and confident enough to stand up to Moscow and Russians sent from Moscow.

The growing tide of anti-Russian nationalism was not confined to the Baltic Republics, and was particularly evident in the South Caucasus. Despite the claims of the Georgian leaders that everything was under control, there is at least anecdotal evidence that, in the wake of the March 1956 events, Georgian nationalism and anti-Russian feeling affected not just parts of the population but the local administration in at least some regions. Towards the end of April 1956 a group of Russian inhabitants of Tskhaltubo sent a desperate appeal to Voroshilov, claiming they were 'in fear of our lives' as a result of the anti-Russian mood. They had been told to leave immediately or face the consequences, amid rumours that anti-Russian atrocities were being prepared for 1 May. One of the leaders of the anti-Russian campaign was the local procurator, and the authorities were doing nothing to stop the threats or protect the population.[25] In the same week a military officer on the railways complained to Zhukov of the growing nationalism in Tbilisi and that he was treated as a 'a Pariah, an undesirable'. According to this letter, official infection with nationalism was demonstrated by the removal of non-Russians from their positions.[26] Later in the year, a member of the CPSU for 20 years, Boris Belkov, alerted the CPSU Central Committee to the growing number of assaults on Russians in his town of Rustavi, which had grown so regular that Russians could not go out at night. He linked rising nationalism on the part of the authorities to corruption, a link that was to be made with growing regularity.[27] Russians were not the only complainants, with citizens of Abkhazia asserting that Beria's Georgianizing policies were being renewed in the wake of the March events.[28]

But the most flagrant official nationalism at the highest republican level was evident in Azerbaijan, where the Supreme Soviet, without consulting the centre, passed a law on 21 August 1956 making Azerbaijani the sole state language in the republic. Although this brought Azerbaijan into line with the constitutional position in Armenia and Georgia, critics argued that in Azerbaijan it was unnecessary and harmful given the large Russian and Armenian populations in Nagorno Karabakh and industrial centres like Baku, Kirovabad, Sumgait and Mingechaur. The law was already being abused to effectively exclude minorities from discussion, while it had also given rise to nationalistic demands for Russians and Armenians to be dismissed from senior posts.[29] While the law was most closely associated with the Chairman of the Supreme Soviet, Mirza Ibragimov, the first secretary of the Communist Party of Azerbaijan (CPAz) Imam Mustafaev, appointed only in 1954, had already been in some trouble in 1955 and was to take political responsibility for the growing tendencies of Azerbaijani nationalism.

In spite of the growing number of reports of local nationalism reaching the CC Secretariat in Moscow, the CPSU leaders took no action at this point. By then a separate department of the CC Secretariat for dealing with the affairs of the Union republics was in existence and its head, Evgenii Gromov, energetically followed up the cases that crossed his desk. But he could not move against senior republic figures without the support of the Presidium. The latter's members were, presumably, anxious not to make enemies in the republics as long as the balance of power in Moscow remained unclear, and may have pigeonholed developments in the republics alongside the host of other unwelcome developments arising from the Secret Speech and the invasion of Hungary. What they were facing bore remarkable parallels to the situation in the 1920s. The return to Leninist policies after Stalin's death was evident not just in cadre policies, but in published pronouncements. As an article published in *Kommunist* in September 1953 put it:

> Our task is to continue diligently to advance and promote local cadres, who know their peoples' languages, life styles and traditions; to develop schools and theatres ... to raise the material and cultural standard of the broad working masses in all republics and territories.[30]

These tasks reflected directly the programmatic statements of the early 1920s, many made by Stalin himself when he was Commissar of Nationalities. Then, the policies of *korenizatsiya* had been intended to promote the integration of the non-Russian regions into the Soviet Union. But this process relied on the promotion of non-Russians who were willing to brand themselves as communists but whose background was, in many cases, in left-wing nationalist organizations. The party leadership recognized early on the dangers of giving a relatively free rein to nationalists. The emergence of a conspiracy to promote pan-Turkic aims led by the Tatar communist Mirsaid Sultan Galiev in 1923, however, did not lead immediately to a crackdown on such manifestations of 'national communism'. Instead, the Bolshevik leaders indicated that they were ready to tolerate this lesser evil rather than antagonize the non-Russians. At this juncture, as in 1956, the question of political succession was unresolved and may have contributed to a more *laissez-faire* policy, with both Trotsky and Stalin being able to claim significant support in the republics. Measures such as the delimitation of new republics in Central Asia contributed to growing anti-Russian and anti-minorities measures on the part of local leaders, until local nationalism began to infect not just those with a nationalist past but many who had been regarded as reliable communists. The climax of the Shumsky affair and show trials of nationalist intellectuals in Ukraine and Belorussia signalled the beginning of a clampdown in 1927, backed up by extensive purges, which was to embrace almost all Union republics as well as the more significant autonomous republics over the next few years.[31]

A similar situation appeared to be developing in the mid-1950s. Beria's courting of the republics had sent out signals that, at least in the areas of cadres and language, the republics would now have a freer hand. But reports of nationalist activity both at the grass roots and official levels mushroomed in the wake of the Secret Speech. For a while, Khrushchev seemed keen to encourage more localized decision-making, but as this took on an increasingly anti-Russian character in several cases it was unlikely that the nationalist direction being taken by republic leaders could be tolerated for long. Having secured his own authority at the centre in 1957, not least because of the support of the republics, an opportunity was soon to present itself to tackle head-on the most recalcitrant of these leaders.

Khrushchev's reform of the Soviet system of education was motivated by his recognition of the importance of schooling a new generation of citizens fit for work in a modern, technology-driven economy, and by his own vision of the future Soviet man and woman.[32] But the reform handed him the opportunity to confront the rise of nationalism in the republics and the excesses of a number of republican leaders. The key paragraph of Khrushchev's theses with regard to native language education was the notorious Article 19. This article states that

> The question ought to be considered of giving parents the right to send their children to a school where the language of their choice is used. If a child attends a school where instruction is conducted in the language of one of the Union or autonomous republics, he may, if he wishes, take up the Russian language. And vice versa, if a child attends a Russian school, he may, if he so desires, study the language of one of the Union or autonomous republics.[33]

While apparently innocuous, this provision threatened to undermine a key principle of nationality policy since Lenin's time, one that provided a secure prop for the national cultural and linguistic policies of the republics. Although the study of Russian as a second language had been almost universal in Soviet schools since the early years of Soviet power, and had been made compulsory by a law of 1938, for non-Russians in the Union Republics of the USSR the overall language of instruction had been their mother tongue since the Revolution. In the immediate post-war years, this principle came under some strain, especially in the autonomous republics of the RSFSR. But proposals to extend the use of the Russian language more broadly were defeating, and the principle of native-language education in the republics was reaffirmed in the last years of Stalin's life.[34] What concerned national leaders was both the threat posed to national schools (should parents choose in large numbers to send their children to Russian schools, even if they were of the local nationality) and the likelihood that Russians would elect to stop learning the main language of the Republic, thus undermining its status.

The highest-level discussions over Khrushchev's education reforms, and Article 19 in particular, are most likely recorded in the closed files of the CC CPSU Secretariat. But from more open discussions of the reform prior to the publication of Khrushchev's theses it can be deduced that Khrushchev introduced the new proposal on the language of instruction at the last minute. A report signed jointly by the RSFSR Minister of Education and the President of the Academy of Pedagogical Science as late as September 1958 envisaged the possibility of extending the period of school study in non-Russian republics by a year in response to the extra burdens of language learning placed on non-Russian children.[35] This option was enthusiastically endorsed in the responses to the proposals from a number of autonomous republics of the RSFSR as well as Union republics.[36] In the latter cases, the extension of school study by a year was to apply to Russian as well as non-Russian children. The emphasis in these discussions was on the practical requirements of learning Russian alongside foreign languages (or for Russian children to learn the local language). These documents suggest that there was no inkling at this stage that the principle of native-language education would be challenged.

Article 19 provoked an outburst of anger on the part of numerous republican delegates at the session of the Supreme Soviet of the USSR, which discussed the proposals in December 1958. Representatives of the three Baltic Republics, Georgia and Armenia opposed the measure in no uncertain terms. Individual delegates from Ukraine and Belorussia also expressed opposition. Only the Central Asian republics and Moldova appeared to accept the need to reform the basis of the language of education.[37]

It was this issue that brought things to a head with the Communist Party of Latvia. The law passed by the Supreme Soviet of Latvia on 17 March 1958 left the question of the total length of school attendance open, nor did it include the principle of parental choice in the study of a second language. By the end of May the Latvian leaders had been forced to amend their education law to bring it in line with Khrushchev's theses. However, in explaining this change of tack, education minister Samsons insisted that

> refusal of compulsion in the teaching of languages does not mean that we should diminish our attention to these languages. In fact, they should be taught even better and more than previously, because everybody needs knowledge of Latvian and Russian for their work and studies in our Republic.[38]

This was an open and brazen declaration of intent that, in practice, nothing would change with regard to the compulsory study of languages.

A commission was sent to Riga to deal with the recalcitrant leaders there on 21 June 1959, and returned with a verdict on 4 July after reporting back to the Presidium in between. In addition to the sluggishness over implementing the school reform, the charges went back to 1956 and Berklāvs' proposals to force officials to learn Latvian; provision of free school books only in Latvian; a discriminatory passport regime in Riga; reduction of Russian-speaking places in higher education; and insufficient attention in propaganda to the friendship of peoples and the role of the Russian people in the Great Patriotic War. Economic nationalism was also uncovered in the form of the republic's objections to doctors being trained in Latvian medical schools who would then go on to work in other republics.[39] Berklāvs was immediately dismissed, and in December 1959 First Secretary Kalnberzin and others were removed from their posts.[40]

The Presidium of the CC CPSU discussed the situation in Azerbaijan in the same session at which it dealt with Latvia, on 1 July 1959. Here the most prominent charge was related to the school reform – the Azerbaijan SSR had passed an additional law that 'in essence undermines the principle of choice in the study of languages and can be interpreted as a demand for the compulsory study of Azerbaijani in all schools'.[41] But the report also discussed economic mismanagement and went back to the 1956 language law, citing a whole series of subsequent abuses and discriminatory practices against Russians and Armenians living in the republic. Economic nationalism was manifested in the refusal to provide Georgia with gas and the declaration by Chair of the Azerbaijan Council of Ministers Sadikh Ragimov that 'the gas is ours, Azerbaijani, and we cannot give it to the Georgians'.[42] Cultural and historical nationalism was equally evident, not just in the field of language, but in propaganda and official versions of history which, for example, more or less ignored the role of the Armenian Stepan Shaumian in Baku during the Revolution.[43] Although in these discussions the main culprits were identified as Ragimov and Ibragimov, in its decision the next day the Presidium laid responsibility firmly at the feet of the republican first secretary, Imam Mustafaev.[44] The latter protested his personal innocence but was the most prominent name on a long list of senior officials to be purged in the republic over the course of the remainder of the year.

The fate of Kalnberzins, Berklāvs and others in Latvia can be compared to that of the leadership in Estonia. Although there was no equivalent history of official nationalism in 1956 or before, in late 1958 and early 1959 the Estonian leaders adopted an almost identical position to the Latvians on the school reform. More important though than the lack of prior history was the readiness of the Estonian first secretary Ivan Käbin to acquiesce to Moscow: having had a polite request to extend the period of school study in Estonia by a year turned down, the proposal was dropped from legislation and Estonia made no formal attempts to resist the reform. Käbin went on to remain one of the longest-serving first secretaries in the history of the USSR.

More political purges, although less extensive, followed in other republics. The First Secretary of the Turkmen Republic, S. Babaev, had already been removed in December 1958. In March 1959, Uzbek first secretary S. Kamalov followed. In April–May 1961 Moldavian First Secretary Z.T. Serdyuk, Kyrgyz First Secretary I. Razzakov and Tajik First Secretary T. Uldzhabaev were all replaced. Only in the Tajik republic did the purge extend to other leading party and state officials. What these other cases suggest is that disobedience over the education reform was not the sole issue to determine the fate of the Republican leaders. The Central Asian republics and Moldavia had accepted Khrushchev's proposals without complaint, while by contrast Arvīds Pelše, promoted to First Secretary of the Latvian Party following the 1959 purge, had been among those openly opposing Article 19 at the December 1958 session of the USSR Supreme Soviet. In Tajikistan and Moldavia accusations of nationalism were listed among the charges against the leaders including, in a clear official repudiation of the post-Stalin policy, the charge that cadres had been promoted on the basis of their nationality.[45] But these slightly later purges occurred just as the consequences of the *Sovnarkhoz* reform were becoming apparent and at the same time the fallout of the Ryazan affair was turning attention towards corruption and deceit in the regions, as discussed elsewhere in this volume.

These purges differed, then, from those in Latvia and Azerbaijan both in terms of their scale and the object of the centre's attacks. While economic distortions were evident in Azerbaijan as well, this was not the case in Latvia and it was in these two republics that leadership nationalism had been most apparent since the mid-1950s. The pattern was remarkably similar to that of the 1920s: *korenizatsiya* had encouraged the activities of republican leaders to pursue the national interest. Where the situation differed radically was that, whereas the 'nationalist deviation' of the 1920s emerged from an alliance of a nationalist intelligentsia with local communist leaders, and was of relatively little concern to the predominantly rural national population, in the 1950s – especially in the Baltics and the South Caucasus – national feeling had much deeper popular roots. Consequently the purges were focused at the top, with the intelligentsia remaining largely untouched. Nor could such administrative measures do much to affect the mood of the population as a whole.

An important change at the highest levels of the Communist Party of the Soviet Union in 1960 underlined the extent to which the policies adopted in 1953 had shifted. Alexei Kirichenko had been appointed First Secretary in Ukraine in line with the new cadres policy in 1953, and was simultaneously promoted to become a candidate member of the CC CPSU Presidium. He was then brought to Moscow permanently as CC Secretary responsible for cadre policies in December 1957. Over the next two and a half years he used this position to energetically implement a national cadres policy in the republics, but was abruptly removed from his post in May 1960. From this point on, the promotion of non-Russians in the central party apparatus slowed down.[46]

The *sovnarkhoz* reform was also reversed eventually, but this did not, however, signal a complete recentralization of control. The republics still enjoyed considerable latitude in decision-making, particularly in areas connected to culture. Language practices varied in schools, but the legal provisions passed in 1959 were often ignored in practice, while republics continued to celebrate national anniversaries and develop monuments and architecture that added national flavour to the major cities. The markers laid down in 1958–59 sent a clear warning against the kind of excesses that had appeared in the republics in response to Beria's measures and, especially, the Secret Speech. Republican leaders, however, still had to contend with the pressures of national feeling coming from below, and were soon, after Khrushchev's ouster in 1964, able to take advantage of the promise of stability of cadres to consolidate their own power bases (and personal fortunes) in part through the pursuit of nationally based cultural and cadre policies. This went for republican leaders including those, like Sharof Rashidov in Uzbekistan, who had risen to the leading republican positions as a result of the Khrushchev purges. While the dismissal of Petr Shelest as first secretary of the Communist Party of Ukraine in 1972 showed that there were still limits that could not be crossed, the longevity in office of leaders such as Rashidov and Käbin, and the further success of Eduard Shevardnadze and Heydar Aliev in gaining promotion to important posts in Moscow – all having built local power bases founded on the pursuit of nationally orientated policies – illustrate that, ultimately, the 1959 purges did little to change the behaviour of republican leaders.

It was in the 1950s that the rural part of the population of the Soviet Union became a minority for the first time. While most of the borderland republics were somewhat behind Russia in urbanization, they were rapidly catching up.[47] For this reason if nothing else, the national question had shifted profoundly since the early 1920s. By the 1950s the non-Russian nations were built on an educated urban population. With this changing social structure, the federal structures of the USSR also took on new significance. Nationality was not just a signifier based on language, but an object of personal identity, loyalty, pride and ultimately mobilization. The tendency of republican leaders to develop as national leaders allowed nationhood to flourish and for infrastructures to develop in the manner characteristic of European nation-states. As this process deepened in the Brezhnev years, we can see in retrospect how the Soviet republics became embryonic nation-states. And while the break-up of the Soviet Union in 1991 still needs some explaining, the way in which the USSR dissolved along federal lines and the character of the new states that emerged from it should not be so surprising.

Notes

1 J. Ducoli, 'The Georgian Purges 1951–53', *Caucasian Review*, 6, 1958, pp. 54–61; A. Knight, *Beria: Stalin's First Lieutenant*, Princeton, NJ: Princeton University Press, 1993, pp. 159–64; Y. Gorlizki and O. Khlevniuk, *Cold Peace: Stalin and the Soviet Ruling Circle, 1945–1953*, Oxford: Oxford University Press, 2005, pp. 109–13.

2 Politburo proceedings of 29 October 1951, RGASPI f. 17, op. 3, d. 1091

3 R. Pikoya, *Moskva. Kreml'. Vlast': Sorok let posle voiny 1945–1985*, Moscow: Rus'-Olimp, Astrel'AST, 2007, pp. 241–2.

4 Pikoya, *Moskva. Kreml'. Vlast'*, p. 243.

5 Pikoya, *Moskva. Kreml'. Vlast'*, pp. 244–5.

6 T. Martin, *The Affirmative Action Empire: Nations and Nationalism in the Soviet Union, 1923–1939*, Ithaca and London: Cornell University Press, 2001.

7 Pikoya, *Moskva. Kreml'. Vlast'*, pp. 245–6.

8 E. Ismailov, *Azerbaidzhan: 1953–1956; pervye gody "ottepeli"*, Baku: Adil'olgy, 2006, p. 19.

9 A. Titov, 'The 1961 Party Programme and the Fate of Khrushchev's Reforms', in M. Ilic and J. Smith (eds), *Soviet State and Society Under Nikita Khrushchev*, London: Routledge, 2009, pp. 14–15.

10 See Chapter 7 of this volume.

11 G. Simon, *Nationalism and Policy Toward the Nationalities in the Soviet Union: From Totalitarian Dictatorship to Post-Stalinist Society*, Boulder, CO: Westview Press, 1991, pp. 234–8.

12 N. Mitrokhin, *Russkaya partiya: Dvizhenie russkikh natsionalistov v v SSSR 1953–1985 gody*, Moscow: Novoe literaturnoe obozrenie, 2003, pp. 78–83.

13 Ismailov, *Azerbaidzhan: 1953–1956*, p. 48.

14 For an accurate account, see V.A. Kozlov, *Mass Uprisings in the USSR: Protest and Rebellion in the Post-Stalin Years*, Armonk and London: M.E. Sharpe, 2002, pp. 112–35.

15 Presidium Protokol no. 17, meeting of 23 May 1956, in A. Fursenko (ed.), *Arkhivy Kremlya: Prezidium TsK KPSS 1954–1964, tom 1; Chernovye protokol'nye zapisi zasedanii Stenogrammy*, Moscow: Rosspen, 2004, pp. 133–4.

16 'Report of the Bureau of the CC CP of Georgia on mass disturbances of the population of Tbilisi, Gori, Kutais, Sukhumi and Batumi on 4–9 March 1956 in connection with the condemnation of the cult of personality of Stalin' [no later than 23 May 1956] in A. Fursenko (ed.), *Arkhivy Kremlya: Prezidium TsK KPSS 1954–1964, tom 2; Postanovleniya 1954–1958,* Moscow: Rosspen, 2006, pp. 289–95.

17 'Report of the Bureau of the CC CP Georgia ...', *Arkhivy Kremlya tom 2,* pp. 296–7.

18 T. Blauvelt, 'Status Shift and Ethnic Mobilisation in the March 1956 Events in Georgia', *Europe-Asia Studies,* 61, 4, June 2009, pp. 651–68.

19 'Report of the Bureau of the CC CP Georgia ...', *Arkhivy Kremlya tom 2,* pp. 298–303.

20 A. Shchegolev and M. Karpov, 'Report to the CC of the CPSU on Shortcomings in the Training of Cadres in Lithuanian Party Organisations', RGANI f. 5, op. 31, d. 59, ll. 101–7.

21 RGANI f. 5, op. 31, d. 59, l. 211.

22 I. Gavrilov, 'Report on Recent Nationalist and Anti-Soviet Displays in Estonia, Latvia and Lithuania', 28 November 1956, RGANI f. 5, op. 31, d. 59, ll. 203–12.

23 E.I. Gromov, 'Report on the XIV Congress of the Communist Party of Latvia', 17–19 January 1956, RGANI f. 5, op. 31, d. 59, l. 41.

24 Letter of Gromov to the CC CPSU, RGANI f. 5, op. 31, d. 59, ll. 58–9.

25 Letter from Russian Inhabitants of Tskhaltubo to Voroshilov, 27 April 1956, RGANI f. 5, op. 31, d. 60, ll. 75–6.

26 Letter from a railway forces officer to Zhukov, 22 April 1956, RGANI f. 5, op. 31, d. 60, ll. 79–84.

27 Letter from Boris Gavriovich Belkov, 28 August 1956, RGANI f. 5, op. 31, d. 60, ll. 93–4.

28 RGANI f. 5, op. 31, d. 60, ll. 121–3, 137–49.

29 Gromov and Lebedev, 'Report on Ibragimov's Language Law', 16 November 1956, RGANI f. 5, op. 31, d. 60, ll. 10–12.

30 Cited in Simon, *Nationalism and Policy ...,* pp. 229–30.

31 Smith, 'National Conflict in the USSR in the 1920s', *Ab Imperio* No. 3, August 2001, pp. 221–65.

32 Smith, 'Khrushchev and the Path to Modernisation through Education', in M. Kangaspuro and J. Smith (eds), *Modernisation in Russia since 1900,* Helsinki: SKS, 2006, pp. 221–36.

33 G.S. Counts, *Khrushchev and the Central Committee Speak on Education,* Pittsburgh: University of Pittsburgh Press, 1959, pp. 45–6.

34 P.A. Blitstein, 'Nation-building or Russification? Obligatory Russian Instruction in the Soviet Non-Russian School, 1938–1953', in Ronald Grigor Suny and Terry Martin (eds), *A State of Nations: Empire and Nation-making in the Age of Lenin and Stalin,* Oxford: Oxford University Press, 2001, pp. 253–74.

35 E. Afanasenko and I. Kairov, Report to the CC CPSU on proposals on Education, 8 September 1958, RGASPI f. 556, op. 16, d. 38, l. 39.

36 RGASPI, f. 556, op. 16, d. 38, ll. 15–16; d. 44, ll. 172–4; d. 46, ll. 22, 27–30, 36, 58, 62, 87; RGANI f. 5, op. 35, d. 96, ll. 12–49.

37 Yaroslav Bilinsky, 'The Soviet Education Laws of 1958–9 and Soviet Nationality Policy', in *Soviet Studies,* 14, 2, October 1962, 138–57, pp. 143–4. Bilinsky and others assume that the fact that individual republics were left to pass their own education laws was a result of the opposition expressed to Khrushchev's theses at the Supreme Soviet in December 1958. However, since February 1957 it had become standard practice for republics to determine their own legal codes and no especial significance should be attached to this procedure; Simon, *Nationalism and Policy ...,* p. 237. I am grateful to Yoram Gorlizki for clarifying this point.

38 *Skolotāju Avīze,* 29 May 1959.

39 Archive of Communist Party of Latvia f. 101, p. 22, d. 48a, ll. 20–7.

40 Archive of Communist Party of Latvia f. 101, op. 22, d. 48a, l. 50.
41 *Arkhivy Kremlya*, tom. 1, p. 366.
42 *Arkhivy Kremlya*, tom. 1, p. 363.
43 *Arkhivy Kremlya*, tom. 1, pp. 357–87.
44 *Arkhivy Kremlya*, tom. 1, p. 387.
45 Simon, *Nationalism and Policy ...*, pp. 253–4.
46 Simon, *Nationalism and Policy ...*, p. 232.
47 R.J. Kaiser, *The Geography of Nationalism in Russia and the USSR*, Princeton, NJ: Princeton University Press, 1994, pp. 201–4.

7 Moscow–Kiev relations and the *Sovnarkhoz* reform

Nataliya Kibita

By 1957 Khrushchev was convinced of the necessity of reorganizing the economic administration. Since the beginning of the 1930s, when the skeleton of the economic administrative system was set, Soviet industry had grown considerably. Yet Stalin's reorganizations of the economic administrative system generally never went beyond merging or dividing the existing People's Commissariats, later renamed into ministries. How to reorganize the economic administration seemed to be clear to Khrushchev as well: along with preserving centralized planning, to abolish the industrial ministries and devolve their managerial authority to the regional economic councils and the republican governments.

Technically, Khrushchev's idea was not original. The regional economic councils, known as *sovnarkhozy* (*sovety narodnogo khozyaistva*), emerged during the years of 'war-communism' and lasted until the beginning of the 1930s; and, as a rule, they administered only the light and food industries.[1] Khrushchev's novelty was in entrusting the republican governments with managing almost all the industry on their territories, even under the centralized planning system.

Having considerably promoted the republican governments in running industry in 1957, Moscow, intentionally or not, had brought 'the nationality policy' motive into economic relations between the republics, just as it provoked the spontaneous formation of interest groups at all levels of republican economic administrations. Yet, whereas the national character of the reform had not been and could not have been perceived by either the central leadership or the republican governments, interest groups pursuing their local economic interests took shape as soon as the reform entered into force. The *sovnarkhoz* reform had provided a new context for the relationship between central and republican/regional political and managerial elites. It pushed the elites to work on new ways of decision-making.

It is with the managerial thinking in Moscow and in Kiev during the years of the reform that the present chapter is concerned. It shows that the new Soviet leadership invited the republics into a dialogue on the ways of improving the economic administration once Stalin died. This dialogue started de-Stalinization in the economic administration that reached its apogee in 1957. Yet, just as de-Stalinization was rather partial and inconsistent in other aspects of Soviet life, it was partial and inconsistent in the economic administration as well. The chapter shows inconsistencies in the reorganization of the ministries after the XX

Congress of the Communist Party of the Soviet Union (CPSU) in introducing and developing the reform. At the same time, it shows that the Ukrainian response to the central reorganization initiatives was, on the contrary, quite consistent. The republic showed interest in running its economy and protecting its economic interests after Stalin died, whether through the republican ministries or through the *sovnarkhozy*.

Khrushchev's note and its discussion in the CC CPSU Presidium (January 1957)

On 28 January 1957, 14 members of the CPSU Central Committee (CC CPSU) Presidium discussed a 'Note of Khrushchev' that revealed his vision of the future economic administration: 'Certain considerations on the improvement of the organisation of the administration of industry and construction'. Khrushchev's idea of improving the economic administration was to get rid of ministerial departmentalism by demolishing the ministries and devolving their managerial authority to the regional councils of national economy. The regional councils would thus cure the economic administration of the shortcomings of the excessive centralization created by the ministries. They would ensure cooperation and specialization between enterprises that were located near each other but previously belonged to different ministries. Cooperation and specialization would prevent wasteful expenditure on transportation and provide the rational usage of production capacities and the development of new products. Moreover, and this was an issue of most major importance for Khrushchev, the regional councils would keep the governmental *apparat* from growing.[2]

The only weak point that Khrushchev anticipated in his note was a possible tendency of certain industrial regions to prioritize the development of their regional economies, so-called localism or *mestnichestvo*. Therefore the USSR Council of Ministers (CM) and republican Councils of Ministers should control the situation and be responsible for it. The USSR Gosplan (the CM USSR State committee for perspective planning) and *Gosekonomkomissiya* (the CM USSR State economic commission for current planning), in their turn, should study the economy of the regions and avoid planning financial investments in the development of economically unprofitable production.[3]

Despite the radicalism of the idea, the majority of the CC CPSU Presidium members supported it, including Bulganin, the Chairman of the government, who was supposed to protect the system he was heading. Among 13 participants in the discussion, there were only three who expressed concerns regarding the reorganization: Pervukhin, Molotov and Voroshilov. Pervukhin considered that with the territorial administration the advantages of the ministerial administration, such as concentration of the branch administration, centralization and specialization, would be lost. He proposed reducing the number of ministries and organizing territorial commissions in order to provide horizontal cooperation.[4] Molotov, in his turn, agreed that the role of ministries should be diminished and the local authorities should be more involved in industrial administration, but it was too early to talk about abolishing the ministries.[5] Finally Voroshilov did not believe that breaking up the centralized administration was the solution. But, as he said, if it was going to pursue the project, the Presidium should turn to specialists in industry.[6]

Regardless, on 13–14 February 1957, Khrushchev's note was submitted to the CC CPSU Plenum.[7] After a countrywide discussion, on 10 May 1957 the USSR Supreme Soviet adopted a law outlining a new institutional model, which became known as the *sovnarkhoz* reform. The model was based on the premise that republican governments would be in charge of administering their economies through the republican Gosplans and the *sovnarkhozy*. USSR Gosplan would provide centralized planning and coordination between the regions.[8]

In 1957, Khrushchev took the reorganization of the ministerial system to an extreme by abolishing the industrial and construction ministries. With the loss of the ministerial system the Kremlin was losing control over the execution of plans, which were still being composed centrally. Was this obvious to the Presidium members or were they confident in the proper leadership of the republican governments? Perhaps this was obvious, but only Molotov showed openly that he did not rely on the republican pro-Union leadership and opposed such a radical reorganization. However, his March suggestion to proceed slowly and less radically when abolishing the system did not get any support in the Presidium.[9] Why not? What were the reasons for the reorganization of the ministerial system? And what was the republican, particularly Ukrainian, contribution to the reform?

Roots of the 1957 administrative reorganization

After Stalin's death, while preserving the system's skeleton and yet realizing that the ministries could not run their industries efficiently with the limited managerial authority that Stalin allowed them to exercise, the first post-Stalin leadership adopted a new approach to the reorganization of the system. Stalinist practices, that is merging and dividing the economic administrations, were combined with showing the ministers more trust. Thus in March–April 1953, a number of ministries were merged.[10] The rights of the USSR ministers were expanded and their responsibilities were increased. The ministers were allowed more latitude in the distribution of investments, salaries and rewards among their enterprises.[11]

The merged ministries lasted for about a year; at the beginning of 1954, the ministries were unmerged.[12] However, the rights and responsibilities accorded to the ministers in April 1953 were not reviewed. The first post-Stalin ministerial reorganization was attributed to Beria and Malenkov, particularly to Beria's desire to consolidate authority in the hands of fewer ministers and deprive the Party of any involvement in the administration of the economy.[13] With Beria's removal, a further step departing from Stalinist practices was made: the approach towards improving administration of the economy changed from expanding the rights of the all-Union ministries to expanding the executive rights of the republican governments.

Involving the republics in management of the economy

The rivalry between Malenkov and Khrushchev had an impact on the *apparat*. In November 1953, Malenkov insisted on a leading role for the government in the economic administration. He accused highly positioned Party leaders of corruption and threatened to prevent them from controlling the governmental agencies.[14] Khrushchev, in turn, attacked the governmental *apparat*. He sought support in the fight against all types of manifested bureaucratism in the work of governmental and soviet institutions.[15] Malenkov's incapacity to protect the governmental *apparat* from criticism in 1954 had, to a large extent, contributed to it becoming the scapegoat for the economic troubles. As of 1954, the Party had been gradually consolidating its role as leader of the economic administration, first by blaming the economic problems on bureaucratism in the governmental *apparat*, and then by gradually interfering with the monopoly of all-Union ministries (by devolving certain executive rights to the republican ministries and governments).

In August 1954, the CC CPSU and the central government revealed financial malpractices exercised by the ministries. For example, by withdrawing above-plan profits and a share of circulation funds[16] from successful enterprises, the ministries[17] covered up evidence of mismanagement. These manipulations allowed the ministries to pass poorly managed enterprises off as successful.[18] At the same time, the ministries were not taking any measures to hold the directors of enterprises accountable for the overuse of material and financial resources.[19] Two months later, the ministries were blamed for certain administrative shortcomings,[20] such as excessive bureaucratic correspondence, 'freezing' material resources and making materials rise in price, or excessive centralization of industrial management, 'because only a small number of enterprises and organizations were within the republican jurisdiction'.[21] As a result, the USSR ministries were instructed to continue transferring administration of enterprises and organizations to the republican ministries.[22]

Ukraine had been participating in the reorganization process since 1953. In 1953, the republic reduced the number of its governmental institutions[23] and, at the same time, actively criticized the style of work of the all-Union ministries, their ignorance of Ukrainian needs and economic administrative suggestions. The criticism of the all-Union ministries might have had a certain impact. At the beginning of 1954, Moscow approved the Ukrainian proposals to transfer some all-Union enterprises to republican control. As a result, two Ukrainian industrial ministries were created: the Ukrainian Ministry of Ferrous Metallurgy in February 1954 and the Ukrainian Ministry of Coal in April.[24]

The Ukrainian authorities were also motivated to ask for the extension of the rights and responsibilities of the republican governmental organs.[25] In addition to the two ministries, they asked to create a few more, and to transfer to them, from the all-Union ministries, administration of all enterprises, organizations, trusts and educational institutes. Regarding the previously existing ministries, Ukraine requested expansion of their authority, as well as transfer of all the enterprises, administrations, trusts, *sovkhozy* (*sovetskie khozyaistva* or state farms) and other organizations from the jurisdiction of all-Union ministries to the jurisdiction of Ukrainian Union-republican ministries. It was at this time that the Ukrainian authorities brought up one most sensitive aspect of Moscow–Kiev economic relations: control of information. Ukrainians requested the transfer of the Statistical Agency of Ukraine, thus far under the sole jurisdiction of the USSR CM Central Statistical Agency, to the joint jurisdiction of the CM of Ukraine and the USSR CM Central Statistical Agency.[26]

Although the jurisdiction of the Statistical Agency of Ukraine was not changed, some other Ukrainian propositions were accepted.[27] In 1954 the central leadership started to concede authority to Ukraine, as well as to other republics.[28] At the end of 1954, Khrushchev initiated the replacement of Malenkov by Bulganin. It was no surprise that Bulganin guaranteed continuation of the currently defined ministerial policy.

At the beginning of 1955, the republics prepared their proposals revising their economic involvement. A governmental commission that included representatives from the RSFSR, Belorussia, Kazakhstan, Georgia, Uzbekistan and Ukraine with Kosygin at the head studied the republican propositions.[29] Based on the materials of the Kosygin commission, on 4 May 1955, the USSR CM issued a resolution that proposed modifications to certain planning and financing procedures. First, the CM of a republic could now approve production and supply plans for managing production in its jurisdiction.[30] Second, in the USSR State plan, the targets of the volumes of gross output and commodity output for the republics were fixed in total:[31] the republican governments would then assign the targets to the enterprises. Finally, the republics obtained the right to keep 50 per cent of construction materials, agricultural products and even of agricultural equipment that were manufactured by the enterprises of all-Union and Union-republican jurisdiction in order to sell it inside the republic.[32] They could also now distribute construction materials manufactured in their territory by all-Union ministries, after the latter retained what they needed.

All in all, the republican ministers got almost all the rights granted to the USSR Ministers in 1953,[33] except the right to manage either financing of construction or distribution of credits.[34] The critical remarks addressed to the all-Union ministries in 1955 were mainly the same as in 1954: the need to eliminate excessive centralization in the administration by abolishing useless organizations that were positioned between the ministries and enterprises. This would bring the administrative *apparat* nearer to production.[35]

Soon after the XX Party Congress, the Presidium initiated the reorganization of 13 ministries.[36] As inspiring as the idea of transferring enterprises of 13 ministries to the republics was, the adopted approach to the transfer was inconsistent. For example, the enterprises of the Ministry of Construction Materials were transferred to Ukraine, but enterprises of cement, slate and asbestos industries were kept under all-Union management; or light, foodstuffs, meat and dairy industries were to be managed by the republic, but enterprises that manufactured machines and spare parts were under the jurisdiction of Moscow.[37] The Ukrainian government's complaints of inconsistencies in the reorganization[38] were not taken into account.

Indeed, the central government had not intended to give up total control over *any* industry.

After his summer 1956 visits to the regions of Urals, Siberia, Kazakhstan and the Ukrainian Donbas, Khrushchev focused on criticizing the planning organs, USSR Gosplan and *Gosekonomkomissiya*. Previously, the planning organs had been criticized less than the ministries. In 1955, they were just advised to focus more on efficient location of production, on bringing industry nearer to sources of raw materials, fuel and consumers, and on the *complex development* of economic regions.[39] In November 1956, however, Khrushchev claimed that the sixth five-year plan and the plan for 1957 had to be reviewed. His criticism of the planning organs was the focus of the forthcoming 20–24 December CC CPSU Plenum. Bulganin then called for vigorous eradication of the shortcomings in planning that were slowing down the implementation of economic programmes initiated by the Party.[40] He reminded the audience of the consequences of excessive centralization and confirmed the direction approved by the XX Party Congress towards the increase of republican role in economic administration. However, he also ensured that the rights and responsibilities of ministers and heads of main departments of ministries as well as of directors of enterprises were to be considerably extended.[41]

In the Party Presidium it was agreed that the system of economic management, as it was, was no longer acceptable. It seemed obvious that the solution for everyone was to continue transferring certain authority and responsibility for managing the economy from the USSR ministries to the republican ministries. However, to continue modification of the ministerial system was not an option for Khrushchev. He did not trust the ministries. Moreover, extension of ministerial authority and multiple reorganizations of ministries did not improve the economic administration as much, or as fast, as Khrushchev had hoped. In addition, there was also Khrushchev's love of radical solutions. Following the general return to Leninist principles in the leadership of the country, Khrushchev appealed to the principle of regional production management.

Moscow–Kiev: developing cooperation in planning and supply

When it launched the reform, the Soviet leadership made at least two assumptions. First, that both the *sovnarkhozy* and republican Gosplans would put all-Union interests first, and republican and regional interests second, as they cooperated to plan and manage the economy efficiently. Second, that the *sovnarkhozy/* republics would produce more at lower cost. At the same time, the *sovnarkhozy* and the republican governments assumed that the reform was designed to provide conditions for the development of their regional economies. The Regulations on the *Sovnarkhoz*, which granted vast planning and management authorities to the regions, demonstrated the confidence of the central leadership in its assumptions while it reassured the regions of the veracity of their assumptions.

The assumptions of both sides appeared to be wrong. From the examples of the 1958 production plan and the management of inter-republic deliveries and resources during the first and last reform years, we will show that, although the administrative system seemed to have changed and the republics had a most important role to play in economic development, the central government failed to create the necessary incentives to induce the republics to pursue all-Union interests. Instead, the incentives of the *sovnarkhozy* and the Ukrainian government were basically similar to those of the abolished ministries.

Plan for 1958

The draft of the 1958 production plan started with the enterprises, continued in the *sovnarkhozy*/preserved ministries, then moved to the republican Gosplans and finished in USSR Gosplan.[42] This procedure was supposed to identify the needs and economic potential of the enterprises and regions more effectively than that used by the ministries.[43] When approved, the plan would reach the enterprises first through the republican Gosplans, then the *sovnarkhozy* or ministries. But the plan drafts composed by the Ukrainian *sovnarkhozy* were found unacceptable by Gosplan of Ukraine. Among others, the *sovnarkhozy* asked for considerable increases in investment. Gosplan of Ukraine had to draft a proposal almost from scratch. Yet in the draft submitted by Gosplan of Ukraine, USSR Gosplan had still to modify 300 plan indices. For 120 types of production, targets were higher than those in 1957. For more than 100 types of production, targets were lower than those in 1957.[44]

USSR Gosplan explained the reasons for these specifications:

> While estimating the 1957 plan fulfilment of gross output, the CM of Ukraine had apparently not taken everything into consideration and the number was underestimated by 4.7 billion roubles. In the submitted 1958 plan draft, gross output also happened to be understated by six billion roubles. Consequently commodity output was underestimated by 5.3 billion roubles in 1957 and 5.7 billion roubles in 1958.[45]

Having compared its own calculations with those of Ukraine and other republics, USSR Gosplan concluded that the expectation of the central government that the regions/the republic(s) would reveal hidden production capacities if the composition of the plan started at the enterprises was not confirmed. At the same time, with respect to the correlation of 'production-consumption-deliveries outside the republic', USSR Gosplan concluded:

> When planning an inconsiderable increase of the production of some products that are important for the entire economy, the Council of Ministers of Ukraine had envisaged consumption of these products bigger than their production.[46]

As Table 7.1 shows, Ukraine indeed planned to increase consumption of the production of heavy industry inside the republic at the expense of exports to other republics.

Table 7.1 Summary of some 1958 plan indices presented by Ukraine and rejected by USSR Gosplan

	Planned increase of production in comparison to 1957		Planned increase of consumption in comparison to 1957	
	Thousand tonnes	%	Thousand tonnes	%
Cast iron	747	4.0	1457	10.0
Rolled metal	785	4.9	1810	21.5
Fertilizer	269	8.4	Not available	200
Cement	629	Not available	2850	500

Source: based on RGAE, f. 4372, op. 57, d. 203, ll. 61–2

According to USSR Gosplan, if these numbers were accepted, the export of rolled metal to other republics would decrease by more than one million tonnes, of cement by almost two million tonnes and of fertilizer by more than one million tonnes.[47] Of course, the fact that Ukraine needed these tonnes was not an issue.

Along with the production of heavy industry, the Ukrainian government intended to increase the consumption of consumer goods in exactly the same way (see Table 7.2).

Compared to 1957, Gosplan of Ukraine planned to increase the republican turnover of consumer goods by 12.4 per cent.[48] For USSR Gosplan, an increase in the consumption of consumer goods at the expense of exports was even less acceptable.[49] Ukraine's plan to keep its internal production was called 'unhealthy and parasitical' by USSR Gosplan.[50] The Ukrainian 1958 plan draft and its discussion in USSR Gosplan revealed that Ukraine perceived the reform as a chance to improve the economic situation in the republic. That perception was wrong. The republican economies were, by 1957, interrelated to such an extent that, if allowed, Ukraine with its 19 per cent of the entire Soviet gross product would destroy established and expedient inter-republic economic relationships, and 'that would provoke other republics to set up production which is more profitable in Ukraine'.[51] USSR Gosplan could not accept this. Allowing the republics to decide which industries to develop would undermine the Soviet economy as well as the coordination/regulation function of USSR Gosplan.

The situation with the Ukrainian 1958 plan draft was not exceptional.[52] Increasing investments while decreasing production targets was typical of the drafts of all republics and of all economic regions.[53] The 1958 plan was the first, and the last, composed by the republics without Moscow's guidance. The next plan they would draft was based on the production volumes and investments[54] determined by USSR Gosplan.

Table 7.2 Planned domestic supply and export of certain consumer goods in 1958

	Increase of domestic supply relative to 1957 (%)	Decrease of deliveries outside the republic relative to 1957 (%)
Motorcycles	10	30
Electric irons	47	51
Electric lamps	9.5	32
High quality china	5	23
Enamel and iron cooking ware	7	23
Vacuum cleaners	10	13

Source: RGAE, f. 4372, op. 57, d. 167, l. 10; RGAE, f. 4372, op. 57, d. 203, l. 62

The supply system: a test for the reform

From the very beginning of the reform, it became obvious that the administrative and economic mechanisms that would allow the supply system to work smoothly were not in place. The advantages of not fulfilling the supply contracts exceeded the disadvantages. This encouraged the non-fulfilment of cooperative deliveries along with the concealment of material and financial resources by the Ukrainian *sovnarkhozy*, further straining relations between Moscow and Kiev.

One purpose of the reform was to eliminate unnecessary transportation expenses by encouraging economic relations either within a region or within neighbouring regions.[55] When proposing this idea to the regions, the central leadership had overlooked the impact of interdependence between regional economies that were at different levels of industrialization. As a result, some regions that were less dependent on others and had the capacity to replace remotely located suppliers, focused rather on developing their own economies at the expense of remotely located customers.

The decision to give priority to local needs over supplying remote regions, made by Stalino and Dnepropetrovsk *sovnarkhozy* already in 1957, was considered scandalous in Moscow for at least one reason: the plan for 1957 was worked out not by the *sovnarkhozy* but by the ministries and could only be roughly adjusted to the *sovnarkhozy* system. The full revision of cooperative relations could not be reviewed in the middle of the year. Thereby customers from outside a republic could count only on suppliers that they 'inherited' from ministries. The reaction of the authorities to the wilfulness of the Ukrainian *sovnarkhozy* was natural. On 3 January 1958, the CC CPU Presidium issued a resolution 'On the responsibility of the chairmen of the *sovnarkhozy*, enterprises and *snabsbyty* (supply agencies) for opportune deliveries of production'.

By April 1958, the USSR CM and the CC CPSU had issued a resolution to modify paragraph 26 of The Regulations on the *Sovnarkhoz*. In the original paragraph 26, it was stated that the economic plan of an enterprise or organization was considered unfulfilled if the cooperative deliveries were not carried out: 'The bonus established for the fulfilment of the plan is not paid in this case.'[56] The modification stipulated that inter-republic deliveries and deliveries into all-Union funds had to be ensured regardless of the impact on the production plan.[57] On 24 April, the USSR Supreme Soviet Presidium issued a resolution stating that if the enterprises, economic organizations, *sovnarkhozy*, ministries or departments failed to meet the planned targets of deliveries to other economic administrative regions or into the all-Union fund by enterprises they would be in serious violation of state discipline, which entailed disciplinary action, financial penalties or criminal charges.[58]

The resolutions had a very limited impact. The *sovnarkhozy* could get away with non-deliveries for at least two reasons. First, because adjusting production plans to match actual production was a typical practice. This made it difficult to relate the realization of deliveries to plan fulfilment. Second, because the system that would enforce disciplinary, financial or criminal penalties did not function properly.[59] Violation of delivery contracts was, in the end, not so dangerous for careers. In fact, the *sovnarkhozy* could sometimes profit from not delivering by, for example, minimizing long-distance transportation costs. For inter-republic deliveries, USSR Gosplan provided resources. Consequently, the *sovnarkhozy* received funds that they might not receive otherwise. So, if rationally spent, the funds could be used for above-plan production, which was better remunerated.[60] Along with this, the *sovnarkhozy* were still judged primarily on production volumes. Delivering to another enterprise of the same *sovnarkhoz* (or not delivering at all) could contribute to fulfilment (or over-fulfilment) of the plan by the neighbouring enterprise and/or contribute to accumulation of excessive material stocks of usually deficit production. Gosplan of Ukraine was supposed to ensure that the non-Ukrainian customers received their deliveries. However, Gosplan of Ukraine was also judged according to production volumes. In the end, its incentives to force the *sovnarkhozy* to realize the deliveries (by, for example, firing the chairmen of the *sovnarkhozy* or directors of enterprises) were not strong enough.

In order to ensure the inter-republic deliveries, *Glavsnabsbyty* of USSR Gosplan, (*Glavnye upravleniya po snabzheniyu i sbytu,* or the main supply and disposal administrations) started violating procedures and establishing direct relations with large Ukrainian exporters. Several tactics were adopted when bypassing Gosplan of Ukraine and the *sovnarkhozy*. On some occasions, USSR Gosplan issued orders for the production for inter-republic deliveries directly to the production sites instead of to Gosplan of Ukraine. *Snabsbyty,* or supply and disposal departments, of Gosplan of Ukraine had no authority to reassign these orders to another plant. Thus they could control neither production nor inter-republic deliveries.[61]

Sometimes, the USSR Gosplan did not provide Gosplan of Ukraine with the plan for inter-republic deliveries and consumption across the Soviet Union.[62] The central government probably hoped to avoid republican judgement about who produced and who consumed how much. Ukrainian Gosplan was thus managing deliveries in the dark.[63] It is important to note that even though Gosplan of Ukraine was bypassed, it was held responsible for meeting manufacturing and delivering commitments. In October 1958 Podgornyi, First secretary of the CC CPU, tried to persuade the CC CPSU to decentralize administration of the statistical system.[64] The Ukrainian government would then be better informed of republican economic performance and be better able to influence the *sovnarkhozy*. However in 1958, as well as in 1959, Ukraine was denied in its request. Only in 1960 would the Statistical Agency of Ukraine be transferred to the direct jurisdiction of the CM of Ukraine.[65]

Such a policy, obviously directed at minimizing the authority of Gosplan of Ukraine over the Ukrainian economy, could not but provoke frustration in Gosplan of Ukraine with the growing power of USSR Gosplan and justify questioning the fairness of inter-republic economic relations. The fact is, when it bypassed Gosplan of Ukraine in inter-republic deliveries, USSR Gosplan violated official procedures. In May 1959, the Ukrainian government suggested that USSR Gosplan share responsibility for orchestrating deliveries with the republican Gosplans.[66] Since the beginning of 1960, the CM of Ukraine questioned the priority of inter-republic deliveries pointing at the same time to a political factor involved in production distribution:

> In certain cases, deliveries inside the republic or inside the region are of particularly important significance and are due priority fulfilment. Considering this, it is expedient to grant the right to USSR Gosplan to allow republics to deliver first where it is needed the most, before fulfilling inter-republic deliveries.[67]

Ukrsovnarkhoz: managing inter-republic deliveries after the November 1962 CC CPSU Plenum

In 1960, the central government decided to impose an additional economic administrative institution on Ukraine, the Ukrainian Republican *Sovnarkhoz* or *Ukrsovnarkhoz*. During the first two years of its existence *Ukrsovnarkhoz* was in an ambiguous position: it was instructed to fulfil important administrative functions – to coordinate the work of the *sovnarkhozy* and ensure that all plans were fulfilled and deliveries realized – yet it was entrusted with very limited authority. Its position started to consolidate only after November 1962. At the 19–23 November 1962 CC CPSU Plenum, Khrushchev announced a change in the approach to the economic administration: along with territorial administration of industrial production, scientific and engineering work had to be administered according to the branch principle.[68] The new administrative policy was reflected in the institutional rearrangement. USSR *Sovnarkhoz* appeared on 24 November 1962. The same day, *Gosekonomsovet*[69] was abolished and USSR Gosplan took over the perspective planning previously performed by *Gosekonomsovet*. Current planning was taken from USSR Gosplan and given to USSR *Sovnarkhoz*.[70] USSR *Gosstroi*,[71] which took over the administration of construction, had to become, as Khrushchev said, 'very centralized, Union-republican, with strict centralization'.[72] After the Plenum, a number of State branch committees were created in order to orchestrate the development of the industries centrally. All in all, the complex of the institutional rearrangements was providing the Moscow governmental functionaries with greatly improved conditions for developing the industrial branches and thus for gradually transferring the administration of the economy back to the centre.

It was due to the rearrangement of the institutional system in Moscow that the authority of *Ukrsovnarkhoz* was secured and stabilized. *Ukrsovnarkhoz* was taking over Gosplan of Ukraine. It was assigned an important role to play in the recentralization process conducted by Moscow. In particular, *Ukrsovnarkhoz* was instructed to increase the *nomenklatura* for the centralized distribution, to establish control over the realization of important deliveries and to draft detailed reports on all deliveries. Yet, being put in charge of Ukrainian industry, *Ukrsovnarkhoz* became as pro-republican as other Ukrainian republican institutions and it cooperated with them when lobbying for Ukrainian interests in fixing planning indices or analysing interruptions of inter-republic deliveries.

By the time of the November 1962 Plenum, the situation with mutual deliveries between the republics as well as between the economic regions had stabilized: the deliveries were irregular and lots of lobbying was necessary to get what was needed, when it was needed. However, after the 1962 November Plenum, the Ukrainian authorities adopted a new methodology for analysing the failures of inter-republic deliveries. Along with instructing the *sovnarkhozy* to correct the situation, they analysed the reasons why the *sovnarkhozy* failed to deliver. The contribution of the central organs was obvious to the Ukrainians. The central planning organs failed to correlate supply plans with production plans, thus the *sovnarkhozy* failed to produce for inter-republic deliveries. The central planning organs failed to accurately estimate the necessary supplies, so the *sovnarkhozy* failed to produce.

From as early as 1959, when the inter-republic deliveries were mostly transferred to the control of *Glavsnabsbyty* of USSR Gosplan, but mainly after the November 1962 Plenum, the Ukrainian central authorities were insisting on holding the central planning organs responsible for the malfunctioning of the supply system. The Ukrainian authorities, while staying loyal to the system, allowed themselves to challenge it. The *sovnarkhozy* did not deliver, but they could not always be held accountable for this. This was the main economic and political accent that the Ukrainian authorities adopted when talking about inter-republic deliveries.

Ukrsovnarkhoz: managing resources

The Ukrainian administrative authorities took a similar approach when it came to analysing the problem of the concealment of material resources by the *sovnarkhozy*, which, along with disruptions of inter-republic deliveries, caused a lot of damage to the supply system. They studied the causes of the problem. At the end of 1963, *Ukrsovnarkhoz* claimed that the reasons for accumulating above-norm stocks of raw materials and other materials were non-fulfilment of the production plan by certain enterprises; multiple modifications in the planned assortment of goods; design modifications introduced in the middle of production cycles; modifications in construction projects, which led to cancellations of orders and to rejections of previously ordered equipment; deliveries of raw materials and materials ahead of schedule; plus some other reasons.[73] An important conclusion of *Ukrsovnarkhoz* was that, in the majority of cases, the reasons that led enterprises to accumulate materials and raw materials above the norms 'did not depend on the enterprises'.[74]

It is interesting to note that the conclusions about the reasons for the accumulating stocks made by *Ukrsovnarkhoz* in 1963 were different to the conclusions it had made a year earlier. In September 1962, *Ukrsovnarkhoz* mentioned two things that allowed the enterprises to accumulate material: obsolete limits on the allowances for storing volumes of production and deliberate over-supply of resources. According to *Ukrsovnarkhoz*, the enterprises *chose* to hoard (that is, to store above the allowed norms); the hoarding was deliberate. Typically, the purpose of hoarding – by applying out-of-date norms or by deliberately supplying too much – was to have material available either to exchange with other enterprises or to handle changes in the production plans, urgent orders, interruptions of supplies, etc. An important factor that encouraged hoarding was that the functionaries responsible for the accumulation of resources bore ethical responsibility only[75] – in other words, the worst that might happen was that they would be denounced by society which would have a negative impact on their reputation, but this cost them very little.

In 1962, when reporting to the USSR CM on the issue, the Ukrainian authorities did not develop the logic of the enterprises further. They did not analyse why the enterprises had to consider 'just in case' situations and why they had to count on exchange with other enterprises or *sovnarkhozy*. Only a year later would the reasons for *mestnichestvo* in accumulating stocks be analysed. Instead, in 1962 *Ukrsovnarkhoz* had used this example of *mestnichestvo* tendencies developed among directors of enterprises and in the *sovnarkhozy* to explain how economic stimulation and material incentives for workers, functionaries and directors could reduce the accumulation of materials and improve the fulfilment of various types of plans in general (i.e. production, production cost, profit, and so on).

The presentation of the issue to the USSR CM in 1962 and in 1963 depended on the general context. When trying to get more financial freedom for the Ukrainian enterprises (through extending their rights) just before the November 1962 Plenum, *Ukrsovnarkhoz* preferred not to bring up the contribution of the central planning organs to the management of resources. *Ukrsovnarkhoz* tried to make Moscow reconsider the importance of the human factor in the management of enterprises and to accept the fact that financial rewards would be a strong incentive for improving the economic situation in general; whereas after the November 1962 Plenum, when the institutional organization in Moscow had been changed, the Ukrainian authorities had to make sure that the balance of authority between republican and central economic institutions in administrating the republican economy was preserved. Therefore, by explaining that the causes for accumulating stocks lay outside Ukraine, Ukrainian authorities tried to prevent further centralization of economic administration in general and of the management of supplies in particular.

At the beginning of 1964 the USSR Supreme *Sovnarkhoz* came up with a draft of a resolution on reduction of the above-norm accumulations of materials and products stocked in enterprises. At this point, the CM of Ukraine had openly stated that 'in a number of cases, formation of above-norms stocks of unfinished and finished products was through the fault of all-Union planning organizations'.[76] Therefore, the Ukrainians proposed that in its resolution the USSR Supreme *Sovnarkhoz* should include paragraphs on the responsibility of central organizations for all changes in the production process that induced the enterprises to accumulate excessive material stocks.

Ukrsovnarkhoz did not and could not meet Moscow's expectations. Having practically replaced Gosplan of Ukraine, *Ukrsovnarkhoz* had itself turned into Gosplan of Ukraine.[77] Along with the functions of the current planning and current administration of the Ukrainian industry previously performed by Gosplan of Ukraine, *Ukrsovnarkhoz* adopted the attitude and understanding that the latter had about the importance of Ukrainian industry to the country. Thereby, being placed by Moscow as an intermediary link between the central institutions and the Ukrainian economic regions, *Ukrsovnarkhoz* could not play a major role in the recentralization processes and lobbied for Ukrainian interests in Moscow rather than for all-Union interests in Ukraine.

Conclusion

The random attempts of the USSR Gosplan to consolidate its controlling and coordinating role during the first years of the reform became systematic with the appointment of Kosygin as its chairman in 1959. Yet the recentralization of the economic administration had not become an official policy until Khrushchev made it official at the November 1962 Plenum, having admitted that the reform had to be curtailed. The new hybrid administrative system, in addition to Khrushchev's other novelties such as the bifurcation of the Party organs, created most cumbersome conditions for the bureaucracy. Yet, at least for the Ukrainian authorities, the recentralization launched by Khrushchev himself meant, first and foremost, his personal withdrawal from his pro-republican policy in the economic administration. This was not forgotten in October 1964.[78]

Despite the fact that the republics did not manage to gain more latitude in running industry throughout the reform years, and that they rather lost some of the economic rights given to them in 1957, the reform marked a major breakthrough in centre–republic relations. Having faced the scale of *mestnichestvo* among the republican Party and governmental authorities, Moscow understood that Stalin had not succeeded in suppressing their will for power and that they had a clear vision of the economic development of their republic, which did not always coincide with the USSR Gosplan's projects. Relations between the republics and Moscow had and could never reach the point of partnership. However, they had matured enough so that the dismissal of Khrushchev did not automatically lead to the abandonment of the *sovnarkhoz* system. It still took Brezhnev almost a year to bring the ministerial system back, which in fact was not a restoration of the pre-1957 system. The experience the Ukrainian politicians and managers had gained since 1957 gave them confidence in their ability to run industry, and allowed them to insist on organizing their own ministries for Ukraine's leading industries.[79] How well they ran the economy after 1965 is a subject for further research, but once again, the ministerial system that emerged after the *sovnarkhozy* was just right for them to continue pursuing their local interests. For the centre, the problem of local interest groups, or clans, thus never ceased to exist.

Notes

1 On the *sovnarkhozy* in the 1920s see, for example, T.P. Korzhikhina, *Sovetskoe gosudarstvo i ego uchrezhdeniya. Noyabr' 1917-dekabr' 1991*, Moscow: RGGU, 1995, pp. 58, 59, 132, 137, 138.

2 On the problems caused by ministerial departmentalism and which Khrushchev targeted in his note, see Note of N.S. Khrushchev 'Certain considerations on improvement of the organisation of administration of industry and construction', presented on 27 January 1957 to the CC CPSU Presidium. A.A. Fursenko (ed.) *Prezidium TsK KPSS 1954–1964 vol. 2: Postanovleniya. 1954–1958*, Moscow: ROSSPEN, 2006, pp. 522–39.

3 *Prezidium TsK KPSS*, vol. 2, pp. 535–6.

4 A.A. Fursenko (ed.), *Prezidium TsK KPSS. 1954–1964. Chernovye protokol'nye zapisi zasedanii. Stenogrammy. Postanovleniya. vol. 1.*, Moscow: ROSSPEN, 2004, p. 221.

5 *Prezidium TsK KPSS*, vol. 1, p. 221.

6 *Prezidium TsK KPSS*, vol. 1, p. 222.

7 *KPSS v rezolyutsiyakh i resheniyakh s"ezdov, konferentsii i plenumov TsK (1898–1986)*, Izd. 9-oe dop. i ispr., vol. 9, 1956–1960, Moscow: Izd-vo Politizdat, 1986, pp. 167–74.

8 *Gosekonomkomissiya* was abolished.

9 Molotov's note, dated 24 March 1957. *Prezidium TsK KPSS*, vol. 2, pp. 613–15.

10 Out of about 50 ministries in Moscow, on 5 March 1953 the first 15 ministries were amalgamated into five. Ten days later, 25 ministries were amalgamated into nine. Our calculations are based on the information provided by Ivkin. See V.I. Ivkin, *Gosudarstvennaya vlast' SSSR. Vysshie organy vlasti i upravleniya i ikh rukovoditeli. 1923–1991 gg. Istoriko-biograficheskii spravochnik*, Moscow: ROSSPEN, 1999, pp. 42–187.

11 Resolution from 11 April 1953. *Resheniya Partii i Pravitel'stva po khozyaistvennym voprosam (1917–1967)*, vol. 4, Moscow: Izdatel'stvo politicheskoi literatury, 1968, pp. 5–14.

12 According to Korzhikhina, once amalgamated the new gigantic ministries were not more efficient than before the amalgamation. According to Pikhoya, the reduction of the number of the ministries held at the beginning of 1953 provoked dissatisfaction among the Moscow governmental bureaucracy. See Korzhikhina, *Sovetskoe gosudarstvo*, p. 231; R.G. Pikhoya, *Sovetskii Soyuz: istoriya vlasti 1945–1991*, Moscow: RAGS, 1998, p. 236.

13 Mentioned at the CC CPSU Plenum held in June 1957, 'On the Anti-Party Group'. See, A.N. Yakovleva et al. (eds), *Molotov, Malenkov, Kaganovich. 1957. Stenogramma iyun'skogo plenuma TsK KPSS i drugie dokumenty*, Moscow: MFD, 1998, p. 45.

At the July 1953 CC CPSU Plenum dedicated to the crimes of Beria, Khrushchev talked about Beria's negative attitude to the involvement of the Party in governing the country. A.N. Yakovleva (ed.), *Lavrentii Beria. 1953. Stenogramma iyul'skogo plenuma TsK KPSS i drugie dokumenty*, Moscow: MFD, 1999, pp. 91–92.

14 W. Taubman, *Khrushchev: The Man and His Era*, New York and London: W.W. Norton & Company, 2003, p. 261.

15 A.V. Pyzhikov, *Khrushchevskaya 'ottepel'*, Moscow: Olma-Press, 2002, p. 90.

16 Circulation funds are funds at the disposal of enterprises.

17 The document does not specify the status of the ministries: all-Union, Union-republican or republican. But we would say that it concerns all-Union and Union-republican ministries: the ministries of republican status did not yet possess such authority. The resolution on the republican ministries would be issued in 1955. See *Resheniya*, vol. 4, pp. 129–44, 213–17.

18 *Resheniya*, vol. 4, p. 129.

19 *Resheniya*, vol. 4, p. 131.

20 Resolution from 14 October 1954, 'On considerable shortcomings in the structure of the USSR ministries and departments and measures on the improvement of the functioning of the governmental *apparat*'.

21 *Resheniya*, vol. 4, pp. 144, 145, 146.

22 *Resheniya*, vol. 4, p. 148.

23 Ukraine intended to reduce the number of governmental institutions from 66 to 32. TsDAVO, f. 2, op. 8, d. 10234, l. 59.

24 TsDAVO, f. 2, op. 9, d. 3088, l. 99 *oborot*.

25 On 8 May 1954, the central government and the CC CPSU received 38 pages of 'The propositions on the extension of the rights and responsibilities of the Ukrainian republican administrative organs and the administrative organs of *oblasti* and *rayony*', TsDAVO, f. 2, op. 9, d. 3088, ll. 98–136.

26 TsDAVO, F. 2, op. 9, d. 3088, l. 99.

27 Three out of four ministries that the Ukrainians were asking for were created in 1954: on 4 August, the Ministry of Rural and Urban Construction; on 28 December, the Ministry of Means of Communication; on 29 December, the Ministry of Higher Education. Moreover, during 1954 a certain number of enterprises, administrations and various organizations were transferred from all-Union to Union-republican jurisdiction. TsDAVO, f. 2, op. 9, d. 3088, l. 99.

28 According to I. Koropeckyj, the concession started in 1955. I.S. Koropeckyj, 'Economic Prerogatives', I.S. Koropeckyj (ed.), *The Ukraine Within the USSR: An Economic Balance Sheet*, New York: Praeger Publishers, 1977, p. 26.

According to Gerhard Simon, the period of decentralization was in the period between 1954 and 1958. See, G. Simon, *Nationalism and Policy toward the Nationalities in the Soviet Union: from Totalitarian Dictatorship to Post-Stalinist Society*, transl. by Karen Forster and Oswald Forster, Boulder, CO; Westview Press, 1991, p. 233.

29 TsDAVO, f. 2, op. 9, d. 329, ll. 97–126.

30 The share of production under republican jurisdiction was growing. In the years 1954–55, more than 11,000 enterprises were transferred to the jurisdiction of republics. While, in 1950, 67 per cent of enterprises were all-Union and 33 per cent were under republican or local jurisdiction, in 1955, the share of all-Union enterprises had dropped to 53 per cent. Korzhikhina, *Sovetskoe gosudarstvo*, pp. 231–2; *Resheniya*, vol. 4, p. 200.

31 *Resheniya*, vol. 4, p. 200.

32 TsDAGO, f. 1, op. 24, d. 4489, l. 139.

33 32 out of 36 paragraphs on the rights of republican ministers in 1955 repeated 32 out of 47 paragraphs of the rights of the USSR Ministers in 1953. *Resheniya*, vol. 4, pp. 213–17.

34 Simon, *Nationalism and Policy toward the Nationalities in the Soviet Union*, p. 235.
35 *Resheniya*, vol. 4, p. 237.
36 *Prezidium TsK KPSS*, vol. 1, p. 109; *Prezidium TsK KPSS*, t. 2, p. 208.
37 *Prezidium TsK KPSS*, vol. 2, p. 304.
38 TsDAGO, f. 1, op. 24, d. 4257, ll. 215–18.
39 For the first time since 1943 the CC talked about the economic regions and about their development. *Resheniya*, vol. 4, p. 241.
40 Mentioning both USSR Gosplan and *Gosekonomkomissiya*, Bulganin focused on the mistakes of the latter. As a result, from 25 December, Saburov was no longer the Chairman of *Gosekonomkomissiya*, although he kept his position of the USSR CM First Vice-chairman. Saburov was replaced by Pervukhin, another USSR CM Vice-chairman. RGANI, f. 2, op. 1, d. 203, l. 69, annex in Pyzhikov, *Khrushchevskaya 'ottepel'*, p. 372; Ivkin, *Gosudarstvennaya vlast' SSSR*, pp. 143, 466, 509.
41 RGANI, f. 2, op.1, d. 203, l. 69, annex in Pyzhikov, *Khrushchevskaya 'ottepel'*, pp. 370, 371.
42 From May 1957, 'Gosplan' stood for *Gosudarstvennyi planovyi komitet*, or the state planning committee. Gosplan was now in charge of both perspective and current planning.
43 'Zadachi Gosplana SSSR v novykh usloviyakh upravleniya promyshlennost'yu i stroitel'stvom', *Planovoe khozyaistvo*, 1957, no. 7, July, p. 7; 'Perestroika upravleniya promyshlennost'yu i stroitel'stvom i zadachi planovykh organov', *Planovoe khozyaistvo*, 1957, no. 5, May, p. 9.
44 RGAE, f. 4372, op. 57, d. 167, ll. 8, 9.
45 RGAE, f. 4372, op. 57, d. 167, l. 8.
46 RGAE, f. 4372, op. 57, d. 167, l. 9.
47 RGAE, f. 4372, op. 57, d. 203, l. 62.
48 RGAE, f. 4372, op. 57, d. 203, l. 62.
49 RGAE, f. 4372, op. 57, d. 167, l. 10.
50 RGAE, f. 4372, op. 57, d. 203, l. 61.
51 RGAE, f. 4372, op. 57, d. 167, l. 10.
52 Even the proposal of the plan composed by RSFSR reflected the same tendencies as the Ukrainian proposal. RGAE, f. 4372, op. 56, d. 1, ll. 43, 45.
53 RGAE, f. 4372, op. 57, d. 167, l. 49.
54 RGAE, f. 4372, op. 57, d. 167, l. 49.
55 In 1957, Ukraine, for example, had incurred 350 million roubles of unproductive expenditures for the transportation of imported equipment from other republics and from abroad – equipment that, according to Gosplan of Ukraine, could have been produced in the republic. TsDAVO, f. 337, op. 2, d. 719, l. 486.
56 RGAE, f. 4372, op. 76s, d. 110, l. 160, § 26.
57 TsDAVO, f. 2, op. 9, d. 7206, l. 43.
58 TsDAGO, f. 1, op. 31, d. 1117, l. 136; I. Evenko, 'Proverka i ekonomicheskii analiz vypolneniya planov', *Planovoe khozyaistvo*, 1958, no. 10, p. 11.
59 In 1959, A. Stoyantsev, the chairman of the Commission of Soviet control of the Ukrainian CM informed the CC CPU that in L'vov *sovnarkhoz*, 14 directors and main engineers were fired for not fulfilling their management tasks. At the same time, some of those who were fired were assigned at a new enterprise to perform the same functions. TsDAGO, f. 1, op. 31, d. 1403, l. 25.
60 V. Andriyanov, *Kosygin*, Moscow: Molodaya gvardiya, 2004, pp. 137, 138.
61 TsDAVO, f. 337, op. 2, d. 874, ll. 112, 123; TsDAVO, f. 337, op. 2, d. 876, ll. 249, 250; TsDAVO, f. 337, op. 2, d. 987, ll. 89–90; TsDAVO, f. 2, op. 9, d. 7201, ll. 74, 77–8.
62 TsDAVO, f. 337, op. 2, d. 874, l. 112; TsDAVO, f. 337, op. 2, d. 876, l. 257.
63 According to Slysh from *Ukrglavmetallosnabsbyt* (*Glavnoe upravlenie po snabzheniyu i sbytu metalloproduktsii pri Gosplane Ukrainy*, or the Gosplan of Ukraine main

administration for supply and disposal of metal products). TsDAVO, f. 337, op. 2, d. 876, l. 257.

64 TsDAGO, f. 1, op. 24, d. 4687, l. 155. Since 1954, Ukraine had tried to get the Statistical Agency under its jurisdiction. TsDAVO, f. 2, op. 9, d. 3088, l. 99.

65 The department for industrial statistics was transferred under Ukraine's jurisdiction sooner than the department for general reports: industrial in February 1960 and general in May. From the introduction to fond 582, op. 2, 3.

66 TsDAVO, f. 337, op. 2, d. 874, l. 124.

67 TsDAVO, f. 2, op. 9, d. 7201, l. 84.

68 *Prezidium TsK KPSS*, vol. 1, pp. 584, 649.

69 *Gosekonomsovet* (*Gosudarstvennyi nauchno-ekonomicheskii sovet Soveta Ministrov SSSR*, or the State scientific and economic council of the USSR CM) was created in 1959 to work on perspective planning while leaving current planning to USSR Gosplan.

70 *Prezidium TsK KPSS*, vol. 1, p. 1104; Ivkin, *Gosudarstvennaya vlast' SSSR*, pp. 147, 148.

71 *Gosstroi* (*Gosudarstvennyi komitet Soveta Ministrov SSSR po stroitel'stvu*, or the USSR CM State committee for construction) was formed on 9 May 1950. When the *Sovnarkhoz* reform was launched in 1957, it was not abolished. Even the head of the committee was not replaced. Ivkin, *Gosudarstvennaya vlast' SSSR*, pp. 92, 93.

72 *Prezidium TsK KPSS*, vol. 1, p. 649.

73 TsDAVO, f. 4820, op. 1, d. 1762, ll. 107, 108.

74 TsDAVO, f. 4820, op. 1, d. 1762, l. 108.

75 TsDAVO, f. 4820, op. 1, d. 1629, ll. 198, 199.

76 TsDAVO, f. 4820, op. 1, d. 1762, l. 105.

77 At the beginning of 1963, after the November 1962 Plenum, the staff of Gosplan of Ukraine was to a large extent transferred to *Ukrsovnarkhoz*, including the Chairman of Gosplan, Rozenko, who was assigned as Chairman of *Ukrsovnarkhoz*. TsDAVO, f. 4820, op. 1, d. 736a, l. 166; TsDAVO, f. 2, op. 10, d. 2524, l. 3; TsDAGO, f. 1, op. 6, d. 3517, ll. 47, 108.

78 Shelest blamed Khrushchev that his leadership did not give the republics enough rights: responsibility was there but the rights were not. *Prezidium TsK KPSS*, vol. 1, p. 863.

79 TsDAGO, f. 1, op. 6, d. 3870, ll. 36, 37, 38, 39, 40, 41, 42.

8   Failings of the *Sovnarkhoz* reform

The Ukrainian experience

Valery Vasiliev

On 26 December 1957, eight months after the decision had been taken to create *sovnarkhozy* (decentralized economic councils), Nikita Khrushchev gave a speech at the plenum of the Central Committee of the Communist Party of Ukraine (CC CPU) in which he gave a positive evaluation of the start of their work. As First Secretary of the Communist Party of the Soviet Union (CPSU), he declared that their work was enabling people's initiatives to unfold. Khrushchev formulated his vision of the problem of governing society as follows:

> Comrades, you know we can't think that under communism we shall have some kind of single administrative centre from where we will govern and say which machines will go where. Above all, we have to be communists. Therefore we must govern in such a way that nobody governs for society, but society itself does the governing. Therefore we need decentralization; this means attracting a wider circle into governing the economy; so that we as an organization actually move in the right direction, towards a deepening of socialist construction in our state. This is good, but we need to go further, and we shall, and we shall broaden the attraction of the general public for governing and for the self-servicing of society.[1]

This idea of governing society in many ways differed from Stalin's approach, which focused on strengthening the role of the state and on centralization of power and government.[2]

Khrushchev's precepts, however, were rather contradictory. Speaking of the defects that had appeared in the activities of the *sovnarkhozy*, chiefly the late deliveries of raw materials and products, Khrushchev suggested that leaders of *sovnarkhozy* who had allowed the non-fulfilment of obligations should be put on trial, at first in communal and later in state courts. In his opinion, if communal government was not backed by firm discipline, then 'that is decomposition' like the Yugoslav variant of reform, signifying the collapse of the local Communist Party and of its attempts to build socialism. The Soviet leader emphasized that the Yugoslav government had been compelled to return to centralized government whereas the USSR had proceeded by a different route, preserving centralized planning and decentralized government.[3]

So what was the main aim in the attempts at decentralization? The Soviet government talked a great deal about departmental impenetrability and of the barriers that hindered the effective development of production in the economic areas of the country. However, analysing speeches by Khrushchev and other Soviet leaders enables the basic reason for attempts at decentralization to be explained on another plane: the Communist Party leaders, just as in Stalin's time, were trying to catch up and overtake the production of the world's most developed market economies, making use of people's initiative and drawing on additional resources that were known to local governing structures.[4]

Ukraine and the *sovnarkhoz* reform

Let us look at the speech of the first secretary of the CPU, Nikolai Podgornyi, at the XX conference of the CPU in January 1959. Podgornyi placed his main emphasis on the tasks of the seven-year plan: the all-round creation of the material and technical base of communism, the further strengthening of the economic and defensive might of the USSR, the more complete satisfaction of the growing material and spiritual requirements of the Soviet people. Podgornyi declared: 'This will be the decisive stage in the competition with the capitalist world, when the historic task will be achieved of catching up and overtaking the most developed capitalist countries in production ...'. It was assumed that the tempo of industrial production in 1959–65 would be extremely high. If, according to figures both of the absolute volume of production and of production per capita, the country emerged as first in the world, securing the highest standard of living in the world, then that would be a historic victory for socialism in the peaceful competition with capitalism.[5]

Ukraine, with its huge industrial base, would have to make a serious contribution to the realization of those ambitious plans. The control figures for the seven-year plan of the republic provided for an average 8.5 per cent annual growth of gross industrial production and a 55.5 per cent growth of labour productivity. Over the seven years, gross production would grow by 77 per cent, the means of production by 82 per cent and the production of consumer items by 67.5 per cent. The building of new and the extension of existing coal and iron ore mining and enrichment combines was assumed, as well as the completion of a cascade of hydroelectric stations on the River Dnieper, and the creation of a series of big thermal power stations, coal mines, oil refineries and chemical factories, and mainline pipelines. The production of consumer items was to rise by 2–2.5 times.[6] Even taking into account the fact that economic plans were generally related more to propaganda than to reality, we can note the scale of the projects, demanding great capital investment, resources and a precise organization of production.

There are a number of reasons for directing attention to Ukraine. About 19 per cent of Soviet gross industrial production originated there. In 1956 the republic produced 48.2 per cent of the Soviet Union's pig iron output, 37.7 per cent of its steel, 39 per cent of its rolled metal, 56 per cent of its iron ore, 32 per cent of its coal, 39 per cent of its tractors, 81.4 per cent of its mainline diesel locomotives,

25 per cent of its cement and 72 per cent of its sugar. Governing the republic's powerful industry were 34 all-union, 18 union-republic, and 15 republic ministries and departments. About 11,000 industrial enterprises were subordinated to these ministries and departments. Construction works were carried out by the 1,712 construction and erection organizations of 50 ministries and departments.[7]

The political leadership of the republic had close ties with Khrushchev, who had led the CPU from 1938 to 1949.[8] He was personally acquainted with many Ukrainian leaders right up to regional secretaries, through chairmen of regional executive committees, factory directors and kolkhoz chairmen, encountering them during his many trips to the republic with frequent face-to-face meetings in which he asked about the local situation. At work, and later during his trips in Ukraine, Khrushchev demonstratively went around in a Ukrainian embroidered shirt (*vyshivanka*) and a straw hat (*bryl'*), these being typical Ukrainian items of dress and headgear. In return, the first secretaries of the CC CPU (A. Kirichenko 1953–57, N. Podgornyi 1957–63 and P. Shelest 1963–72) supported Khrushchev up to 1964 in the Kremlin power struggle.[9]

Many leaders in all-union structures had substantial clienteles in the republic. Khrushchev devoted particular attention to the leaders of the Donetsk region (where he came from, and which he held in particular respect). Podgornyi, from 1963 secretary of the CC CPSU, relied in Kiev and Moscow on promotees from the Kharkov region, where he worked in the late 1940s and early 1950s as first secretary of the regional party committee. The CC CPSU secretary L. Brezhnev was patron of the flourishing Dnepropetrovsk group of leaders.[10]

The reorganization of central ministries in 1953 and 1954[11] was mirrored in Ukraine.[12] During the first half of 1954 union-republic ministries were established for ferrous metallurgy, coal, the fish industry, and the meat and milk products industry. In 1955 ministries of light industry, and of the communications, textile and wood-processing industries were set up.[13]

In July 1955 the plenum of the CPSU CC listed the causes of negative processes in the Soviet economy: excessive centralization of government, its bulkiness and multi-layered nature, and the existence of superfluous links.[14] The conclusion, the need for a certain decentralization of government, followed logically. It was obtained by means of delegating part of the rights, functions and responsibilities from the top to the lower levels. This transfer took place both in individual branches of government and in the main state institutions, from the USSR Council of Ministers to the governments of the union republics. From 1955 the ministers of industrial ministries received the right to keep at their disposal up to 5 per cent of the capital investment allocated to the branch, and also to create a 5 per cent reserve of material-technical resources. The rights of directors of enterprises in planning, capital construction and reconstruction of enterprises were somewhat extended.[15]

Serious changes took place in the functioning of republic organs of government. In accordance with the decree of the USSR Council of Ministers of 4 May 1955, 'On changes in the pattern of state planning and the financing of union-republic economies', extra income, with the fulfilment of a republic's budget, could be spent on financing the residential/communal sector and on social/cultural measures. The rights of the finance ministry of a republic, and of regional, urban and district financial departments of the executive committees of local soviets, were significantly widened. A republic was given the right to keep at its disposal 25–50 per cent of consumer goods produced in excess of the quarterly plan by enterprises of the all-union and union-republic ministries. The goods were sent for sale to the population. Significant sums of money, which could be spent independently, came to a republic government.

In 1956 the council of ministers of the Ukrainian Soviet Socialist Republic (and to a large degree the departments of the CC CPU) coordinated the activity of 12 union-republic and republic ministries: ferrous metallurgy, coal, light industry, timber, paper and wood processing, construction materials, meat and milk products, food products, fish, and textile industries, the Ukraine finance ministry and the republic ministry of local and fuel industry. During 1953–56, in Ukraine, more than 10,000 industrial enterprises moved from Soviet to republic subordination, and the share of the gross production of industries under republic or local supervision rose to 78 per cent.[16]

The Ukrainian government was in a subordinate position relative to the politburo/presidium of the CC CPU. The degree of delegation of authority from the all-union power centre to the sub-centre of power in Ukraine varied by period, but the CC CPU politburo always decided a host of administrative and economic questions. The Ukraine council of ministers had limited functions, and concentrated on the fulfilment of the republic budget and the activity of republic ministries. When in the mid-1950s the process began of transferring part of the authority and functions from the all-union to the republic level the government, having a small *apparat* in which there were no experienced administrative cadres, proved to be unprepared for it. Archival research has shown that there were many discussions devoted to the search for answers to questions of how to govern an enormous mass of industry, coordinating the activity of ministries in the republic and beyond its borders, and effectively bringing the new rights into action.

At the beginning of 1957 a new task came before the republic government: the move from the branch to the territorial principle of administering industry, in accordance with the initiatives of Khrushchev and of the CC CPSU plenum of February 1957. In the republic a commission of representatives of various republic structures was at work, and in April it sent its suggestions to the CC CPU. Their essence was to reform the *apparat* of the CC CPU, creating territorial departments, with branch sectors, to replace the branch departments. Insofar as it was precisely CC CPU departments that occupied themselves with administrative problems in the functioning of the republic's industry, it was not possible to lose control from their side over the situation in industry. They were called on to coordinate the activity of the *sovnarkhozy* of eight economic districts. At the same time the authors of the suggestions noted that they still could not understand who exactly would coordinate the development of industrial branches within the borders of the whole republic. It is indicative that nothing was said of the role of the republic government in guiding the industry that was being transferred to its management. The Ukraine council of ministers had to lead the regional executive committees and regional administrations, and likewise the district executive committees in the sphere of activity of local industry transferred to the local administrative organs. The second administrative 'vertical' could be the Ukrainian SSR Gosplan, which led the regional planning administrations, and likewise acted with the administrations of local industry that were subordinated to the regional executive committees.[17]

On 25 April 1957, at the CC CPU plenum, Kirichenko underlined the basic feature of the transformation (agreed with the CC CPSU and the USSR Council of Ministers): the liquidation of 12 union-republic and republic ministries; the transformation of the Ukraine state planning commission (the Ukraine Gosplan) into an organ of general state planning and material-technical supply in the republic, subordinated to the Ukraine council of ministers; and the creation of 11 *sovnarkhozy* in the economic administrative districts: Stalinsk (Donetsk), Lugansk (Voroshilovgrad), Dnepropetrovsk, Zaporozhe, Odessa, Kiev, Kharkov, Vinnitsa, Lvov, Stanislav (Ivano-Frankovsk) and Kherson.

'The *sovnarkhozy* will bear full responsibility for fulfilling production plans and for the results of the economic activity of enterprises subordinated to them in the economic district,' Kirichenko asserted, '[they] work out and carry out perspective and current plans of production, specialization and cooperation, material-technical supply, [and] decide current tasks in the development of industry and construction of the district arising from general state interests'. Responsibility for selection of cadres, the regions' economic development and raising people's participation was placed on party organizations at various levels. This meant that party committees had to keep a check on the activity of *sovnarkhoz* leaders.

From the point of view of the leaders of the CPU the role of the party organs in the new system of government did not diminish. Kirichenko's speeches appeared, at first sight, out of tune with the declaration by Khrushchev, cited above, about the need for a decentralization of the economy's administration:

> It is possible that somebody might mistakenly view the liquidation of ministries and creation of *sovnarkhozy* as a step towards weakening centralized leadership. The manifestation of parochial feelings among a few economic and political workers can lead to a weakening of state discipline in enterprises, and to the interests of separate economic districts standing in opposition to general state interests. Obviously, such harmful local manifestations must be strangled at birth.[18]

Yet Kirichenko well knew what he was talking about. Khrushchev, at a session of the USSR Supreme Soviet in May 1957, had pronounced on the danger of the emergence of a 'closed in' economy inside a district or republic under the new system of government, and the appearance of parochial positions on the part of the economic leaders of lower administrative levels. Such formulations allowed party organs to focus on checking *sovnarkhoz* activity and also on the struggle with subjectively interpreted 'local manifestations', although in real life it was hard to determine where to draw the line between the latter and the development of local initiative. On the one hand there were calls for using the governmental reform in the struggle against bureaucratic *apparats*, sources of conservative political tendencies of Stalinist bent. On the other hand party organs, as a constituent part of the bureaucratic *apparat*, had to keep track of the *sovnarkhozy*. The Soviet leadership's approaches to the creation of *sovnarkhozy* brought up a complex of intellectual and logical contradictions that led to certain governmental problems and political consequences. We will look at their basic features, taking Ukraine as an example.

Beginnings of the reform

In Ukraine, at the same time as the *sovnarkhozy* were formed, the ministries were liquidated (apart from those for the coal and ferrous metallurgy industries, which became republic ministries). Several republic committees and administrations were created, subordinated to the Ukraine council of ministers.[19] The reform was accompanied by the transfer of extra authority to the republic governments, into whose management came the setting of standards for industrial and food goods, retail prices, reorganizations of enterprises of republican or local subordination, allocation of individual material stocks, redistribution of some capital investment, and quarterly planning.[20] How did these components change with the *sovnarkhozy* in operation?

In this case it was a question of the particular features of current planning, because five- and seven-year plans were concerned more with ideological functions than with scientific planning and forecasting.[21] Up to 1954 the USSR Gosplan in its annual development plans laid down the basic tasks for all republic ministries and departments. From 1955 it laid down for the republics the indicators of gross output, size of labour force and wage fund. The USSR Gosplan formulated tasks for republic ministries on the basis of figures declared by the republic Gosplans.[22]

From 1956 the USSR Gosplan began to determine for republic ministries the tasks for output of gross and traded production, general volume of capital works, number of workers, size of wage fund and labour productivity increases. Planning of production and the allocation of all types of production from republic industry was transferred to the union republics. As a result there was an increase in the volume of Ukraine's Gosplan work.

In 1957 the Ukrainian leadership formulated the role of the Ukraine Gosplan as having 'the basic task to secure in the republic, on the basis of the state plan, the carrying through of the single centralized policy in development of the USSR economy as a whole and of its most important branches'.[23] In the ideal model (as conceived by the Ukrainian leaders) the Ukraine Gosplan, on the basis of the *sovnarkhoz* plans, coordinated planning in the economic administrative districts, and also coordinated the activity of the regional planning commissions for the development of local industry. Composing on this basis the general republic plans, the Ukraine Gosplan was to take an active part in working out that part of the general all-union plans that applied to Ukraine. Initially, in July 1957, to fulfil these functions the deputy chairmen of the Ukraine Gosplan entered the Ukraine Council of Ministers as ministers, and subsequently heads of departments came in as their deputies.[24] A certain merging of government and Gosplan functions occurred, with Gosplan accorded broad rights: fixing material balances and production reserve norms; inserting partial changes in the tasks of the annual plans; confirming and changing wholesale and retail prices. Its decisions were obligatory for *sovnarkhozy*, ministries, departments and executive committees of local soviets. It is indicative that in April 1957 Kirichenko declared that the Ukraine Gosplan had to keep a check on the fulfilment of current and prospective plans.[25]

In fact, an instruction of 4 May 1958 of the CC CPSU and the USSR Council of Ministers established that the basis of the planning system were plans composed by enterprises, construction sites, *sovnarkhozy*, local soviets of deputies of workers, ministries and departments, that were derived from the control figures of the perspective plans. It was assumed that the plans would have to be discussed by the collectives of enterprises and sites, and come into force after confirmation by *sovnarkhozy*, regional executive committees and ministries. The indicators and patterns from which the plans would be composed were determined annually by the USSR Council of Ministers; so until these patterns were received in the localities the appropriate work was not done. Thus it was only in the middle of May 1960 that the Ukraine Gosplan received instructions for composing the 1961 plan, sending directions to the localities at the end of that month. *Sovnarkhozy* and regional planning commissions had to present their projects by 25 June 1960, so had less than a month to compose the plan for the following year.

It is only with great caution that one can speak of real current plans, since the Ukraine Gosplan workers had only a restricted period in which they could make use of the plans of the enterprises of the *sovnarkhozy* and regional planning commissions. It appears (although this problem needs further research) that the political component of current plans, and also the control figures of the USSR Gosplan, had more meaning for republican planning staff than 'suggestions from below' of enterprises and *sovnarkhozy*.[26]

Taking into account the enormous scale of republic industry, was it possible in principle to compose effective current plans? At various party forums of those years the inability of the Ukraine Gosplan to get plans for bringing new production on-stream to match with the plans for material-technical supply was noted, as was the fact that the planning of volume produced lagged behind the growth of production capacity, and that the plans for applying new technology were unreal, and so on.[27] *Sovnarkhozy*, and the Ukraine Gosplan, were criticized for the frequent changes of plan, which had a negative effect on the activity of enterprises. Special resolutions of the USSR Council of Ministers took up this question and obliged republic governments to make sure that quarterly plans were not altered later than 45 days before the end of the period, and monthly plans not later than 20 days before. However, plans continued to change with kaleidoscopic rapidity.[28]

Distribution of resources

To understand the role of changes in the administration of industry in the course of the Khrushchev reforms, an analysis of the distribution of material resources is very helpful. As well as causing tensions between Ukraine and the USSR, this issue created difficulties between the Ukrainian institutions. Up to 1957 the union and union-republic ministries had powerful subdivisions that were responsible for the so-called '*snabzheniye i sbyt*' (supply and marketing). In 1957 Khrushchev declared that the *sovnarkhozy* would have a certain independence in the planning and disposition of material resources. But the situation turned out differently. After the liquidation of ministries in Ukraine, the supply functions were transferred to the Ukraine Gosplan. Three departments were organized in it: material stocks; equipment and fitting out; and marketing. The latter was responsible for the planning and distribution of material resources and also coordinated the activity of 17 republic structures responsible for supply to *sovnarkhozy* of individual types of resources and products. In these conditions the *sovnarkhozy* and the departments of material-technical supply in the regional executive committees, which carried out deliveries to local industrial enterprises, fulfilled only a technical function: they controlled the distribution of material stocks that had come into the network of bases (organizations that had stores in regional and district centres).[29]

The *sovnarkhozy* leaders asked for the distribution of material resources to be transferred to their management. But the Ukraine Gosplan actively opposed this. In December 1957 there was a conference in the CC CPU of *sovnarkhozy* chairmen on the question of material-technical supply. It became clear that the *sovnarkhozy* had serious problems with supply and marketing. The Ukraine Gosplan, by concentrating the republic supply/marketing structures within itself, had made possible their liquidation in the localities at the price of staff increases in Kiev. The *sovnarkhozy* chairmen had to solve the problem of insufficient resources at the enterprises within their purview even though such problems were under the management of the republic marketing and delivery departments. Thus the union marketing structure (under the USSR Gosplan) in many cases carried out delivery of resources and marketing of products, bypassing the republic structures and working directly with factories on Ukrainian territory.[30]

Two points of view were expressed at the conference. Some *sovknarkhozy* chairmen proposed that the republic structures should be eliminated, concentrating distribution of material resources in the union distribution agencies. The secretary of CC CPU, O. Ivashenko, who led the conference, did not agree with them.[31] Others stood for the transfer to the *sovnarkhozy* of responsibility for production, supply and marketing on the territories of the economic administrative districts. Representatives of the republic structures ignored this suggestion. In reply, the *sovnarkhozy* chairmen declared that they were in an ambiguous situation since, in developing production, they ran into accusations of parochialism. They could not guarantee deliveries of raw materials, component parts or market production of enterprises in their domain, and therefore had sent a large number of '*tolkachi*' ('pushers', recruited from staff of the *sovnarkhozy* supply and marketing departments) to the republic and union organs, and to big industrial centres of the USSR. That is, the previous working practice of union and republic ministries was being repeated.[32]

Organization of the *sovnarkhozy*

Having briefly analysed the impact of changes in the planning and supply of material resources we will take a look at the most characteristic problems in *sovnarkhoz* activity. Several months were needed to organize the *sovnarkhozy*, during which questions of cadre appointments were solved, premises were sought and material provision for the work of the *apparats* of the *sovnarkhozy* was investigated. Each of these numbered 300–500 people, some of whom were staff from the liquidated ministries. They were appointed to their new place of work by decisions of the Ukraine SSR council of ministers. But to go off from Kiev to regional centres was not to the liking of the bureaucrats; *sovnarkhozy* leaders, who were confirmed by the CC CPU politburo, had to seek specialists on the spot. Staff complements filled up slowly and the majority of *sovnarkhozy* continually had vacancies for which specialists could not be recruited. To all this were added problems with the transfers of enterprises and organizations from subordination to ministries to management by *sovnarkhozy*, which sent numerous staff groups to Kiev for information about work plans, about indicators of enterprise activity, and also about the volumes of material resource deliveries.[33]

Towards the end of 1957 the Ukraine SSR *sovnarkhozy* had a practically uniform structure, determined by resolutions of the Ukraine and Soviet councils of ministers. They were led by chairmen and deputies. With the agreement of chairmen of the councils of ministers of the republics and of the supreme soviets, *sovnarkhoz* chairmen could be appointed by ministers in the governments. Each *sovnarkhoz* had branch and functional departments and administrations. Thus, the Vinnitsa *sovnarkhoz* had 12 functional and branch departments and also six administrations, of which two combined the basic branches of industry in this agricultural region – the fish and sugar industry administrations. Specialized trusts went into these, combining enterprises making their own particular kinds of product. The structure of the Stalinsk *sovnarkhoz* was more complex, having nine administrations and four coal and mine-building combines.[34]

As can be seen, *sovnarkhozy* preserved the branch structure for administering enterprises. At the same time they could carry out a series of measures for production cooperation. After the formation of *sovnarkhozy*, most of the big industrial enterprises fell into their jurisdiction. Up to 1962 their number diminished as, within the framework of the policy of inter-*sovnarkhoz* cooperation, more than 1,000 enterprises were merged, with 400 combines being formed from them. In most cases the enterprises that were combined were those producing similar ranges, and also organizations of special factories for equipment repair, semi-finished industrial products, and so on.[35] Such reorganizations helped develop the production infrastructure, facilitating the growth of the prevalent production in the economic districts. At the end of 1962, enterprises of local industry that had hitherto been managed by regional executive committees were subordinated to the Ukraine *sovnarkhozy*.

The independence of *sovnarkhozy* in deciding many economic problems was evaluated by the central committees of the CPSU and CPU, and regional party committees, which fixedly followed the activity of *sovnarkhoz* leaders and factory directors (although often maintaining close informal connections with them), on the alert for violations of state discipline. In accordance with the decisions of the May 1958 plenum of the CC CPSU, party organs checked the correct utilization by the Ukrainian *sovnarkhozy* of planned capital investments and material resources. It became clear that the Dnepropetrovsk, Stalinsk, Lugansk, Lvov and other *sovnarkhozy* wilfully reduced capital investment in the ferrous metallurgy, coal, chemical, mineral extraction and other branches of industry, redirecting 118 million roubles to the development of other branches. Redistribution of resources without the agreement of the Ukraine Gosplan happened in practically all branches of industry. At a session of the presidium of the CC CPU, four *sovnarkhoz* chairmen were reprimanded for such conduct, obliging all *sovnarkhozy* to restore capital investment in the leading branches of heavy industry.[36]

However, redistribution of capital investment persisted. *Sovnarkhozy* chairmen under various pretexts convinced the Ukraine Gosplan of the need for this. Since leaders of regional committees and regional executive committees had an interest in the attraction of big investments in regional development, they lobbied for them in the CC CPU and CC CPSU. Deficiency of a certain production in a region, the lagging of a branch or region (behind, as a rule, a frivolously chosen analogue), a rise in the anticipated economic effect of a project, the existence of opportunities to start work, and especially natural or industrial potential, all served as arguments for getting capital investments. Study of the interaction of party leaders, economic managers, and planners during the preparation of investment projects carried out in Ukraine has shown that selection of the actual location for carrying out a project depended on how strong the support was of the regional leadership. Investments were distributed on the basis of an already-achieved level, while success in utilizing them improved the chances of regional leaders in the following period. From this point of view investment activity for Ukraine inside the USSR borders was eased, while in the republic the best conditions came to be in the eastern and central regions.

Repeated checks by party organs revealed a series of other contradictions and problems in the work of *sovnarkhozy* and enterprises under the conditions of governmental reform. Making use of the non-agreement of governmental resolutions and directives in the sphere of price regulation, and of the weakened control over financial discipline, enterprise directors strove to increase assignments to enterprise funds. At that time a turnover tax was imposed on enterprises. Having fulfilled production and output plans and paid the tax, an enterprise that over-fulfilled the plan and raised wholesale prices of products that had been deliberately lowered in calculating production costs, received a profit. Profit was assigned in two parts: the first stayed with the enterprises' '*spetsfond*' (special fund) for meeting their needs (capital investment, liquidity, additions to wages in the form of money prizes and rewards), while the second part went to the state. The actual profit received by the enterprise was the difference between its wholesale prices and the total production cost. Enterprise directors did not have the right to do so, but they established factory-gate prices, which were later sanctioned by the *sovnarkhoz* leadership. Sometimes prices were not artificially increased but on the contrary were lowered, which permitted over-fulfilment of gross production indicators and likewise led to profit, which the book-keepers and financial departments of *sovnarkhozy* preferred not to notice (although they had to do so, and do a recount, when violations were revealed in relation to money going to the state).

In all cases, to receive a profit it was necessary to obtain reduced planning tasks, and an increase of capital investment, to have unaccounted stocks of material resources and productive capacity, and to present untrustworthy accounts. Surviving accounts of checks by party organs on the activities of *sovnarkhozy* show that such 'rules of the game' were maintained by enterprise directors, leaders of *sovnarkhozy* and local party/soviet bureaucrats.[37] Despite the fact that a large part of enterprise profits was taken by the state (which lowered the incentive for successful production), the remaining part allowed directors to receive substantial monetary resources that were used not only in the interest of production and the development of the social sphere, but also for 'lining the pockets of essential people' in the *sovnarkhozy*, and party/soviet organs of a different level. It was not for no reason that in those years many leaders began to build their own houses, and purchase private cars, furniture and valuables. On the basis of such criminal connections, groups and clans were formed of local leaders, and there were clienteles having patrons in the top echelons of Soviet power.

To a certain degree such a situation was facilitated by the operating conditions of the *sovnarkhozy*. Whereas staff of their functional departments received pay from the republic budget, the branch departments received theirs on account of the official rise in the cost of products of subordinate enterprises. Therefore it was very hard to control the legality of extra costs, outgoings, wages and rewards in kind. Moreover, production plans for the *sovnarkhozy*, which the Ukraine Gosplan changed scores of times (46 times, for example, in 1959) left a vast field for manoeuvring resources, finances and productive capacities. A great flow of administrative questions and economic problems compelled *sovnarkhoz* leaders to send to Kiev and Moscow hundreds of staff, who for weeks and months lived in the capitals, taking no part in the work of the *sovnarkhozy*. The leaders themselves were in no position to carry out the avalanche of governmental resolutions and directives from the Ukraine Gosplan.[38]

Non-fulfilment of the 'cooperated deliveries' plans led to feverish production in the *sovnarkhozy*. When specialization of production was high enough, tens of thousands of Soviet enterprises were interconnected through raw material and product deliveries. With the formation of *sovnarkhozy* it became evident that it was easier to develop many kinds of production locally rather than bring the planned production from 1,000 kilometres away. In the absence of market relationships, and with the attempt to dismember the administrative vertical management connections ('locked' on to union structures), the *sovnarkhozy* within the republic and also those within the USSR had difficulty in establishing relationships with each other. As delivery shortfalls ran counter to the principles of the so-called 'single economic complex' and 'fraternal relationships of the Soviet peoples', they had not only an economic significance but also bore an additional political burden. Party leaders used this to place blame on the *sovnarkhozy*. At the XXI CPU conference in February 1960, Podgornyi declared:

> Unfortunately, quite a few *sovnarkhoz* workers brought with them the obsolete, condemned, working methods of the former ministries. So in the activities of individual *sovnarkhozy*, where such workers are to be found, there are some very serious faults. The effective solution of questions is undermined by changes wrought by numerous resolutions and directives, and their fulfilment is not properly organized and controlled. Leading *sovnarkhoz* workers are still rarely present at enterprises and construction sites and know of their situation only on paper or by what workers say, and so they often do not have their own opinions when correct decisions need to be made.[39]

Podgornyi's conclusion had not only a subjective subtext, but reflected an objective reality: staff and leaders of the *sovnarkhozy* were reproducing the working practices of the liquidated ministries.

Under the slogans of struggle against bureaucratism and inadequacies but, in the main, to create an extra channel of supervision, the Communist Party leaders formed commissions for controlling the activities of industrial enterprise administrations, of trade and of *sovnarkhozy*. More than 80,000 communists were selected for these. About 7,000 members served as external inspectors of party district and urban committees. The Ukraine council of ministers at the same time created a commission of soviet control and this drew in over 15,000 people who, for example, in 1959 took part in inspections of the work of about 14,000 enterprises, kolkhozes and construction sites in the republic.[40] The letters of ordinary participants in these commissions are an interesting source for studying administrative relationships in practice and a broad spectrum of problems at the micro level of the Soviet economy, but that is a subject for a separate investigation. The commissions did not, of course, have any real significance for perfecting the managing practice of enterprise and *sovnarkhozy* administrations.

The all-Ukraine *sovnarkhoz*

Deficiencies and problems (real, or imagined out of fear of losing control over managing the economy) were met through the creation by the communist party leaders of universal republic *sovnarkhozy*, which had to help the regional all-Ukraine *sovnarkhoz* (USNKh) with rapid solutions for their problems. In June 1960 was created, headed by M. Sobol' and later by A. Kuz'mich.[41] This was substantially a parallel Ukraine council of ministers organ, occupying itself with particular aspects of planning and also with coordinating the activities of the Ukrainian *sovnarkhozy*. In less than a year the staff of this governing organ comprised about 1,000 people, who did not go to *sovnarkhozy* or enterprises but occupied themselves with paperwork. In 1961 more than 137,000 letters, telegrams, reports, information items and other documents came to the USNKh, while more than 58,000 letters and telegrams were despatched to subordinate organizations. Drafts of documents sent to the leadership were scrutinized by no fewer than three leading staff members, taking two to three days. Instructions and requests of the CC CPU and Ukraine council of ministers were not carried out. Thus, in 1961, out of the 2,590 normative state acts, fulfilment of which was compulsory, the USNKh dealt punctually with only 69 per cent. The USNKh was given the near impossible task of solving the problems already encountered in distribution and marketing, but in contrast to the republic *sovnarkhozy* of the RSFSR and Kazakhstan, USNKh did not have the right to distribute material resources. Answering for the fulfilment of production and delivery plans, USNKh was in no condition to effectively decide problems of supply and marketing, which in non-market conditions lacked the natural character determined by supply and demand.[42]

The questions that arose in coordinating the activity of *sovnarkhozy* and assisting them were decided extremely slowly. To decide any question it was necessary to get representatives of the *sovnarkhoz*, USNKh and the Ukraine Gosplan to agree the actions. When Gosplan worked out annual republic plans, more than 50 per cent of the USNKh *apparat* were sent to its departments to get an idea of the control figures. In the early 1960s the *sovnarkhozy* continually turned to USNKh and the Ukraine Gosplan with requests for corrections of the plans. USNKh, without real authority, had to turn to Gosplan with these questions.

At the same time, the USNKh leaders spent huge sums on prize giving, payment of 'raises', and personal additions to wages. On the whole, the work of USNKh bred those 'bureaucratic distortions' condemned by the party. And in many cases USNKh duplicated the activity of the Ukraine council of ministers, arousing perplexity and often dissatisfaction on the part of leaders and staff in the government *apparat*.

Sovnarkhoz leaders expressed their indignation at evolving practices in economic and managerial activity. The chairman of the Donetsk *sovnarkhoz*, I. Dyadyk, sent Khrushchev (with a copy to Podgornyi) a frank letter in April 1961 about the problems that existed in *sovnarkhoz* activity. In particular, he noted that *sovnarkhozy*, the basic state form of administering industry and construction, were becoming remote from active influence on the economy[43]. Increasingly, many questions were being decided by republic and all-union organizations. *Sovnarkhoz* leaders had expected that the role of *sovnarkhozy* would increase on account of an extension of rights, the transfer to their authority of new enterprises, and that the new organs of administration would have equal weight in the administrative system as ministries. But in reality they were not regarded as administrative organs directly subordinate to the Ukraine council of ministers; between them and the republic government were numerous departments. The *sovnarkhozy*, therefore, 'had got closer to the enterprises but further from the government'. Many points in the *sovnarkhoz* statute (accepted by the USSR Council of Ministers in 1957) were not being observed; in fact, government resolutions had been removing them. The suggestions that *sovnarkhozy* should have the right to reallocate up to 5 per cent of capital investment within enterprises of the heavy industry branch had not been backed up. The *sovnarkhozy* had no formal influence on the process of planning and distribution of material resources.

As an example, Dyadyk dealt with how union and republic structures limited the rights of the *sovnarkhozy*. Despite the right of *sovnarkhozy* to affirm the financial plans of enterprises and to change the rates of planned profit, at the beginning of 1961 the Ukraine finance ministry instructed its regional offices not to accept changes in profit rates introduced by *sovnarkhozy* without the official approval of regional finance departments of the regional executive committees.[44] In addition, a petty tutelage over the activity of enterprise leaders had been preserved. It was not accidental that the draft law on the state enterprise, which the *sovnarkhoz* leaders learned about at the beginning of 1958 during the one conference in which they took part in the CC CPSU, had never been published. Instead, USNKh had been created, frankly appraised by Dyadyk as an extra link in administration[45]. He suggested merging *sovnarkhozy* with the aim of gaining a more effective activity in the enterprises of various branches, to which the secretary of the CC CPU, I. Kazanets (formerly first secretary of the Donetsk party executive committee), reacted with a pencilled note on the text of the letter: 'The question of unifying the *sovnarkhozy* is also centralization but on a smaller scale. This is an attempt to move out from under the control of party and soviet organs.' Kazanets had precisely grasped the meaning of Dyadyk's letter to Khrushchev. The chairman of the Donetsk *sovnarkhoz* was writing that departmentalism (as a basic defect of the branch system of administration) 'had been quickly forgotten but nevertheless the label of parochialism was automatically stuck on to many phenomena irrespective of whether they had any connection to parochialism or not'.[46]

At a May 1961 conference of *sovnarkhoz* chairmen, meeting under the auspices of the CC CPU, Dyadyk was supported by his colleagues. They said, bluntly, that USNKh was a structure parallel to the republic government. It was the leadership of USNKh that summoned enterprise directors and decided essential questions, without informing the *sovnarkhozy* (the all-union structures did the same). But as USNKh lacked extensive rights, the *sovnarkhoz* leaders (often at the request of enterprise directors) phoned the chairman of the Ukraine council of ministers or the secretaries of the CC CPU, trying to quickly resolve problems that arose. In their turn, Ukraine's leaders spent a lot of time talking and exchanging telegrams with the party central committees of union republics, with their party executive committees, with republic governments, and with *sovnarkhoz* chairmen, in an attempt to influence delivery of products needed by the republic's enterprises, and justifying their own delivery shortcomings.

Aware of this practice, Ukrainian *sovnarkhoz* leaders suggested conferring republic government rights on USNKh, which would ease the operation of all administrative links. In reply, Kazanets declared: 'The *sovnarkhoz* and the economic district were not created to create autonomy, but to more effectively lead enterprises and to better utilize the resources and opportunities that are there.' That is, all the suggestions were judged to be assaults on the administrative prerogatives of republic structures.

It must be understood that in Soviet political language and culture the terms 'autonomism' and 'parochialism' could be regarded as describing something resembling political separatism or 'national tendencies' (terms of the 1920s and 1930s). So Kazanets' expression 'to create autonomy' implied a lot to experienced bureaucrats; at any moment the political leadership could hold forth about incorrect political activities. How such accusations would end was well known from Stalinist times. Not for nothing did the secretary of the CC CPU confidently declare that the rights of *sovnarkhozy* were perhaps being constrained but republic organizations were doing this out of concern for the republic's interests and not for individual regions or enterprises.[47]

Further reorganizations

This meant that the position of the republic government in relation to the *sovnarkhozy* remained rigid and would become more rigid in the future, matching the general direction of union structures' activity. In November 1962 the National Economic Council (SNKh) of the USSR was created; this bore responsibility for the realization of plans for economic development. In March 1963 the Supreme Council of the National Economy of the USSR of the Council of Ministers was formed, to which Gosplan, SNKh, and individual union state committees were subordinated. Simultaneously USNKh and SNKh gained the status of union-republic organs.[48] All these 'reformations' were called on to decide not only the problems of coordinating the activities of USSR *sovnarkhozy* but, as a main task, to keep a check on their work.

In this context the division of party and soviet organs into industrial and agricultural branches, initiated by Khrushchev, looks completely logical and in no way resembles 'voluntaristic improvization'.[49] Whatever the official phraseology of the Soviet leadership might have been, the industrial regional committees of the party were aimed at a strict control over *sovnarkhoz* activity. However, the secretaries of regional committees, one of the main systemic supports of the communist party, found themselves in an ambiguous situation. By the decision of the CC CPU plenum (December 1962) seven instead of 14 economic districts were created: the Donetsk, Kiev, Lvov, Podolsk, Pridneprovsk, Kharkov and Chernomorsk districts.[50] Enterprises of local industry were transferred to them,[51] which had a negative effect on the activities of regional and district soviets, and also on the state of the social infrastructure, supported by a portion of the turnover tax of such enterprises.

In the autumn of 1963, three instead of seven economic districts were formed in Ukraine: Donetsk-Pridneprovsk, South-Western and Southern. For the party regional committees such changes were very unpleasant. Instead of *sovnarkhozy*, which were formally equal to former ministries, the majority of industrial regional committees could have control only over the administration of enterprises.

In the Soviet administrative system this meant that the authority and influence of regional committees in fact was reduced (in the sense of influence in the taking of administrative decisions on the scale of the republic and of the entire USSR). Party workers lost confidence in their future. At party conferences and meetings of active party members 'critical remarks' were continually heard about *sovnarkhozy*, republic and union organs (the CC CPSU and the Ukraine Council of Ministers excepted). These complaints had objective causes: defects in the activity of administrative structures, their leaders and their co-workers. But the 'critical remarks', carefully collected by the party regional committees who sent them on to the CC CPU, indirectly testified to the serious discontent with the permanent reforms felt by party bureaucrats at a different level.

Apparently, this discontent was one of the reasons why Shelest supported L. Brezhnev and Podgornyi in the struggle against Khrushchev.[52] It is revealing that not a single party leader at republic or regional level in Ukraine officially came out in defence of Khrushchev when the latter was pensioned off in October 1964. A year later, in September 1965, at a plenum of the CC CPSU, Brezhnev declared that the system of administration through *sovnarkhozy* was coming into contradiction with the trend of branch development and it was, therefore, necessary to go back to the ministries. At the CC CPU plenum of October 1965 Shelest expressed his attitude to *sovnarkhozy*:

> One of the basic defects of administering industry through *sovnarkhozy* is that enterprises of a branch were too disconnected to fit economic districts, and this complicated the organized direction both of them and of the whole branch, and held back technical progress. Multi-level administration was leading to endless questions of coordination and often gave rise to irresponsible work. The role of planning was debased. Plans were not always based on calculation, and after confirmation were often changed. This disturbed the normal work of enterprises and lowered production effectiveness. An enormous number of plan indicators that were established for enterprises limited their rights and initiative in economic activity.

Ten ministries were restored in Ukraine.[53]

Conclusion

The *sovnarkhoz* epoch had ended. What were the results of the attempt by the Soviet leadership to reform the administration of the economy from the late 1950s into the first half of the 1960s? It can be noted that the aims of the reform were utopian: to catch up within a few years the world's developed countries' per capita production, and to draw large groups of society into administrative work. By 1965 the USSR had been able to surpass developed countries in the gross output of some individual items of production, but not in terms of output per head. Instead of a widening participation of society in administrative work there was a growth of the administrative *apparat*, which in the Ukraine alone exceeded a million people in the first half of the 1960s.[54]

The Soviet leaders lacked a precise programme for reform. Attempts to reform the administrative system were of a chaotic character. In the mid-1950s there had been a significant redistribution of administrative functions in favour of the union republics. In the Ukraine SSR the share of industry moved into the authority of republic organs was more than 95 per cent. The Ukraine government and other republic organs were not ready for a serious widening of administrative functions; experience was lacking, as were trained cadres and professional structural subdivisions. Creation of the *sovnarkhozy* complicated the situation. A number of trained administrators did not want to exchange the 'Kiev pavements' for the 'greasy dirt' of the fertile Ukraine Black Earth, even if they were to work in regional centres.[55] Republic leaders in the framework of existing administrative/bureaucratic relationships asked what to them was the logical question: who, what and how are we going to lead, if the *sovnarkhozy*, not only *de jure* but also *de facto* in terms of the scope of their rights, turn into the former ministries? Therefore highly placed officials in Kiev, just like their opposite numbers in Moscow, applied numerous stratagems and models of behaviour to limit and liquidate the authority of the *sovnarkhozy*.

But there were even deeper problems. What Soviet leaders called 'centralized planning' was never there, especially as regards the annual plans that were changed over and over again. In addition, the planning indicators for output in kind were unconnected with plans for the supply of materials to enterprises. The breach was not only technical (arising from the impossibility of realistically coordinating plans) but was also institutional.

After the liquidation of ministries in 1957 the republic supply-and-marketing organizations came under the Ukraine Gosplan, but the principal questions of supply and deliveries between *sovnarkhozy* of the union republics were decided by secretaries of the CC CPU, and the chairman and his deputies of the Ukraine council of ministers. CPU leaders clearly understood that the distribution of resources was one of the 'commanding heights' of the command-administrative system, securing control over the work of the *sovnarkhoz* leadership and enterprise administrations, signifying conversion or transformation into real power-administrative functions. The communist leaders of Ukraine preferred to create extra controlling structures, USNKh for example, rather than allow the *sovnarkhozy* a wide spectrum of independent activity. With the transfer of many administrative functions from the union to republic level it became clear that *sovnarkhoz* leaders were under the strict tutelage of CC CPU secretaries and party executive committee secretaries, without which there could be no action either on the republic or general union level. Our conclusion is that despite the declarations of Khrushchev, there operated in the USSR in the 1950s to 1960s a system not of centralized (or, rather, compromising) planning, but rather one of centralized administration having a highly personalized nature (as in the close relationships of Podgornyi and Shelest with Khrushchev). In this sense the formal transfer of enterprises from union ministries to republic governments and *sovnarkhozy* did not bring the anticipated decentralization of the administrative system.

No less paradoxical was the administrative and economic practice of the *sovnarkhozy* and the enterprises under them. Fixed taxes on turnover and part of the profits, with the so-called centralized distribution of materials and centralized marketing, condemned enterprises to 'the realm of the bed', deterring the quest for profit growth. At the same time factory directors and enterprise administrations were obliged to fulfil the often-changing output plans. For this, a variety of strategies were used which, while fulfilling the plans, permitted escape from the control of party/soviet organs (in many instances corrupting them), and allowed accumulation of material and financial resources. The *sovnarkhoz* workers and leaders knew these 'rules of the game', taking part in what was known as 'deception of the state'. But the economy, based on anti-market ideology and administrative allocation of resources, could not function otherwise. This leads to questions not only of possibility and practicability but also of the potentiality in principle of reforming the Soviet command-administrative system.

The attempt by CPSU leaders to arouse a new wave of popular labour enthusiasm[56] did not lead to a serious improvement of society's standards of living. Widely circulating talk about how Soviet people for the first time had been moved from hovels and semi-hovels into the new apartments of the '*khrushchevki*',[57] and how peasants had finally received passports and could leave their villages, do not give an entirely correct view of the trends of Soviet economic development. In Ukraine, from the late 1950s to the mid-1960s, the average annual growth rates of industrial production was a little more than 9 per cent. But these rates were achieved by the construction of new enterprises and output from newly built enterprises. During the seven-year plan alone, more than 900 big enterprises were brought into use in Ukraine. A diminution of national income resulted (capital investment was directed to new industrial construction) as well as a drop in labour productivity and a slow-down of the development of branches producing consumer goods.[58]

In March 1966 at the XXII conference of the CPU, Shelest made the ritual statement that the tasks of the seven-year plan had been achieved in Ukraine in October 1965. The CPU leader, amid thunderous applause, enthusiastically affirmed:

> The republic is now producing more pig-iron, steel, rolled metal, iron and manganese ore and gas than any other capitalist state in Europe, and more coal is extracted than in the developed capitalist countries France, Belgium and Japan taken together. This is a great victory for the Communist Party of Ukraine and all the republic's workers.[59]

Naturally, the promise 'to catch up and overtake' the developed countries in per capita output and to secure the world's highest standard of living, implying the victory of communism over capitalism, did not get a mention.

Notes

1 Tsentral'nii derzhavnii arkhiv gromads'kikh ob'ednan' Ukraini (Tsentral'nyi gosudarstevennyi arkhiv obshchestvennykh ob'edinenii Ukrainy (-henceforth TsGAOOU), f. 1, op. 1, d. 1445, l. 27.
2 After the publication of a series of hitherto unknown documents of the highest organs of power and of the USSR administration, the opinion of academics that only Stalin had an exact picture of how the Soviet power system should be built appears fully well founded. See O.V. Khlevnyuk, *Politbyuro: mekhanizmy politicheskoi vlasti v 1930-e gody*, Moscow: Rosspen, 1996; P. Gregori, *Politicheskaya ekonomiya stalinizma*, Moscow: Rosspen, 2006.
3 TsGAOOU, f. 1, op. 1, d. 1445, l. 28. At the VI conference of communists of Yugoslavia (1952), Iosip Broz Tito declared that the USSR had built state capitalism rather than socialism, which was directed by 'a Soviet counterrevolutionary bureaucracy', a new group of exploiters. An alternative could be seen in an extension of independent workers' collectives in the administration of production. Quests for a 'market socialism' different from the Soviet model unfolded in Poland and Hungary.
4 At the end of 1929 the USSR leadership declared that industrial production in the country, amounting to 5 per cent of that in the USA, would exceed the US level by the beginning of 1940. See R.W. Davies, *The Soviet Economy in Turmoil, 1929–1930*, Cambridge, MA: Harvard University Press, 1989, p. 68.
5 TsGAOOU, f. 1, op. 1, d. 1486, ll. 12, 23.
6 TsGAOOU, ll. 28–9. *Resheniya partii i pravitel'stva po khozyaistvennym voprosam*, Moscow: Politizdat, 1968, vol. 4, p. 531.
7 TsGAOOU, f. 1, op. 1, d. 1409, l. 7; op. 31, d. 2894, l. 1.
8 For more about the activities of Khrushchev and other Ukraine leaders, see *Politicheskoe rukovodstvo Ukraina, 1938–1989gg*, Moscow: Rosspen, 2006.
9 P.E. Shelest, *Da ne sudimy budete. Dnevnikovye zapisi, vospominaniya chlena Politbyuro TsK KPSS*, Moscow: Edition q, 1992, pp. 172–242.
10 For more details, see Nikolai Mitrokhin's chapter in this volume (Chapter 3).
11 See Nataliya Kibita's chapter in this volume (Chapter 7).
12 See 'O preobrazovanii Ministerstv SSSR', *Vedomosti Verkhovnogo Sovieta SSSR*, 20 March 1953; 'O rashirenii prav ministrov SSSR', *Postanovlenie SM SSSR*, 11 April 1953; *Resheniya partii i pravitel'stva po khozyaistvennym voprosam*, Moscow, 1968, vol. 4, pp. 5–14; 'Pro peretvoreniya soyuzno-respublikans'kikh ministerstv Ukrains'koi RSR; Ukaz Presidii Verkhovnoi Radi Ukrainskoi RSR, 10 Kvitnya 1953 r', *Vidomosti Verkhovnoi Radi URSR*, no. 1, 1953, pp. 14–15; 'Pro peretvoreniya respublikans'kikh ministerstv Ukrainskoi RSR; Ukaz Presidii Verkhovnoi Radi RSR, 25 travnya 1953 r', *Vidomosti Verkhovnoi Radi URSR 1953*, no. 1, pp. 28–9; 'Pro utvoreniya soyuzno-respublikans'kikh Ministerstva promislovosti prodovol'chikh tovariv Ukrains'koi RSR; Ukaz Presidii VP URSR 7 zovtnya 1953 r', *Vidomosti Verkhovnoi Radi URSR*, no. 3, 1953, p. 22; 'Pro utvoreniya soyuzno-respublikans'kogo Ministerstva promislovosti tovariv shirokogo vzhitku Ukrains'koi RSR; Ukaz Presidii VP URSR, 7 zhovtnya 1953 r', *Vidomosti Verkhovnoi Radi URSR*, no. 3 1953, p. 23; 'O ser'eznykh nedostakkakh v rabote partiinogo i gosudarstvennogo *apparata*, Postanovlenie TsK KPSS, 2 January 1954g', *Izvestiya*, 26 January 1954. T.I. Korzhikhina, *Istoriya gosudarstvennykh uchrezhdenii SSSR*, Moscow: Vyshaya shkola, 1986, p. 203.
13 TsGAOOU, f. 1, op. 46, d. 6971, ll. 5–6; *Vidomosti Verkhovnoi Radi URSR*, no. 2, 1954, p. 16; no. 3, pp. 41, 44, 48.
14 'O zadachakh po dal'neishemu pod'emu promyshlennosti, tekhnicheskomu progressu i uluchsheniyu organizatsii proizvodstva: Postanovlenie Plenuma TsK KPSS, 4–12 iyulya 1955g', *Resheniya partii*, vol. 4, pp. 225–43.

15 'O nekotorykh dopolnitel'nykh pravakh ministrov SSSR: Postanovlenie SM SSSR, 4 maya 1955', *Resheniya partii*, vol. 4, pp. 19–20; 'O rasshirenii prav direktorov predpriyatii: Postnovlenie SM USSR, 9 avgusta 1955', *Resheniya partii*, vol. 4, pp. 244–50.

16 Tsentral'nii derzhavnii arkhiv vishchikh organiv vdali ta upravlinnya Ukraini (henceforth TsDAVOVU), f. 582, op. 3, d. 3339, pp. 3, 6.

17 TsGAOOU, f. 1, op. 46, d. 7168, l. 1; op. 31, d. 2894, l. 1.

18 TsGAOOU, f. 1, d. 1409, ll. 5–6, 8, 23–4, 26, 28, 31.

19 On 31 May 1957 a session of the Ukraine supreme soviet accepted the law 'On the further perfection of the organization of industry and construction administration in Ukraine SSR', legislatively strengthening the administrative reform in the republic.

20 See *Sobranie postanovlenii Pravitel'stva SSSR*, no. 4, 1957, pp. 39–41; no. 7, 1958, p. 66; no. 16, 1958, p. 161; *Resheniya partii*, vol. 4, pp. 370–5.

21 As we shall see, the current plans were no less of 'an ideological burden'.

22 *Resheniya partii*, vol. 4, pp. 200–17.

23 TsGAOOU, f. 1, op. 31, d. 2932, l. 3; d. 2894, l. 13.

24 *Vidomosti Verkhovnoi Radi URSR*, no. 6, 1957, pp. 99, 102

25 TsGAOOU, f. 1, op. 1, d. 1409, l. 21.

26 For more detail, see TsGAOOU, f. 1, op. 31, d. 1873, l. 15.

27 TsGAOOU, f. 1, op. 46, d. 7388, ll. 3–4.

28 TsGAOOU, f. 1, op. 31, d. 1371, l. 119.

29 TsGAOOU, f. 1, op. 1, d. 1409, l. 23; op. 31, d. 2894, l. 15.

30 TsGAOOU, f. 1, op. 76, d. 830, ll. 4, 5.

31 The memoirs of Ukraine leaders of that time talk about how O. Ivashenko maintained friendly relations with Khrushchev's kinsfolk. See A.P. Dashko, *Gruz pamyati: Trilogiya: Vospominaniya*, Kiev: Delovaya Ukraina Publishing House, 1997, book 3, ch. 1: 'Na stupenyakh vlasti', pp. 45–6.

32 TsGAOOU, f. 1, op. 76, d. 830, ll. 21, 31, 39, 44.

33 TsGAOOU, f. 1, op. 46, d. 7173, ll. 31–6; d. 7197, ll. 25–9.

34 TsGAOOU, f. 1, op. 24, d. 4475, ll. 134, 297.

35 TsGAOOU, f. 1, op. 46, d. 7391, ll. 18–19; TsDAVOVU, f. 582, op. 3, d. 3872, l. 3.

36 TsGAOOU, f. 1, op. 24, d. 4687, l. 13.

37 TsGAOOU, f. 1, op. 46, d. 7337, ll. 53–4, 77, 79, 83.

38 TsGAOOU, f. 1, op. 31, d. 1403, ll. 15–25.

39 TsGAOOU, f. 1, op. 1, d. 1588, ll. 46.

40 TsGAOOU, f. 1, op. 1, d. 1588, ll. 102–4.

41 See *Sobranie postanovlenii Pravitel'stva USSR*, no. 11, 1960, p. 82; *Vidomosti Verkhovnoi Radi URSR*, no. 15, 1961, p. 189.

42 TsGAOOU, f. 1, op. 34, d. 1817, ll. 321–4, 327.

43 In this case the letter writer's term is used.

44 This can be evaluated from several points of view, but Dyadyk presented the actions of the Ukraine finance ministry as violating normative state documents.

45 Podgornyi sent the letter on to members of the CC CPU presidium.

46 TsGAOOU, f. 1, op. 31, d. 1848, ll. 47, 55–6, 62, 68, 71, 78, 81.

47 TsGAOOU, f. 1, op. 31, d. 1880, ll. 146–7, 160, 164.

48 *Vidomosti Verkhovnoi Radi URSR*, no. 47, 1962, p. 587; no. 17, 1963, pp. 208, 210, 309

49 Khrushchev has been blamed for this ever since the end of 1964.

50 In May 1960, with persistent support from regional party committees, three economic administrative districts were formed, the enterprises of which were administered by *sovnarkhozy*: Crimea, Poltava and Cherkassy. See *Vidomosti Verkhovnoi Radi URSR*, no. 17, 1960, p. 116. Later there were 14 *sovnarkhozy* operating in the republic, which were refashioned in accordance with the instruction of the Ukraine supreme soviet of

26 December 1962 into seven *sovnarkhozy*. See *Vidomosti Verkhovnoi Radi URSR*, no.1, 1963, p. 26.

51   TsGAOOU, f. 1, op. 1, d. 1876, ll. 1–12.

52   For more detail, see Petro Shelest, *Spravzhnii sud istorii shche poperedu. Spogadi, shchodenniki, dokumenti, materiali*, Kiev: Geneza, 2003.

53   TsGAOOU, f. 1, op. 1, d. 1927, ll. 13–16.

54   TsGAOOU, f. 1, op. 1, d. 1813, l. 51.

55   Here, common slang expressions of the time are used.

56   In August 1964 Shelest at the CC CPU declared that about five million people in the republic were competing for the honorary titles of collectives and shock-workers of communist labour. See TsGAOOU, f. 1, op. 1, d. 1887, l. 6.

57   Houses of panel construction, at that time built with accelerated methods. The new inhabitants used their own resources to finish off what the builders had left undone.

58   TsGAOOU, f. 1, op. 1, d. 1927, ll. 5, 9.

59   TsGAOOU, f. 1, op. 1, d. 1936, l. 22.

9   Khrushchev and the challenge of technological progress

Sari Autio-Sarasmo

Introduction

When Nikita Sergeevich Khrushchev came to power, he became the leader of a new era. The Soviet Union was now the other pole of a bipolar world, as well as leader of the Socialist bloc, surrounded by newly developed people's democracies. The clear aim of the Khrushchev leadership was not only to strengthen the Soviet Union's status as a superpower in world politics, but also to boost its economic and ideological role as leader of the Eastern bloc. In the new structure of world politics, this was no longer a question of the Soviet Union alone; it now had ramifications for the entire bloc, and for the historically determined victory of communism over capitalism. The Second World War and rapid technological progress in the West had created a technological gap between East and West. Khrushchev was aware of the gap but did not give it too much weight, since he was convinced that it was only a matter of time before the main capitalist countries would fall behind the Soviet Union technologically.[1] After the launch of Sputnik in 1957 and subsequent successes in the Soviet space programme, this seemed entirely plausible. During his visit to the United States in 1959, Khrushchev took every chance to propagate the superiority of the Socialist system over the West.[2]

Throughout the history of the Soviet Union, 'overcoming backwardness' was a determining factor in the policy-making of Soviet leaders. Backwardness was treated as a direct threat to Russia's sovereignty and geopolitical interests, especially during the closed period of late Stalinism. While Lenin wanted Russia to become part of Europe and Stalin wanted to catch up with Europe, Khrushchev wanted to catch up and then overtake America.[3] During the Khrushchev leadership, Bolshevik utopian beliefs in the omnipotence of technology from the 1920s converged with the Cosmic utopianism created by the 'space age' of the 1960s. Technological progress became one of the main features of the Khrushchevian vision of modernization, and technological development became one of the tools that was to be used in order to catch up with the West and overtake America.[4] The main obstacle to the creation of domestic technological progress in the Soviet Union was the prevailing technological backwardness of Soviet industry, and the Soviet inability to transform the high technology created in the space programme and military sector for civilian use. In order to take part in the worldwide upsurge in technological progress, the Soviet Union needed foreign technology.

Since Peter the Great the fight against backwardness in Russia had been based on the transfer of foreign technology.[5] Before the Second World War technology was transferred through normal trade and it mainly took the form of machinery based on innovations that were already widely known. However, the war changed the structure of technology transfer and trade. The technological innovations of the Second World War boosted technological progress in the West, and this became a key factor in the subsequent intensive economic growth in that part of the world. After the Cold War began, technological innovations also became connected to military technology and the arms race, which made technology, and especially the transfer of technology, a matter of world politics. The United States wanted to prevent the flow of high technology to the Soviet Union and the Socialist bloc. A high technology embargo was raised against the Socialist states by the United States and the West, and this hampered the Soviet leadership's plans to transfer Western technology to the Soviet Union. Access to high technology and know-how, and especially the dissemination of related information, became the subject of its own Cold War. This 'Information Cold War' established new structures aimed at concealing and collecting information. During the Cold War, Soviet technology transfer demanded new methods of operating.

The main aim of this chapter is to investigate Khrushchev's expedients for overcoming technological backwardness in the Soviet Union, and the aims that he pursued in technology transfer during the Cold War. Why was technology transferred, and how was this executed? What kind of networks were created and how did these networks operate during and after the leadership of Nikita Khrushchev? New networks of technology transfer recast Cold War politics behind the scenes by creating new structures of cooperation in Europe. Archival materials are changing the standard picture of the Cold War as a political confrontation of two poles, East and West, and they also highlight the importance of the figure of Khrushchev in this process.

Escaping backwardness

The transfer of foreign ideas and technology to Russia had a long tradition and the tradition continued after the October revolution, which was, as Sheila Fitzpatrick points out, a means of escaping backwardness.[6] Lenin realized the need for foreign technology and expertise in the development of Soviet Russia. For Stalin, industrialization was equal to modernization and represented a means of strengthening the Soviet Union which was surrounded by 'hostile' capitalist countries. The undeveloped state of heavy industry forced Stalin to import foreign technology to the Soviet Union, mainly in the form of machinery from the West, in order to create a base for domestic heavy industry. By importing machinery and prioritizing heavy industry Stalin managed to industrialize the Soviet Union and to create an immense military-industrial complex before the outbreak of the Second World War.[7] During the Stalin period, technology transfer more or less amounted to the imitation and duplication of foreign advanced equipment, rather than transforming it as a base for domestic high technology production.[8] One reason for this was the complexity of Soviet attitudes towards the West and Western know-how. Especially during the late Stalin era, Soviet scientists were encouraged to copy Western innovations, even though Western scholarship was officially treated as 'idealistic and reactionary'. At the same time, Stalin's anti-cosmopolitan campaign propagated the achievements of Soviet science.[9] Stalin's suspicious attitude and the beginning of the Cold War created a juxtaposition that hindered the Soviet Union in its drive to become part of Western technological progress through normal trade.

During the Khrushchev period the importance of technological progress for intensive economic growth was recognized and reflected in Soviet plans for economic modernization. The importance of automation in intensive economic growth was realized and the need for advanced foreign technology to create domestic automation was accepted.[10] This was a clear continuation of the policies of former Russian leaders who had turned backwardness into an advantage: by borrowing advanced Western technology, it was possible to move forward quickly.[11] Advanced Western technology was needed because the conventional machine-building industry that produced metal-working machinery was the only relatively developed branch of post-war Soviet industry. Connections between the civil and military sectors such as existed in the West were never established in the Soviet Union, and the divide separating the civil and military sectors served to isolate the prioritized and developed military-industrial complex from wider Soviet research and development (R&D).[12]

From the viewpoint of the Soviet Union's role as leader of the Socialist bloc, the need for Western technology was somewhat problematic. The need for capitalist technology clashed with the idea of the 'superiority' of the Socialist system and thus created an ideological problem. This was even more awkward after the glories of the space programme and the intra-bloc problems of the recent past. The United States managed to erode unity within the Soviet bloc by launching the Marshall Plan in the late 1940s.[13] Nor were developments positive in the 1950s: disturbances in the German Democratic Republic in 1953 and Hungary in 1956 caused concern in the Soviet Union.[14] Khrushchev faced a huge challenge: how to transfer technology without losing the credibility achieved with the help of the space victories and without causing ideological problems inside the Socialist bloc?

Scientific-technical revolution

One of the means used in order to address the problems involved in transferring Western technology to the Soviet Union was the adoption of the concept 'scientific-technical revolution' (STR).[15] The STR concept was popularized in the West in the late 1950s to explain the rapid technological progress of that period and the change it created. The STR included the close integration of science, theory, technology and production, and its main elements were various new technological processes: more effective utilization of nuclear energy, computers and electronic data processing, automation and cybernetics, and rocket technology. The economic content of the STR included improvements in the factors of production and products enabling a rapid expansion of production, and also substantial increases in social welfare.[16] The idea fitted well with Khrushchev's aims, and it was a useful concept in propaganda, both outside and inside the Socialist bloc.

Especially after the launch of Sputnik in 1957, the STR offered a useful propaganda tool: according to the STR, future development was to be based on the wide utilization of nuclear energy and the exploitation of outer space (*kosmos*).[17] The STR was also a tool for transforming extensive economic growth in the Soviet Union and the Soviet bloc into intensive growth – and 'to catch up and overtake' America, in accordance with Khrushchev's intentions. In this situation, it was only understandable that the success in the space programme should be effectively exploited in Soviet propaganda: the superiority of Soviet technology and science was strongly emphasized and the Soviet need for Western technology was concealed.[18] This propaganda was directed outside the Socialist bloc in particular, with a view to strengthening the Soviet Union's superpower status. On the other hand, official political pronouncements that increasingly stressed the importance of technological progress in economic growth[19] and the role of the STR were earmarked for audiences within the Soviet bloc itself.

The starting point for the STR campaign was very promising. There were remarkable scientific and technological breakthroughs in the Soviet Union in the late 1950s and early 1960s.[20] Sputnik was followed by several Soviet victories over the United States, the most important of which was the first manned space flight in 1961. Initially, the space programme seemed to be a huge blow to the United States and clear proof of Soviet technological superiority. Another Soviet success of the 1950s was the 'MESM', the first electronic stored-program, digital computer in continental Europe.[21] According to some recent estimates, in the 1950s and early 1960s the Soviet Union was on a level of parity with the United States in computer technology. Moreover, the future looked quite optimistic; because of Khrushchev's technocentric vision of modernization, official attitudes towards cybernetics had become more positive after 1958, and this enabled support for computer technology.

The adoption of the STR was based on Khrushchev's willingness to adopt new scientific innovations. Throughout the whole period of his leadership, Khrushchev lent his support to unprecedented scientific projects including in the field of computer technology, such as a plan to create a broadband connection linking 100 industrial cities across the Soviet Union, ushering in a new era of Soviet computer science. After Khrushchev's ouster, however, all these projects were closed down immediately and 'Soviet cybernetics transformed from a vehicle of reform into a pillar of the status quo'.[22] Khrushchev's methods for escaping backwardness were modern and technologically orientated, but his willingness to support technology without really understanding it ultimately worked against him. His incapacity to learn from practical experience, especially in science, was one of the reasons why he was ousted in 1964.[23]

Although the space programme was a great victory for the Soviet Union and a useful tool for propagating the STR, its practical benefits for Soviet technological progress were slight. The main reason for this was that it was not an exclusively Soviet achievement: major assistance for the technical applications was provided by Germany. Both superpowers – the United States and the Soviet Union – exploited Nazi German expertise and technology after the war.[24] Unlike the United States, the Soviet Union did not manage to reap the full benefits of German technology because of the culture of suspicion prevailing in Soviet science during late Stalinism. In the Soviet Union, German specialists were put to work on secondary tasks, and because the Soviet regime insisted on keeping German and Russian specialists separate, the Soviet Union was not able to transfer the German scientific capacity effectively into wider use.[25] The culture of secrecy that pervaded the Soviet space programme was exemplified by the case of chief engineer Sergei Korolev, whose identity was shrouded in total secrecy. Because knowledge was limited to such a small group of people and the whole space programme effectively connected to a single individual, the death of Korolev led to a downturn in the entire Soviet space programme in 1966.[26]

Information Cold War

While the triumphs of the space programme were widely propagated both inside and outside the Soviet bloc, the real situation in Soviet R&D and the need to acquire advanced Western technology were matters of the strictest secrecy. Selective dissemination of information was an essential part of the Soviet strategy during the Cold War and it engendered a special Information Cold War. Especially after the launch of Sputnik, the West was using all possible means to obtain information on the real technological level of the Soviet Union. The main aim of Western interest in the Socialist bloc was the arms race, which started immediately after the United States' detonation of the atomic bomb in 1945. Four years passed before the Soviet Union tested its own atomic bomb, but the testing of the Soviet hydrogen bomb came only ten months after the detonation of the American hydrogen bomb in October 1952.[27] The space programme was closely entangled with the arms race; it was evident after Sputnik that the Soviet Union possessed rocket technology for ballistic missiles.[28] This caused concern in the West; it was known that there was a technological gap between East and West, but precisely how wide the gap was remained unclear. After Sputnik the direction of the gap could no longer be taken for granted.[29]

Intelligence services were working actively to obtain information on the Soviet Union and the Socialist bloc but other, more official channels for collecting information were also established in the West. In the summer of 1958 the Central Committee of the Communist Party of the Soviet Union (CC CPSU) was concerned by the West's intensified interest in the Soviet economy, science and technology. The United States and Great Britain had established Pergamon Press, whose main target was to publish books based on the content of Soviet scientific and technological journals. At the same time there were plans in the OECD

(Organization for Economic Cooperation and Development) to establish a special institute to spread scientific-technical information originating in the Soviet Union, Eastern Europe and China. In order to strengthen the selective dissemination of information, new systems were established to control the publication of information in the Soviet Union.[30]

In spite of Khrushchev's official proclamations of willingness to take part in international cooperation and to seek out foreign know-how, the opening up of the Soviet Union proved to be rather complicated. In July 1959 the CC CPSU refused to allow a visiting American delegation access to Soviet oil industry installations, because the production units were located in areas that were closed to foreigners.[31] Reciprocal visits would have been useful for the Soviet Union, which needed advanced American oil-drilling technology. The secrecy problem came up not only during the visits of foreign specialists to the Soviet Union, but also on those occasions when Soviet specialists were sent abroad. In 1961 the KGB raised a question in the CC CPSU about the leaking of secret information during Soviet scientific visits to the West. Information on Soviet metallurgy had been published in a British journal after one Soviet specialist's visit to Great Britain. The KGB viewed such leaks as harmful, and demanded that a stricter information policy be put in place.[32] Connections with the Western press and the publication of scientific information inside the Soviet Union were duly tightened up.[33] Attitudes and practices involving the publicizing of Soviet achievements underwent numerous changes as a result.

When direct contacts caused problems, Western scientific-technical knowledge was collected, translated and exploited indirectly. The Soviet Union was concealing information on Soviet science and technology, but meanwhile, it had itself created a massive system for collecting information from the West and translating it into Russian. This information was administered by the All-Union Institute of Scientific-Technical Information (VINITI), which also had a role in the system of controlling information.[34] VINITI was established in 1952 but it started to work most actively during the Khrushchev leadership.[35]

'Mobilization of internal resources'

One of the systems that shaped the Soviet Union's efforts to obtain high technology and know-how was CoCom, a multilateral export and control mechanism established by Western powers in 1949.[36] The Western strategic embargo, in which the US took a leading role, was implemented by NATO members as a response to the Soviet atomic bomb, with the main aim of retarding Soviet technical progress in key strategic areas. The embargo was directed against exports of technology that might contribute to military and civilian economic performance, and it was directed not only against the USSR but against all Warsaw Pact countries.[37] At a time when Khrushchev was promoting the need for a more active technology import policy,[38] then, the Cold War was setting strict restrictions.

One of the clear outcomes of the Western embargo inside the Soviet bloc was Socialist economic integration, which was very close to Stalin's policy of turning inwards and 'mobilizing internal resources'. The Council for Mutual Economic Assistance (CMEA) was established in 1949 inside the Soviet bloc, and it strengthened the Soviet Union's industrial, technical and military grip on the rest of the bloc. Although economic and technological cooperation was the main purpose of CMEA cooperation, it also had significant bearing on the military technology and defensive capabilities of the Warsaw Pact countries. Economic performance within the CMEA was based on the division of labour, which, together with the aims of economic cooperation, was based on priorities set by the Soviet leadership.[39] Adoption of the STR as the main target in technology policy directed the priorities in the CMEA, and during the Khrushchev leadership the CMEA's role in realizing the STR was strengthened. In 1958, the role of scientific-technical cooperation and planning based on intensified utilization of new technology was emphasized in Soviet and CMEA plans. This policy was adopted quickly and by 1959 the Soviet Union was already actively developing cooperation within the bloc.[40]

In the late 1950s Sputnik generated enthusiasm in the CMEA and the main target was to strengthen Socialist technological performance with intra-bloc innovations. One of the main fields of scientific-technical cooperation was automation, which was also a major feature of the STR and the main booster of Western economic growth.[41] The lack of standardization across the CMEA countries caused problems, but as soon as technological progress and the increasing role of the international division of labour started to take effect, the need for standardization began to be taken more seriously.[42] With a view to fostering more connections internationally, discussions regarding the establishment of common standards in all CMEA countries started in 1958 and were soon put into effect.[43] Another aspect of cooperation was the transfer of scientific-technical discoveries and designs from one country to another. This mainly involved cooperation and a division of labour in R&D within the CMEA; one country took care of one process and then sent it on to the Soviet Union, which gathered together the related information.[44] This was intended to work for the common good, but ultimately the main benefit was accumulated in the Soviet Union.

The CMEA and scientific-technical cooperation within the bloc did not solve the problem of the lack of technology. In spite of serious efforts, the main problem plaguing cooperation within the CMEA was that of low quality; only the conventional machine-building industry was relatively developed. Already by the early 1960s it was recognized in the Soviet bloc that advanced Western processes, designs, know-how, machinery and equipment were needed throughout the whole CMEA area. Technology obtainable inside the bloc was no longer sufficient to keep abreast of the STR.[45] This made Khrushchev's networking and cooperation with the West more target-orientated: new opportunities for cooperation were sought, and existing cooperation was strengthened.

In spite of its leading role in the embargo, the United States had only a limited ability to persuade its allies to adhere to a strict embargo policy through CoCom.[46] This caused dissent among the CoCom partners. The Soviet Union was an eligible trade partner; Soviet markets were large, and the Soviet Union had a high credit ratio.[47] Those CoCom partners dependent on foreign trade, especially Great Britain, Italy, Japan and France, were reluctant to support the embargo. This was the case especially when it came to products that had a commercial value to the exporters and could become the subject of commercial trade with the Soviet bloc.[48] Khrushchev's willingness to cooperate was met with growing interest in most of the West European states. The Soviet Union had agreements with France, West Germany, Italy, Japan and Great Britain, all of which were members of CoCom. These were also the countries that were the main targets of the Soviet specialists – a point that was a major cause for concern in the United States.[49] During this period the Soviet Union was concluding not only inter-governmental agreements but also agreements with Western firms. The Soviet State Committee for Science and Technology (the GKNT) concluded agreements on technical and scientific cooperation with leading Western firms such as AEG, Krupp, Mannessman, Fiat and Siemens in the late 1950s, based on Khrushchev's initiative.[50] Connections with Western high-technology firms were of great interest for the Soviet authorities because this represented a means of obtaining tested high technology.

Soviet scientific-technical networking with the West

CoCom restrictions placed limits on normal trade and, as a result, the Soviet Union started to seek alternative ways to obtain technology from the West. A good example of the new types of technology transfer launched in this connection was the push for scientific-technical cooperation (*nauchno tekhnicheskoe sotrudnichestvo*) in the mid-1950s. Scientific-technical cooperation was a non-commercial and officially approved channel for transferring technology. In contrast to inter-war technology transfer, which concerned only machines, scientific-technical cooperation also involved transfers of know-how and expertise. The Soviet Union promoted scientific-technical cooperation with Western countries especially actively in the early 1960s, and concluded numerous inter-governmental agreements on cooperation with Western governments.[51]

The Soviet Union engaged in scientific-technical cooperation with the neutral states of Austria, Finland and Sweden in particular. The importance of the neutral states was emphasized by Khrushchev on several occasions, especially in connection with his policy of peaceful coexistence, which fitted well with the aims of technology transfer and networking. The group of neutral states was defined as a 'zone of peace' between the two camps. Finland was described as a neutral state, comparable to Sweden and Austria, which defined Finland's new role between the East and West.[52] The role of neutral and un-allied states in Scandinavia was strengthened in Soviet policies during the 1960s. According to Soviet archival materials, Finland and Sweden were the main targets of the Soviet Union and were also of great interest in the field of scientific-technical cooperation. In Europe, the Soviet Union's interest was focused mainly on West Germany, especially during the Khrushchev leadership.

Finland was the first market economy to conclude a five-year trade agreement with the Soviet Union, in 1950.[53] The Soviet–Finnish agreement on scientific-technical cooperation was signed in 1955,[54] and was the first treaty between any two states with different economic systems agreeing upon scientific-technical cooperation.[55] Initial cooperation was established at the commission level. The partners involved were the Finnish Ministry of Foreign Affairs, the Finnish Academy of Sciences, and the GKNT and the Soviet Academy of Sciences respectively. Relatively quickly, most of the technologically orientated Finnish firms became involved in scientific-technical cooperation and direct connections were established linking Finnish enterprises and Soviet partners, mainly ministries and state enterprises.[56]

In the early stages, Finnish–Soviet scientific-technical cooperation was confined mainly to visits to factories and production units. The themes of these visits were set by meetings of the scientific-technical commission but the best targets were chosen by the Soviet embassy in Finland, which did all of the background work.[57] Soviet embassies drew up a multitude of reports about the development of the target country. For example, the Finnish economy and technology were analysed monthly, especially whenever there were changes in the leadership or economic policy of Finland.[58]

Soviet specialists, who came mainly from ministries and state enterprises, started to collect information from those branches of the economy that were most useful for the Soviet partners in the late 1950s. In the initial stages the information collected during their visits to Finland was very practical. After visiting Finnish production units, Soviet specialists wrote up hundreds of pages of practical suggestions for action based on what they had experienced and observed during their visit. Such visits invariably had to involve a clear benefit to the Soviet side: when there was no such benefit, no visitors were sent.[59] Soviet specialists took their task seriously but their report-writing skills were also the result of long years of experience. From the early 1930s, Soviet specialists had been sent on missions (*komandirovka*) aimed at transferring knowledge, and in this way they became experienced report writers.[60] In the 1960s these visits became more target-orientated and more focused on technological observations. Soviet specialists travelling abroad produced reports with descriptions of the relevant technology, illustrated with dozens of photos and constructional drawings, which were then distributed for the benefit of Soviet industrial designers.[61]

Initially the main interest of the Soviet partners was focused on those production units that utilized Western technology in fields that were backward or not prioritized in the Soviet Union and the CMEA countries. The Finnish oil enterprise Neste, which utilized Western technology, regularly received visitors from Soviet ministries and state enterprises from the late 1950s.[62] These visits were generally viewed quite positively on the Finnish side, but sometimes the obviously target-orientated behaviour of the Soviet delegates was seen as annoying. The main reason for the negative attitude was the Soviet side's keenness to obtain detailed information on those processes that were forbidden under patent agreement. This was especially irritating in those cases when Finnish reciprocal visits were subsequently blocked by the Soviet side: Neste's proposed visit to the Kuibyshev region to learn about the production of arctic diesel was not approved by the Soviet partners.[63] However, after experiencing this kind of obstructiveness from the Soviet side, the Finns also learned to deny permission for undesirable visits suggested by the Soviet partners by invoking reciprocity.

Reciprocity was not always a practical tool for cooperation. In the early stages of cooperation in particular, Finnish partners were forced to take Soviet offers into account whether they wanted to or not. In the late 1950s the Finnish leadership decided to choose nuclear energy as a source of energy for Finland. The decision to build a nuclear plant based on Western technology had already been taken and the first rounds of competitive bidding launched when it became evident that a Soviet offer would have to be accepted for political reasons. Negotiations with the Soviet side began in the early 1960s. For security reasons the Finnish partners demanded the right to acquaint themselves with Soviet nuclear technology through visits to Soviet nuclear plants and major physics institutes. Although these visits came under the terms of the contract, they would frequently be cancelled or refused without explanation. In the end, the nuclear plant was built and it became a unique example of East–West cooperation based on Soviet nuclear technology and Western security systems with Finnish construction expertise.[64]

In the early stages of the scientific-technical cooperation the Soviet partners viewed Finland as merely a mediator of Western technology. In the late 1950s Western and modern telephone exchange technology utilized in Finland was of great interest to the Soviet Union.[65] In the 1970s Finland was transformed from a mediator into a producer or even a pioneer, and the cooperation thus entered a new phase. The Finnish enterprise Nokia is a good example of the rapid progress characterizing the cooperation during this period. Nokia embarked upon cooperation with the Soviet Union through the scientific-technical commission during the Khrushchev period, in the late 1950s, with basic industrial products such as products of the timber and cable industries. Only two decades later Nokia had widened production into the field of automatic telephone exchanges and wireless communication systems. During the economic recession in the 1980s Nokia could count on a large demand from the Soviet market and thus managed to avoid economic meltdown. Based on the scientific-technical cooperation started during the Khrushchev leadership, Nokia managed to create remarkable growth.[66]

During the 1960s Finland became the Soviet Union's main partner in Scandinavia. However, in the early 1960s the Soviet leadership was also interested in Swedish technology and cooperation with Sweden. In the West, Sweden was seen as one of the countries that might potentially transfer technology to the Soviet Union. This fear was not without reason – the Soviet Union had a strong interest in Swedish technology, especially during the early 1960s. In 1961 the Soviet embassy in Stockholm drew up a report on Swedish scientific-technical achievements. According to the report, the embassy's primary role was to widen scientific-technical connections with Swedish firms and scientists. The Soviet Union's interest in Sweden was quite similar to its interest in Finland. Soviet specialists visited Swedish firms, and Swedish specialists travelled to the Soviet Union. Compared to Finland, Sweden was technologically more developed in the early 1960s and exploited atomic energy, for example. The Soviet Union was interested in exporting uranium to Sweden, which imported most of its nuclear fuel from the West.[67] Yet although it was of great interest to the Soviet Union in the 1960s, scientific cooperation with Sweden never reached the level it did with Finland. There was cooperation but no similar joint projects involving Soviet and Swedish partners. Sweden's more active role in the Western bloc was probably the main reason for this. Furthermore, for the United States, Swedish–Soviet cooperation was far less acceptable than Soviet–Finnish cooperation. In Sweden, active Finnish–Soviet cooperation was seen, especially later in the 1980s, as a security threat.[68]

In Europe, the most attractive partner for the Soviet Union was West Germany. The establishment of technology connections between the Soviet Union and West Germany started in the late 1950s when Soviet specialists were sent to Western scientific conferences and technology exhibitions to collect information and to establish connections with Western enterprises and specialists.[69] One of the key events here was the international congress and exhibition in Düsseldorf, West Germany, in November 1957. During the conference, Soviet specialists visited not only the exhibition stands but also the production units of the West German enterprises in order to acquaint themselves with projects for the automatization of production and development of the machine-building industry. The CC CPSU received a detailed report from the Soviet delegation, which analysed the main fields of German technology. An operations model based on the recommendations given in the report was subsequently created. This visit was the first of a series of visits to German firms including, among others, Siemens. In the late 1950s the Soviet interest was directed in particular at firms like Siemens and AEG, which were the pioneers in control systems.[70] Interest in cooperation was also shown on the German side: Siemens made a proposal to the Soviet delegation on the translation of technological terminology from German and English into Russian.[71] This cooperation bore fruit in the late 1960s when Siemens started to export third-generation computers and components to the CMEA countries and especially to the Soviet Union.[72] Trade was against CoCom recommendations, but these were ignored in West Germany – a matter that caused some anxiety in the United States.[73] In the case of West Germany, Soviet tactics in politics clearly differed from tactics in trade. Negotiations with the Soviet Union and the West German enterprises continued through the 1961 Berlin crisis.[74]

Conclusion

Khrushchev's main aim – to catch up and overtake America – was based on the idea of technological modernization and a strong belief in the superiority of the Socialist system. Khrushchev's methods for escaping backwardness were based on a strong belief in technology, and there was a sense in which this represented a return to the Bolshevik utopia of the 1920s. The Soviet leadership of the 1950s and early 1960s, and the Bolsheviks of the 1920s both shared the same technocentric belief in the omnipotence of technology. Khrushchev's technological utopianism was further strengthened by the cosmic utopianism that was generated by Sputnik,

Laika and Yurii Gagarin in the 1950s and 1960s. Successes in the space programme created an atmosphere of admiration outside the Soviet Union and enthusiasm inside the Soviet bloc, which strengthened belief in the superiority of the Socialist system. After Sputnik, the realization of the scientific-technical revolution (STR), the aim set by the Soviet leadership in 1958, seemed to be an easy task for the Soviet Union. Despite a strong belief in the preconditions created by Socialism and intra-bloc abilities, technological progress did not materialize through CMEA cooperation. The Soviet Union needed foreign technology in order to achieve technological progress capable of transforming extensive economic growth into intensive growth.

The import of foreign technologies had a long tradition in Russia and the Soviet Union. During the Khrushchev leadership the Soviet Union engaged with world politics as a superpower. The new structure of world politics affected the Soviet Union's ability to transfer technology through normal trade. The CoCom embargo hindered technology transfer from the West and the Soviet Union needed to find other channels to become part of the technological progress. Especially after Soviet success in the space programme the need for Western technology proved to be rather problematic for the Soviet leadership. The ideological problem engendered by this need was solved by the Soviet leadership by emphasizing the international propaganda aspect of the STR. The concept was connected to global technological progress and emptied of ideological content. Any information that might have jeopardized realization of the STR was kept secret, while all information that nurtured the aim was collected with great diligence. This created, especially during the Khrushchev leadership, a new dimension to the Cold War – the Information Cold War.

The Soviet leadership created alternative ways to transfer technology and know-how in order to boost the Soviet drive for technological progress and intensive economic growth. The main channel for obtaining technology and know-how from the West during the Khrushchev leadership was scientific-technical cooperation, and this cooperation remained the main feature in the transfer right through until the collapse of the Soviet Union. While most of the reforms launched by Khrushchev were rescinded after his removal from power, scientific-technical cooperation was further developed and intensified. Scientific-technical cooperation proved to be very useful for the Soviet Union all the way through to the 1990s. It offered an approved and official way to transfer technology and know-how, but it also offered a chance to turn cooperation into conventional trade.

Although technology transfer was controlled in the 1960s, through trade agreements based on scientific-technical cooperation with several Western states, Khrushchev proved that not even CoCom was effective in preventing high technology transfer to the East. West Germany was one of Khrushchev's main targets and became also a good illustration of the huge gap between economic aims and political rhetoric. In spite of West Germany's strong role in the Western bloc as a member of NATO and CoCom she remained an important trade partner for the Soviet Union until the end of the Cold War. German machinery constructed the technological basis for Stalin's industrialization in the 1930s. During the

Khrushchev leadership the basis of Soviet–West German trade was created. Under the Brezhnev leadership, this would become a stronghold of Soviet trade with the West and one of the main sources of Western high technology.

Scientific-technical cooperation proved to be valuable for the Soviet Union but it was important for the partners as well. The role of the neutral states in scientific-technical cooperation was emphasized by Khrushchev, and the Soviet Union duly concluded agreements with these states. Finland soon became the most important partner for the Soviet Union. History and geopolitical proximity are the reasons usually mentioned for Finland's special role in Soviet–Western trade. Post-war Finland was a relatively fast-developing and technologically oriented Western state, which proved to be a very useful partner for the Soviet Union during the Cold War. For Finland, especially from the economic point of view, cooperation with the Soviet Union was extremely beneficial. For Finnish enterprises like Nokia, the possibility that scientific-technical cooperation offered for trade with the Soviet Union was key to later development and to global markets. Sweden, another Scandinavian neutral state, was an important target for the Soviet Union in the 1960s. Sweden's stronger position in the Western bloc, however, prevented deeper involvement in joint projects with Soviet partners.

Scientific-technical cooperation was a good example of Soviet networking with the West during the 1960s and the system created during the Khrushchev leadership proved to be successful for the Soviet Union. Although the network of scientific-technical cooperation was widely utilized until the end of the Soviet period, the aims of the Khrushchev leadership, especially the STR, were not realized as such. In spite of wide-reaching scientific-technical cooperation and technology transfer there was no remarkable technological progress that would have transformed extensive economic growth into intensive growth in the Soviet Union and changed the balance of superpower rivalry. The optimism and utopianism of the Khrushchev era began to crumble away bit by bit after Khrushchev's removal from power in 1964. The Soviet invasion of Czechoslovakia in 1968 was the end of optimism, and the decision to return to imitation and duplication of Western technology and to give up Soviet R&D was the end of technological utopianism.

From the point of view of Cold War historiography, Khrushchev's scientific-technical networking in Europe appears to have had a greater impact on Cold War politics than was previously believed. The networks that Khrushchev created with the small states in Europe established a grey area between the two blocs. Because of this grey area, interaction in Cold War Europe took the form of multipolar activity across the Iron Curtain. By utilizing these networks, small European states formed their own relationships with the great powers and their allies. The technology transfer that Khrushchev initiated across the Iron Curtain had a significant impact on Cold War bloc politics.

Notes

1 J.L. Gaddis, *The Cold War*, London: Allen Lane, 2005, p. 84.
2 The maximum benefit was extracted from the visit, with wide use of propaganda made both outside and inside the Soviet Union. See, for example, the 'report written on the meeting of Glavproektmontazmashina 28.10.1959', RGANTD f. 20, op. 4–6, d.194, ll. 1–2; *Face to Face with America. The story of N.S. Krushchev's visit to the USA*, Moscow: Foreign Languages Publishing House, 1960; W. Taubman, *Khrushchev: The Man and His Era*, London: Free Press, 2003.
3 V. Shlapentokh, *A Normal Totalitarian Society*, New York: M.E. Sharpe, 2001, pp. 18–19.
4 See e.g. Taubman, *Khrushchev*, pp. 130, 423.
5 The transfer of technology is the process whereby a technique is substantially moved from one set of users to another, or the process by which innovations made in one country are subsequently brought into use in another country. See E. Nironen, 'Transfer of Technology between Finland and the Soviet Union' in K. Möttölä, O.N. Bykov and I.S. Korolev (eds), *Finnish–Soviet Economic Relations*, London: Macmillan Press, 1983, p. 161.
6 S. Fitzpatrick, *The Russian Revolution*, 2nd edn, Oxford: Oxford University Press, 1994, p. 9.
7 P.R. Gregory and R.C. Stuart, *Soviet and Post-Soviet Economic Structure and Performance*, 5th edn, London: HarperCollins,1994, pp. 15, 30; see also P. Hanson, *The Rise and Fall of the Soviet Economy*, London: Longman, 2003, p. 62; on the connection between science and industry during the Stalin era, see R. Lewis, *Science and Industrialisation in the USSR*, London: Macmillan, 1979.
8 S. Autio-Sarasmo, 'Soviet Economic Modernisation and Transferring Technologies from the West', in M. Kangaspuro and J. Smith (eds), *Modernisation in Russia since 1900*, Studia Fennica Historica 12, Helsinki: SKS, 2006, pp. 108–9.
9 S. Gerovitch, *From Cyberspeak to Newspeak. A History of Soviet Cybernetics*, Cambridge, MA: MIT Press, 2001, pp. 15–16.
10 E.P. Hoffman and R.F. Laird, *'The Scientific-Technological Revolution' and Soviet Foreign Policy*, Oxford: Pergamon Press, 1982, pp. 7–8. See also S. Autio-Sarasmo, 'Soviet Economic Modernisation', pp. 110–11.
11 Fitzpatrick, *The Russian Revolution*, p. 19. See also Gregory and Stuart, *Soviet and Post-Soviet Economic Structure and Performance*, p. 8; *Technology and East–West Trade Report*, Office of the Technology Assessment, US Congress, November 1979, pp. 214–15, 217.
12 I. Susiluoto, *Suuruuden laskuoppi. Venäläisen tietoyhteiskunnan synty ja kehitys*, Helsinki: WSOY, 2006, pp. 172–3. Even the computer systems that were created inside the military complex were incompatible with one another.
13 Gaddis, *The Cold War*, pp. 32–3.
14 See e.g. C. Kennedy-Pipe, *The Origins of the Cold War*, London: Palgrave Macmillan, 2007, pp. 17–18.
15 P. Hanson, *Trade and Technology in Soviet–Western Relations*, London: Macmillan, 1981, p. 87.
16 J. Wilczynski, *Technology in COMECON. Acceleration of Technological Progress through Economic Planning and the Market*, London: Macmillan, 1974, pp. 6–7. According to Wilczynski, the concept is attributed to Bertrand Russell; for more analysis of the STR see e.g. E. Rindzeviciute, *Constructing Soviet Cultural Policy. Cybernetics and Governance in Lithuania after World War II*, Linköping: Linköping University Arts and Science, no. 437, 2008, pp. 192–5.
17 MID f. 135, op. 36, por. 39, papka 128, ll. 18–24. 'The prognosis of the development of the foreign trade of the Soviet Union' was given in 1960.
18 Shlapentokh, *A Normal Totalitarian Society*, pp. 18–19.

19 J. Berliner, *Soviet Industry from Stalin to Gorbachev*, Cheltenham: Edward Elgar, 1985, p. 250.
20 L. Graham, *Science in Russia and the Soviet Union*, Cambridge: Cambridge University Press, 1993, p. 180; see also G. Ofer, *Soviet Economic Growth: 1928–1985*, RAND/UCLA Center for the Study of Soviet International Behaviour, 1988, p. 62.
21 Graham, *Science in Russia and the Soviet Union*, pp. 73, 222–3, 256. MESM was designed by S.A. Lebedev and was developed totally independent of Western efforts.
22 Gerovitch, *From Newspeak to Cyberspeak*, p. 279. The new Soviet leadership decided to copy Western computers rather than to develop domestic computer R&D: Susiluoto, *Suuruuden laskuoppi*, pp. 152–3.
23 Taubman, *Khrushchev*, p. 620. On Khrushchev's attraction to scientists and engineers, see Taubman, *Khrushchev*, p. 130.
24 See e.g. A. Sutton, *Western Technology and Soviet Economic Development 1945 to 1965*. Third volume. Stanford, CA: Hoover Institution Press, 1973, pp. 270–3. During and after the war, among others, two testing sites, technology and some 6,000 technicians were transferred to the Soviet Union.
25 P. Maddrell, *Spying on Science. Western Intelligence in Divided Germany 1946–1961*, Oxford: Oxford University Press, 2006, pp. 17–18, 25–6, 30–1, 206–7.
26 See, e.g., A.A. Aleksandrov, *Put' k zvezdam. Iz istorii Sovetskoi kosmonavtiki*, Moscow: Veche, 2006.
27 Wilczynski, *Technology in COMECON*, p. 335.
28 See, for example, Gaddis, *Cold War*, p. 68–9.
29 Maddrell, *Spying on Science,* pp. 119–20.
30 RGANI f. 5, op. 40, d. 98, ll. 33–5, 36–8.
31 RGANI f. 5, op. 40, d. 121, ll. 60–1.
32 RGANI f. 5, op. 40, d. 157, ll. 52–3.
33 RGANI f. 5, op. 40, d. 98, ll. 36–8.
34 RGANI f. 5, op. 33, d. 46, ll. 15–16, 21. Vsesoyuznii Institut Nauchnoi i Tekhnicheskoi Informatsii (VINITI).
35 J. Seppänen, *Tieteellis-tekninen informaatio Neuvostoliitossa*, Helsinki: Suomen ja Neuvostoliiton välisen tieteellis-teknisen yhteistoimintakomitean julkaisusarja 2, 1978, pp. 9–10; see also Susiluoto, *Suuruuden laskuoppi*, p. 180.
36 For more detail see e.g. Wilczynski, *Technology in COMECON*, p. 331.
37 Hanson, *Trade and Technology in Soviet–Western Relations*, p. 223. Iceland was not a member of CoCom but Japan and Australia were members; G. Bertsch, 'Technology Transfers and Technology Controls: A Synthesis of the Western–Soviet Relationship', in *Technical Progress and Soviet Economic Development*, Oxford: Basil Blackwell, 1986, p. 127–8; Hanson, *The Rise and Fall of Soviet Economy*, p. 161. As Hanson points out, the Volga Automobile plant was reviewed to see whether the Italian-made machine tools could be diverted to tank production.
38 G.D. Holliday, *Technology Transfer to the USSR 1928–1937 and 1966–1975*, Boulder, CO: Westview Press, 1979, pp. 78–9.
39 Wilczynski, *Technology in COMECON*, pp. 77, 110–11, 261–3, 264–6, 341.
40 RGANI f. 5, op. 40, d. 121, ll. 29–30; see e.g. RGANI f. 5, op. 40, d. 121, ll. 142, 153, 154.
41 Wilczynski, *Technology in COMECON*, pp. 122–3. Automation included the most classical examples of automation: the methods and means of steering and controlling technical equipment without direct human participation; see also RGANTD f. r-20, op. 4–6, d. 313.
42 Wilczynski, *Technology in COMECON*, p. 141.
43 RGANI f. 5, op. 40, d. 98, ll. 98–178.

44 See e.g. RGANTD f. r-20, op. 4–6, d. 227, l. 34; RGANTD f. r-20, op. 4–6, d. 312, l. 97. See also Wilczynski, *Technology in COMECON*, p. 272.

45 Wilczynski, *Technology in COMECON*, pp. 14, 145, 185, 275, 296.

46 *Technology and East–West trade*, pp. 14, 155–6, 160. See also E. 'Nironen, 'Lännen embargopolitiikka murrosvaiheessa', *Ulkopolitiikka*, 3, 1990, p. 44.

47 Hanson, *Trade and Technology in Soviet–Western Relations*, p. 123.

48 I. Jackson, *The Economic Cold War: America, Britain and East–West Trade, 1948–63*, London: Palgrave-Macmillan, 2001, pp. 173, 178.

49 See e.g. RGANI f. 5, op. 40, d. 121, ll. 52–3. An interesting detail in the decision to embark upon cultural relationships with the UK was that the KGB coordinated the planning; RGANI f. 5, op. 40, d. 121, l. 153.

50 Wilczynski, *Technology in COMECON*, pp. 297–9, 301; see also RGAE f. 9480, op. 7, d. 949 (2), ll. 162–3, 169–73; RGANI f. 5, op. 40, d. 98, ll. 9–20.

51 Bertsch, 'Technology Transfers and Technology Controls', pp. 117, 120; Holliday, *Technology transfer to the USSR 1928–1937 and 1966–1975*, p. 47.

52 A. Kähönen, *The Soviet Union, Finland and the Cold War. The Finnish Card in the Soviet Foreign Policy*, Helsinki: SKS, 2006, p. 31.

53 Rantanen, 'The Development of the System of Bilateral Agreements between Finland and the Soviet Union', in Möttölä, Bykov and Korolev (eds), *Finnish–Soviet Economic Relations*, pp. 43–4, 52.

54 Sopimus tieteellis-teknillisestä yhteistoiminnasta Suomen tasavallan ja SNTL:n välillä, 16.8. 1955. File Ad 13/3647–55, MFA; see also RGAE fond 9480, op. 7, d. 931, ll. 154–60.

55 A.K. Romanov, 'Suomen ja Neuvostoliiton välisen tieteellis-teknisen yhteistyön tuloksia', in *Suomen ja Neuvostoliiton välinen tieteellis-tekninen yhteistoiminta 30 vuotta*, Helsinki, 1985. The cooperation was based on the treaty of friendship, cooperation and mutual assistance signed in 1948 between Finland and the Soviet Union.

56 Suomen ja Neuvostoliiton välisen tiedeyhteistyön kanavat, 1–2; see also e.g. RGAE 9480, op. 7; RGAE fond 9480, op. 3, d. 1610.

57 For example, theme no. 4223 was oil refining. RGAE fond 9489, op. 7, d. 925, ll. 139–42; see also Pöytäkirja [Protocol] 7.3.1959. File Soviet–Finnish scientific-technical cooperation 1956–1959. Archive of the enterprise UPM-Kymmene.

58 A good example of this is the Soviet embassy's report on the Finnish economy and markets in 1960 and the economic policy of the government of PM Sukselainen. MID 135, op. 42, por 16, p. 89, ll. 5–24.

59 RGAE f. 9480, op. 3, d.1610, l. 42.

60 Reports on *komandirovka* inside the Soviet Union, see e.g. GASO f. 4697, op. 1, d. 9, ll. 4–6.

61 See e.g. RGANTD f. p-18, op. 2–6. d. 204, Samara.

62 RGAE f. 9480, op. 7, d. 925, ll. 139–42; see also RGAE f. 9480, op. 7, d. 925, ll. 146–56. A good example of this was the cable industry in the CMEA countries; on the problems in cable industrial cooperation in the CMEA, see RGANTD f. r-388, op. 3–1, d. 135, l. 2; See also RGANI f. 5, op. 40, d.157, ll. 4–23 Swedish cable industry. In Finland cooperation with Nokia was based on the cable industry in the early years.

63 Neste Oy:n vastaus TT-komission tiedusteluun [Response from Neste to ST-commission] 16.10.1961. File Ad 13/3647–55, FMA. The Kuibyshev (Samara) region was closed until 1991.

64 See e.g. MID f. 135, op. 36, por. 39, p. 128, ll. 18–24. These institutes were closed and highly secretive units; MID f. 135, op. 46, por. 11, p. 98, ll. 6–7, 23–4, 28. The host of the visits to Moscow, Tbilisi and Kiev was A.V. Zakharov. The rector of the Polytechnical Institute in Helsinki wished to visit the institutes in Leningrad and Moscow. In information provided to the Soviet side in the lead-up to the visit it was mentioned that the Helsinki polytechnic had an American-made atomic reactor. In

a letter dated January 1965 he was informed that permission for the visit had been refused. On the construction of the first Finnish atomic plant, see K.E. Mickelsen and T. Särkikoski, *Suomalainen ydinvoimalaitos*, Helsinki: Edita, 2005.

65 RGAE f. 9480, op. 3, d.1610, l. 49.
66 M. Häikiö, *Sturm und Drang. Suurkaupoilla eurooppalaiseksi elektroniikkyritykseksi 1983–1991. Nokia Oyj:n historia*, part 2. Helsinki: Edita, 2001.
67 RGANI f. 5, op. 40, d. 157, ll. 4–23; see also Michelsen and Särkikoski, *Suomalainen ydinvoimalaitos*.
68 See e.g. O. Kronvall, *Den bräckliga barriären: Finland i svensk säkerhetspolitik 1948–1962*, Stockholm: Elander Gotab, 2003; see also Thorsten B. Olesen (ed.), *The Cold War – and the Nordic Countries*, University Press of Southern Denmark, 2004.
69 RGANI f. 5, op. 40, d. 121, ll. 54–5.
70 RGANI f. 5, op. 40, d. 98, ll. 9–20.
71 RGANI f. 5, op. 40, d. 98, ll. 9–20; RGANI f. 5, op. 40, d. 121, l. 47. In Finland there were similar translation projects but these came later, mainly during the 1970s; see e.g. Susiluoto, *Suuruuden laskuoppi*, p. 215.
72 Wilczynski, *Technology in COMECON*, p. 113.
73 Nironen, 'Lännen embargopolitiikka murrosvaiheessa', p. 46.
74 See e.g. RGANI f. 5, op. 40, d. 121, ll. 54–5.

10 Khrushchevism after Khrushchev

The rise of national interest in the Eastern Bloc

Katalin Miklóssy

Khrushchevism versus the Khrushchev period: challenging concepts

Historians often develop strong preferences for thinking in periods: far too often we investigate historical problems in the context of predefined epochs. Nothing is so emblematic of our profession than the interruption of the flow of time according to governing elites that followed one another. 'Periodization' sends an encrypted message of the explicit beginning and end of an era that emphasizes the limited focus of interest. By distinguishing historical periods we anchor the dynamic, social developments to static divisions and hence we downplay the significance of long-term phenomena that overarch our classification. However, by observing the processes that run counter to our perceptions of different periods we could gain new understanding of the significance of these chapters of time.

In this chapter, we shall focus on those political initiatives launched by Nikita Khrushchev that grew into their full importance long after his governance and marked actually the beginning of the end of the entire socialist epoch. In this respect it is inevitable to differentiate between the static period and dynamic afterlife of some political decisions – in other words, between Khrushchev's time and Khrushchevist ideas. Needless to say, in comparison to the preceding and subsequent periods, Khrushchev's era was exciting: in relation to the Stalinist years his line represented a clear-cut and dramatic change that Leonid Brezhnev, despite his relentless attempts, was unable to root out. Hence the basic period-distinguishing characteristics do not need further elaboration. The period-surviving elements, on the other hand, create a different theoretical problem. How do we define what can be justly called Khrushchevist elements if we take into consideration that some major processes, which Khrushchev brought about, led to unintended consequences that Khrushchev himself would never have agreed with. Unintended outcomes alone can also be questioned in relation to the evaluation of processes since these results emerged due, for example, to Khrushchev's successors' policies to discredit his efforts or to other circumstances. Subsequently, we have to explicate the late development of Khrushchev's ideas as processes and pinpoint those that undoubtedly can be counted as being in line with Khrushchev's intentions. We shall relate to his strivings that had long-term consequences as Khrushchevism. The most important question to be answered here is what kind of changes Khrushchev did induce that had significance for the development of the Eastern bloc.

Ironically, the most perceptible conceptual alteration that Khrushchevism brought up actually challenged the conventional image of the Eastern bloc itself. According to the still prevalent understanding of the centre–periphery construction of the bloc, the East European states in the Soviet sphere acquired satellite[1] status and constituted a rather uniform group in a monolithic power structure. In this vision, the Moscow-based centre acted as the ultimate decision-maker and the periphery countries formed a subordinated and loyal constellation. This simplistic view is rooted in the old Cold War paradigm and in the research tradition of Sovietology that either emphasized the bipolar collision and concentrated on major actors that had the power to shape the international system, or explored the Eastern bloc by focusing on the centre and manufactured educated guesses about the rest.[2] The so-called small bloc members attracted scholarly attention only if they proved disobedient by uprisings, revolutionary movements or other political dissent.

Due to the marginalization of interest in the 'small' states of the Eastern bloc the scientific community could not foresee the collapse of the communist system. As became evident later, observing the *small* through the glasses of the *great* produced a distorted, cursory picture that left unnoticed the delicate political mechanisms under the immediate surface of the quasi-coherent community – mechanisms that in the end brought down the whole era.

The science of history is, however, based on a never-ending dialogue between the past and present, where past events get reinterpreted either by the introduction of new evidence or by applying new angles to old sources and by asking new questions of the material at hand. This study represents a new stance: it shall revise the traditional point of view and concentrate on mapping the possibilities of manoeuvring in the context of a constantly changing leverage where we find a fascinating world of multidimensional power-play. We argue that, due to Khrushchev's influence, the notion of *satellite* turned out to be an inaccurate term to describe the post-Stalinist era and remained in scientific language unduly long as a useless relic. Quentin Skinner recalled that the introduction of a new vocabulary is the best sign of a conscious conceptual change.[3] Thus, turning around Skinner's thought concerning scholarly works, we could ponder if the general use of the word 'satellite' would not forestall the spread of a new understanding of that era.

In this chapter, we shall focus first on the loopholes the centre–periphery construction offered by Khrushchev's reforms, then we move on to analyse the nature of national strategies, and finally elaborate on what was the best form of action to achieve different objectives.

Khrushchev's revolution

Khrushchev's period, like any other, was full of contradictory characteristics. He introduced refreshing new ideas concerning intra-bloc relations at the same time as applying old methods to punish renegades in order to secure the Soviet sphere of influence. During the Stalinist era no special theory of international relations was needed for the socialist states since the governing idea, socialist internationalism, relied on the general and oversimplified demand of working-class solidarity without national boundaries.[4] In practice this idea served to justify the extension of power of the Soviet Empire to Eastern Europe. After Stalin's death, Khrushchev replaced 'the informal hegemony' with a formal system,[5] where instead of rigid direct command, the institutional framework became emphasized. Accordingly, the Warsaw Pact and the Council for Mutual Economic Assistance (CMEA) became the dominant means of control through which a new form of intra-bloc relations was introduced. These institutions were restructured to improve efficiency of cooperation, but at the same time these reforms also provided more leverage for the member states.[6] The attitude change became apparent for the first time in the rapprochement to Yugoslavia.[7] In the Belgrade declaration of 1955, representatives of the Soviet Union and Yugoslavia praised the principles of non-interference in domestic affairs.[8] Shortly after this declaration the Cominform, the most rigid instrument of control, was dismantled to emphasize the new course. This was a grand opening where a new political theory for international relations was introduced, which later the so-called 'secret speech'[9] made official. At the 20th Party Congress in 1956 Khrushchev condemned the mistakes of Stalinism and stated that socialism could now be built differently from the Soviet model.[10] The dramatic consequences of this statement in Poland and Hungary later that year were evidence of the general uncertainty of acceptable interpretations of Khrushchev's declaration. As a response, the notion of 'the commonwealth of the socialist nations' was invented on the eve of the suppression of the Hungarian revolution, sending a message that the community was, first and foremost, in the interest of every socialist country.[11] After the intervention, the Soviet government admitted that the equality principle of the socialist countries had failed in practice.[12] In order to avoid future problems and to explain the actions already taken, the principles of intra-bloc relations required more precise definition. From now on, the fraternal countries would avoid interfering in the domestic affairs of others, but would provide mutual aid if needed,[13] in the name of international solidarity.[14]

From the Hungarian perspective we could naturally question whether the famous secret speech was, after all, mere rhetoric. Although historians disagree over the evaluation of those events, there are signs indicating that modest changes or moderate social reform probably would have been accepted,[15] especially in the experimental era of Khrushchev. Zubov and Pleshakov also argued that, due to the improvement of Soviet–Yugoslav relations in the previous year, Khrushchev could have tolerated a diverse line of national communism if it proved otherwise loyal to the centre.[16] However, the fatal mistake was most likely the radio announcement (1 November 1956) of the Prime Minister Imre Nagy (in office 1953–55, 1956) declaring that Hungary would resign from the Warsaw Pact.[17] This statement endangered the Soviet Union's influence over the area and thus questioned the Yalta agreement itself. In the Cold War climate the disintegration of the bloc would have been considered a sign of Soviet weakness.

The next remarkable change in Soviet bloc policy occurred at the 22nd Congress in 1961, where Khrushchev defined the common goal as the economic victory of the socialist world over the capitalist one, and invented the new motto 'Dognaty i peregnaty', 'reaching and surpassing' the West. This would be achieved by a division of labour between the socialist countries where every country would specialize in a sphere of production that best fitted its economic profile and capabilities. Accordingly, the CMEA was to be transformed into a more efficient organization furthering the expansion of cooperation and developing coordination of the national economic plans.[18] Until 1954 the CMEA had only a symbolic significance, completely lacking any institutional framework, functioning on the basis of direct commands from Moscow, and it furthered mostly industrialization projects. In 1960 the organization was reformed to serve the demands of the new economic policy.[19]

The new programme suggested that the Western level should be reached by means provided by the entirely different Eastern model, i.e. communist management of the economy.[20] Hence, it was a 'mission impossible', as the leading economists described it.[21] The irony is that the new CMEA integration was highly reminiscent of the classical capitalist factory model in the Fordist sense with extreme specialization of production sectors coupled with central planning and management. But it gradually introduced a semi-capitalist economic thinking that started to spread especially during the era of the reformist Prime Minister, Aleksei Kosygin. Probably the most interesting paradox of the Cold War was that in spite of the spectacular East–West dichotomy, there emerged a tremendous interdependence where continuous challenges required adequate responses from each side.[22] This situation created an overemphasized demand for competitiveness with the West since success was to provide the ultimate proof that socialism was after all the right societal model, as ideology predicted. Later, however, competition appeared also among socialist countries resulting in increasing economic nationalism and diminishing loyalty to the community. This trend induced a growing awareness of the Western challenge, on a daily basis.

Still there was a more serious issue generated by the new Khrushchevist policy that eventually undermined the ideological basis. The theoretical problem was that the supremacy of the socialist system over capitalism should have been self-evident since it represented an immanent axiom of the ideology. According to the inward logic of ideological argumentation the 'objective' laws of social evolution proved that communism was the highest level of social development.[23] So *dognaty i peregnaty'* revealed that there were serious troubles in the evaluation of basic principles. The new motto of progress indicated that the concept of societal development became stripped of its deeper social-historical evolutionary content and was now conceived more simplistically, only in the economic aspect. It raised plainly the question of Western standards, which were now declared worthy of aspiring to.

Subsequently, the undisguised attraction of Western progress had grave consequences for the magnetism of the shared credo. Identification with the communist cause, in spite of its legitimacy advantages for the communist regimes, came under challenge from the ideal of a faster and more efficient development. What happened was that what Antonio Gramsci called 'the philosophy of practice'[24] took over the role of ideology. Ideology was now replaced by rationality, and the utmost means of persuasion in this hegemonic structure was not a faint image of a future ideal society but the very down-to-earth advantages of present economic growth. The political theory of building socialism differently coupled with the demand of reaching and overhauling the West actually represented a new course of ideological politics where politics was cut off more clearly from the ideological base and was bound emphatically to the economic sphere. From a coercive dominance of ideology on politics, there occurred a transitional situation where this economically orientated ideological mutation started actually to converge with political decision-making and began to serve political goals instead of providing them.

The above mentioned Khrushchevist line can be perceived as an attempt at modernization, especially in the political and economic spheres, after Stalin's reign. From this point of view it is sufficient to examine the Eastern bloc's working mechanisms in terms of a centre–periphery relationship. The centre's purpose of modernization aims either to secure a hegemonic stand in the face of its periphery or to maintain competitiveness in proportion to other centres.[25] In our case, the Soviet centre started to face opposition from its 'satellites' after Stalin had died. The workers' revolt in East Germany (1953), social unrest in Poland and revolution in Hungary (1956) were due to dissatisfaction with everyday living conditions and the economic hardship of the ordinary people. The revitalization of economic production, particularly elementary consumer goods, was also a way to maintain social peace and secure the system. The political disturbances in the member states undermined the credibility of Soviet rule and weakened its superpower status in the Cold War juxtaposition.

The Khrushchevist announcement about the possibility of building socialism differently from the Soviet model made the modernization projects even easier since it released the periphery from its obligations to uniformity. What we can observe is that some 'satellites' (Czechoslovakia, Hungary, Poland and Romania) seized the opportunity and started to experiment on their own. The fact that these countries stopped following the Soviet model meant that the model failed to provide them with working solutions to their problems, hence Moscow began to lose ground and prestige as a respectable *real* centre in front of its minor partners.

Reform projects in chain reaction

This section, based on selectively diverse sources, does not intend to provide a full chronicle of reforms in the East European countries, but will emphasize specific characteristics of political agendas that led to a peculiar development of the entire bloc.

The reformist projects of the small countries were rooted in the Khrushchevist structural change although in practice they started after the Khrushchev period. What is striking in these reforms is the great variety of theoretical perceptions and methods on which these political decisions were implemented. The emerging diversity revealed the different interpretations of the new principles Khrushchev had introduced, and this discrepancy derived from the distinct national interests, dissimilar domestic power equilibriums and divergent political preconditions in terms of dependency levels on Moscow's support.

The gradual disentanglement of the periphery countries was also furthered by the fact that in the late 1950s and early 1960s a new generation of party leaders came to power. The 'national or domestic' communists that replaced the Muscovites had less admiration for the Soviet model and more respect for their countries' past experiences. Hence, it can be argued that the attitude towards common bloc policies and also performance on the intra-bloc arena became subordinated to domestic political agendas. National interest started to dominate all foreign political considerations and coloured not only the enthusiasm for but even the actual engagement in the responsibilities bloc membership required. This development was due to the fact that, despite the differences, the main Khrushchevist message was understood quite similarly. More room for manoeuvre had been allowed for the satellites and they took full advantage of the situation – and, moreover, started to require even more elbow room.

Romania: a paradoxical model

The countries of our analysis were situated very differently within the Soviet sphere of influence, deriving mostly from geopolitical, economic and not the least from historical factors. Romania in this respect was an agricultural country at the time of the communist takeover, even though it was relatively rich in raw materials and energy. Since the country was well equipped with natural resources, it adjusted easily to the forced industrialization of the Stalin period. Thus Romania benefited fully from Soviet economic influence. Consequently, the early period of communist rule in Romania can be evaluated as a historical modernization project.[26] Enthusiasm for cooperation was motivated by the need for industrial machinery and technology to build up a functioning industrial base. As became clear in the 1960s, these supplies could be replaced with Western products. So the Soviet grip on Romania's economy was less extensive than in many other countries, thus allowing Bucharest considerable leverage. Romania was also a special case geopolitically since the Red Army had left Romanian soil in 1958. Therefore, the immediate reminder of the centre's constant presence and the symbol of the threat of intervening in domestic issues disappeared too. However, Romania was surrounded by the socialist community, a fact that evidently determined its elbow room.

The Khrushchevist declaration about building socialism differently gave birth to the formulation of a new political line in Romania. Incidentally, the core message of Khrushchev's thinking about the de-Stalinization campaign had not been heard in Bucharest. On the contrary, the Romanian leadership referred to the same Khrushchev speech about the right to choose the model that suited each country the best. This, on the other hand, was a very important factor of the

Romanian strategy regarding its position on the periphery. The party leadership continued not only its Stalinist methods of rule but also maintained the orthodox idea of building a communist state.

However, a considerable change started to become apparent after 1961. The Khrushchevist plan of division of production sectors ordered Romania to act as the main agricultural producer and supplier of raw materials for the more industrialized CMEA countries.[27] Evidently, this new role would have pushed Romania into a growing dependency on others and would have forced the leadership to abandon its programme of rapid industrialization. Hence, the Plenum of the Central Committee (in November–December 1961) made a decision to resist the unfavourable Soviet plans. From now on the Romanians would aim for economic self-sufficiency in order to become as independent as possible under bloc circumstances. Since the country was endowed with natural resources, the leadership was confident of standing up to Soviet economic pressure. Therefore the fast development of heavy industry was set as the absolute goal for the period 1960–65.[28] To achieve these tremendous objectives and to make itself more independent from the Comecon system, Romania was the first of the Eastern bloc countries that actively searched for Western economic contacts and it soon made lucrative trade agreements with France, England, Italy and Austria. To help the Romanians, Western equipment and technological aid was provided – a unique phenomenon in the interrelations of East and West in the early 1960s.[29] In consequence, in 1964, the outspoken preference for national interest instead of internationalist goals was brought to light in the 'Declaration of Independence' where Romania announced her own road towards communism.[30] To implement the Stalinist line in a period of de-Stalinization ordered by Khrushchev, the Romanians ironically exercised a daring anti-Soviet foreign policy to acquire more leverage.

With the rise of Nicolae Ceausescu (22 March 1965; stayed in office until 25 December 1989),[31] the national emphasis was made official at the IX Party Congress by stressing the concepts of 'sovereign nation' and 'independent state'.[32] A new and radical phase in Romania's foreign policy occurred in 1967[33] when it became gradually clear that the Ceausescu administration chose not only a more independent but also a more active role in world politics. Moreover, it was no longer disguised as being motivated by plain economic interest. This new attitude was rewarded when Corneliu Manescu, Minister of Foreign Affairs was elected as the President of the 22nd Session of the United Nations General Assembly (19 September 1967). Manescu and his deputy, Mircea Malița, later estimated that the international recognition was a result of the Romanian efforts to strengthen the role of small countries on the international stage and their constant fight to destroy the barriers between the political blocs.[34] From that time on, the political language started to change and foreign political argumentation gradually replaced ideological references not only at home but also in the international arena, outside of the bloc.

To evaluate the new Romanian line we have to recall the ideological background the dominant economic thinking relied on. Social evolution depended basically on economic performance, thus the economy was a vital factor and the key to modernization. The undoubted omnipotence of heavy industry was a very typical Leninist–Stalinist concept of progress, which was the guiding star for the Romanian communists. This orthodox model of modernization became a canonized salvation strategy against the Khrushchevist reforms and acted as a strong incentive to revise the country's foreign political line accordingly. In the Cold War context, Romania emerged as a fresh breeze, a true reformist of the international arena. Inside of the Eastern Bloc Romania demonstrated, on the other hand, that Moscow could be successfully resisted even in vitally important questions (like refusal to participate in the Warsaw Pact intervention in Czechoslovakia, 1968) without serious sanctions.

Reshaping theory: Czechoslovakia

A significantly different response to the Khrushchevist call was given in Czechoslovakia. A country with strong social-democratic traditions tried to apply reformism in its totality and build a completely new version of socialism. The Czech theoretical innovations grew out of the need to provide a functioning economic model without overlooking its consequences for society. 'Socialism with a human face' was an innovative way of creating a third model that challenged simultaneously the Western capitalist and the Eastern communist systems.

Czechoslovakia was a unique case in the whole region. In the inter-war period the country was industrialized and prosperous with a stable parliamentary system. After the Second World War, the Communist Party was elected to the Parliament as the biggest political force.[35] In contrast to the Polish and Hungarian sibling parties, the Czechoslovakian communists relied in the beginning on strong popular support, hence they were not dependent to the same extent on welfare policies to feel secure in power. In addition, Czechoslovakia had traditionally friendly relations with the Soviet Union, hence it was spared the stationing of Red Army troops. Nevertheless, the country relied heavily on Soviet supplies for its industry.

Khrushchev's secret speech did not launch an immediate reaction in Czechoslovakia as it did in Poland or in Hungary, because the party elite ardently resisted any questioning of its position. De-Stalinization emerged as a delayed process only after Khrushchev's renewed request at the 22nd Congress of the CPSU, partly a result of domestic economic tension and mounting popular dissatisfaction in the early 1960s. Unlike in the neighbouring countries, here the reformist party opposition started to take shape relatively late in the mid-1960s. Hence, the search for new values and debates influenced by Khrushchevist ideas took place in intellectual circles and focused first on philosophy.[36] The two most prominent philosophers, Karel Kosik and Ivan Svitak, offered a complete reformulation of the Marxist–Leninist foundation.[37] They wanted to set Marxist philosophy free of its dogmatic duties. The new Marxist philosophy was supposed to explore human reality without political constraints and party interference, with particular interest in the working people. The new thoughts of the reinvented Marxist humanism found their way to reformers of political theory who engaged themselves with reintroducing the model of democratic socialism. The chief architect of political reform, Zdenek Mylnar, and his circle demanded in the main theoretical monthly review of the Central Committee a quick transition to a pluralistic socialist system.[38] Pluralism was perceived on a corporatist model where the leading role of the party would be secured but also other key groups of society (like representatives of different sectors of labour) would have access to decision-making through the elevated role of the National Front movement. This model was believed to create a new consensual society.[39] In the economic sphere, theoreticians tried to find a modernization model that would not reintroduce market capitalism but would diverge from rigid central planning and management. The leading economist of the time, Ota Šik, came up with the idea of a socialist market that would work for the benefit of society.[40] According to this model, central planning would be limited to price policy, while enterprises would be granted more autonomy to decide about production. Consumer needs, quality of products, market sensitivity and efficiency were to be the key factors that would determine economic mechanisms. Management of factories was decentralized and delegated to the local actors. Heated discussions on the democratization of society and party structures culminated in April 1968 in the revolutionary Action Programme of the Czechoslovak Communist Party. Based on these thoughts, Alexander Dubcek finally declared the new national goal to establish socialism with a human face.[41] We could argue that the Khrushchevist proposal of building socialism differently was nowhere else taken as seriously as in Czechoslovakia.

Hungary and Poland: pragmatic reformism

Hungary and Poland showed important similarities in their relation to Khrushchev's declaration. Both countries accepted basically all his principles, or at least they carefully avoided collision with them. In contrast to the Czechoslovakian alternative, both administrations limited their interests to practical questions and downplayed the significance of the obvious need for theoretical renewal. Down-to-earth arguments, like referring to economic problems, were also handy to use as justification for daring reform policies that otherwise could have raised fierce opposition from Moscow, especially during the more conservative Brezhnev period. In these countries, economic rhetoric started to serve as a substitute for ideological argumentation in intra-bloc negotiations as well as in the domestic political discourses. The image of the top decision-making process became publicly mundane, 'economicized' and cut off from its ideological background.

Hungary

Hungary was in many respects in a different situation from that of Romania, Czechoslovakia or Poland. It was traditionally an agricultural country that started to develop a modern industrial base which was linked to agricultural production, much before the communist takeover.[42] The Stalinist emphasis on heavy industry in a country that completely lacked coal and iron was, however, an odd foreign idea that was hard to adopt.[43] Thus, Hungary depended entirely on Soviet supplies of raw material and energy for its speedy industrialization programme. This situation could not be changed through development, but was a basic factor that linked Hungary to the Soviet Union in constant subordination. In this way, Hungary's dependency on necessary supplies for industrial production was stronger than that of many other countries. In addition, conversely to Romania and Czechoslovakia, Soviet troops were a significant factor in domestic political life from 1945, but especially after 1956, until 1991.[44]

By the mid-1960s the possibilities for extensive economic growth were exhausted and the exaggerated emphasis on heavy industry led to a dead end. By 1963, it was clear that the centralization of the industrial sector decreased even further the results of production.[45] This situation forced the Hungarian leadership to search for new solutions to improve commercial capabilities and to find ways to stimulate economic growth. As a way out, the Hungarians wished to trade more with non-socialist countries, so they had to take the realities of the world market into consideration and concentrate on improving competitiveness.[46] The orientation towards Western markets also indicated intentions for decreasing trade with Eastern Europe. The administration made serious efforts to diversify trade, and held secret negotiations with the International Monetary Fund and the World Bank.[47] The leadership finally launched a new programme in 1968, the 'New Economic Mechanism', which was in fact an introduction of market mechanisms into a socialist economic model.[48] The primary objectives of the reform were the transformation of the economic structure, increasing efficiency in production, improving the quality of products and a balance in foreign trade.[49] The cornerstone and main reason for the reform was, however, as the Central Committee stated, to secure an uninterrupted growth in the standard of living[50] that would guarantee the support of the people and urge them at least to a minimal cooperation. János Kádár (in office 1956–88), who had cordial personal relations with Khrushchev, was a true follower of his principles, especially those that were based on welfare considerations. The constant increase in living standards was a central element of the consolidation policy of the post-1956 Kádárist regime trying to achieve a 'social agreement' by buying the silent acceptance of the people. This welfare-anchored line was coupled with considerable relaxation of control over society that generated a liberal atmosphere unique in the Eastern bloc.

Poland

The Polish case differed somewhat from its Hungarian counterpart, although it relied on the same premise: revitalizing economic development. Poland, like Romania, was also allocated an unfavourable sector of specialization in the Khrushchevist division of spheres of production, but unlike the Romanians the Polish leadership was unable to resist the Soviet plans, partly because of its different kind of dependency on Soviet supplies. Focusing on energy and material, consuming electro-mechanical engineering did not allow Poland to foster other industrial areas that could have furthered export orientation with more success.

Due to the low quality of the products, the Western markets stayed closed for Polish goods. In the early 1970s, however, attempts were launched to change the export profile through new specialization in high-tech products.

Party leader Edward Gierek's (in office 1970–80) reform started two years later than the Hungarian experiment, in 1970, and only remotely resembled it. This was another example of the differentiated approach to economic development encouraged by Khrushchevist ideas. The new policy was popularized by the slogan 'building a second Poland',[51] which clearly referred to a fresh beginning in contrast to the past. The key aim of the reform was to raise productivity by using energy and raw materials more efficiently by applying new technologies. During the 1970s, the importance of new technologies for industrial production was underlined and was to be acquired by imported licences. The leadership set out to extend the purchase of Western licences as a major element of its economic strategy.[52] Accordingly, from 1972 Gierek sought actively to establish new forms of contact with the West that, besides economic relations, included various areas of scientific and cultural fields, opening up also the possibility for people to travel. By 1975 trade with the West made up almost 50 per cent of Polish trade. Even the obstacles to developing contacts with West Germany[53] diminished due to Willy Brandt's efforts at Ostpolitik in 1970.[54] The reconciliation with the Bundesrepublik lessened the dependence on Soviet security support and opened up new perspectives in economic relations, strengthening the tendency to create distance from Soviet tutelage. This vivid approach towards the West induced a relaxed attitude towards society, and even the top rank of the party became more mundane by shutting out hard-core ideologists.[55] Gierek succeeded in establishing better relations with the Catholic Church, which even improved for a while after the election of Cardinal Karol Wojtyla to the papacy in 1978.

Unlike the Hungarian case, Polish modernization did not reach the level of managing the economy, which remained highly centralized. As Norman Davies put it, Gierek's reform was caught between Kadarization and Sovietization.[56] Out of fear of social unrest the leadership did not cure structural inconsistencies, like the imbalanced industrial sector that favoured heavy industry, or the increasing discrepancy of artificially high wages and low prices of subsidized commodity goods. The reformers were caught in the structural traps that the giant enterprises represented in the industrial sector with a powerful party lobby and aggressive workers' unions. Gierek came to power in 1970, after his predecessor, Wladyslaw Gomulka (in office 1956–70) mishandled a dock workers' riot and was forced to resign. Unsurprisingly, Gierek wanted to strengthen the relations of the party, especially with the industrial workers.[57] In other words, every attempt that could have jeopardized this liaison, like the restructuring of the industrial base or the much needed wage and price reform, met strong opposition and was doomed to failure. This halfway-modernization, based on huge Western loans, turned out to be an evolutionary modification of the existing system and hence could not rehabilitate the economy but caused unforeseeable troubles in society. Mounting economic problems in the late 1970s paired with a relaxed general atmosphere launched domestic instability and increasing unrest that eventually undermined

Edward Gierek's attempt at domestic détente.[58] Still it can be argued that, compared to his predecessor, Wladyslaw Gomulka, who although not a Stalinist was definitely a Soviet-minded leader[59] and to his successor General Jaruzelsky (in office as the first secretary of the party 1981–89), who came to power by a military coup and hence strongly relied on Soviet political support, Gierek's period represented a conscious liberalization and opening towards the West, away from the Soviet grip.

The Brezhnevist backlash against the Khrushchevist reforms

As a consequence of the reformist experiments rooted in Khrushchevist policy, political dissent started to grow within the bloc from the early 1960s onwards. The divergence of opinion was not about the communist faith in principle, since all of the member states were communist regimes and were eager to remain as such. The elementary difference originated from the centre–periphery structure or, more precisely, from the unsatisfactory conditions the periphery status offered. As Khrushchev's declaration of building socialism differently could be interpreted, the centre fell short in providing a flawless model to secure continuous development for the periphery – which is why the periphery was left alone to invent its own strategy. Yet, at the same time, the centre insisted on maintaining the centre–periphery structure and its leading position in it. But because it could not fulfil the centre's primary role of model-providing, it began to use secondary means (sheer military force, economic blackmail and political intrusion) to preserve its power status.

The first major crisis since the Hungarian revolution originated in the reformist movements that spread around the Eastern bloc. Twelve years after the new era had been declared, the Brezhnev leadership, supported by the most orthodox bloc members (the GDR and Bulgaria), decided to strike against the most challenging reformist country, Czechoslovakia. As a result, the principle of inter-state relations was modified again by the so-called 'Brezhnev doctrine' in 1968.[60] It still preserved the principle of sovereignty of the allies, as well as non-interference in domestic matters, but at the same time it declared that the common goals of the Eastern bloc dictated the order of priority. In other words, if domestic or foreign hostile forces tried to change the political system in any member state, it was considered an attack on the whole socialist camp. Therefore, it was an obligation of all to participate in the defence of socialism. Regarding Czechoslovakia, Brezhnev tried to provide a true picture of internationalism, by ensuring that the other member countries were now drawn into the intervention, whereas 1956 Hungarian actions were purely a Red Army operation.

Until the introduction of the Brezhnev doctrine, the limits of de-Stalinization and the range of deviation from the Soviet model were not unambiguous. The line between what could be changed and what could not be touched in any circumstances was unclear. Brezhnev's 'restricted sovereignty' theory marked a radical change from the comparatively liberal alliance policy of Khrushchev. The events in Czechoslovakia showed the vulnerability of the system. If the people's democracies followed their own path in the creation of socialism too independently, it would lead sooner or later to ideological disputes. On the other hand, if socialism could be built in different ways it would result in a situation where nobody could be criticized or condemned because of their choices, nor could any country be forced back on to the 'right track'. Hence the leading position of an authority inside the alliance could be questioned, as well as the reassessment of the relation between the Soviet Union and its allies. To put it simply, the centre–periphery construction was in danger.

The year 1968 was one of constant crisis management for the centre. The Czechoslovakian human-faced socialism differed strikingly from the Soviet example, not to mention the Hungarian experiment with its small-scale market economy or the Romanian challenge on foreign policy. All questioned the leading role of the centre by rejecting the model it provided to be followed. Even though the centre had an undoubted military and economic supremacy over the periphery, nevertheless, as it appeared in 1968, this was not enough to secure a solid political influence on the subordinated minor partners.[61] Although order could seemingly be restored quickly (that is, Czechoslovakia returned to a 'satisfactory' domestic line shortly after the military intervention and Hungary forced to abandon reform policy by 1972;[62] Poland settled down in 1981, while Romania never really gave up her independence-seeking policy), the results were dubious. It seemed that 'straightening the lines' of the mavericks required more and more energy from the centre and produced only a fragile illusion of loyalty, for the time being. The periphery obeyed when there was no other possibility left, but was obviously waiting for the next opportunity. It can be stated that the Khrushchevist line in support of reform in the end eventually undermined the very structure of the bloc's centre–periphery setting in spite of the Brezhnevist efforts to prevent the process of erosion.

Reforms in perspective: concluding remarks

The new context of Khrushchev's dawn generated a major metamorphosis in the bloc countries' behaviour. The re-evaluation of national interest and of attitudes towards common goals induced a wave of reformism that affected the way of thinking as well. Which version of these diverse political choices achieved the most, compared to their purposes, is not only an interesting issue but also a vital matter for understanding the degree of leverage these countries enjoyed. By answering this question we can also measure what form of divergence was acceptable within the framework of the Eastern bloc and thus had the potential to inflict change most efficiently.

For the sake of analysis we differentiate ideology and mode of rule in relation to political practice as the key structural factors of the Soviet-type system.[63] It would be a failure to argue that Marxism–Leninism was the ultimate impulse of political practice or to categorically deny the significance of ideology for politics in state socialism. Ideology also functioned as an instrument of system legitimization since the system was legitimate for those who could identify themselves with

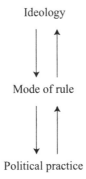

Ideology

Mode of rule

Political practice

Figure 10.1 Ideology and the mode of rule in Soviet-type systems

its ideological basis.[64] The mode of rule was in a sense a political theory that interpreted ideology for daily political decision-making and offered the strategy of how to achieve the ultimate goals that ideology prescribed. Thus, the mode of rule limited the space of political action, since it clarified the line between the possible and the forbidden. There was a two-way interaction between these elements (see Figure 10.1; this downward and upward relation is emphasized with the signs ↓↑). Ideology provided a supply of principles for the ruling mechanism, but also defined the limits of governance. The mode of rule in its turn served as the basic line-drawer for everyday political decision-making (this is marked 'from top to bottom' aspect with ↓). However, the mode of rule can be also perceived as a derivative of political practice because changing realities demanded adjustment of the mode of rule.[65] However, the new mode of governance had to be consistent with the inner logic of the ideology and rely on the storehouse of established principles (this demand from 'below' is displayed with ↑).

Khrushchev's inventions, the de-Stalinization campaign and the idea of building socialism differently, represented alterations in the mode of rule. It changed the direct command system to indirect governance. Since it transferred new responsibilities (of economic development) to the member states it had to extend space for their political practice that was favourable for national experiments.

Now, if we reflect on the types of reform these countries applied, we can see an interesting pattern that explains the prospects for the chosen agenda (see Figure 10.2). It is rather obvious that reforms differed in (a) what idea characterized the key concept of the aimed transformations, (b) how these purposes were manoeuvred on the international arena, and (c) what structures of the communist establishment they challenged. So, according to the analytical framework, the Czechoslovak, Hungarian and Polish models were based on liberal ideas with Western characteristics that were disguised with an accentuated loyalty to the Moscow-dominated community on the international stage. Romania, on the other hand, conducted a neo-Stalinist agenda but turned publicly against the centre, protesting openly against the restrictions the bloc structure inflicted upon her. The reforms sought after dissimilar changes and subsequently they challenged the structure differently. This aspect, reveals for example, the logic and strength of sanctions some of these reforms incurred from the centre.

From the centre's point of view, the Czech reform was the most frightening because it reinterpreted the role of ideology by offering a new version of Marxism without Leninism, and consequently the whole Soviet system came under attack. Actually, the other aspects of the Czech model were insignificant in this respect because according to the inner logic of the system, all the others derived from the perception of ideology. In contrast, the Hungarian and Polish lines modified only the political practice because these changes focused primarily on reforming the economy and did not reach out to the upper levels. In addition, this goal was strongly supported by Moscow because the failing economies of the socialist countries were becoming burdensome to maintain. Interestingly, though, when the Hungarian reform entered a new phase where liberalism started to alter society and affected the governing methods, Brezhnev intervened immediately.

The Romanian case was significantly different. The stubborn foreign line, which questioned directly the mode of rule of the centre in relation to its allies, was tolerated. To solve this puzzle we have to remember three aspects of the Romanian policy. First, it did not threaten the existence of common institutions (as did, for instance, the Hungarian revolution in 1956 when the country decided to leave the Warsaw Pact) that were the operational means of the mode of rule in the Eastern bloc. Second, Romania was situated geographically in the middle of the bloc so there was no security risk threatening the borders of the Soviet sphere. The third and most important factor was that Romanian intentions were intrinsically inconsistent: they challenged the Soviet mode of rule only in the international arena while, on the domestic stage, Romania applied Soviet ruling methods literally. Hence, the ideological sphere and the diversification of central power were not jeopardized. Thus, as far as the Kremlin's interests were concerned, the communist system was stable.

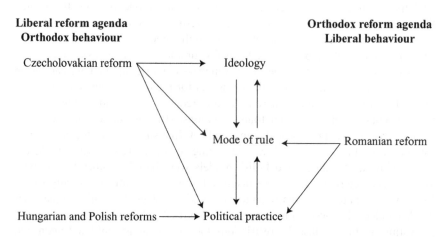

Figure 10.2 Reform programmes in Communist states

In the research literature, reforms have generally been viewed as slight modifications of the existing system that compromise with the basic structures and represent only a controlled amendment from above. Reform movements have been compared to revolutions that are considered as the ultimate form of political action that can bring about radical and comprehensive changes in a relatively short period of time.[66] In this regard, reforms are perceived as failures or the wrong means to inflict considerable change.[67] If we look at the reforms discussed above in this context there arise serious doubts as to whether the sharp division between revolution and reform is useful to describe what happened in the Eastern bloc. We argue that, in some of our cases, the transformation became as extensive, deep and wide structurally as any revolutionary movement, where the only decisive difference is actually the speed of changes.

We have already established that there are consciously defined goals and unintended consequences in every political decision. One of the most significant unanticipated alterations took place in the mindset of the ruling elite. The idea of development and economic thinking became overwhelming in every segment of these societies, including in the ideological sphere, and gave birth to a strong economic-based nationalism and technocratism. Economic progress also provided through the steady rise of living standards a new possibility for seeking legitimacy. As it was conceived, welfare-based legitimacy could have decreased the communist leadership's political dependence on the Kremlin's support for their power. Replacing ideology with the new religion of economy destroyed the last link of cohesion of the Eastern bloc and provided a false belief of legitimacy for the national political elites, especially in Hungary, Poland and Romania. Most of all in the case of Hungary the generational change of party personnel confirmed an accentuated liberalization process of the whole social structure by the mid-1970s. The pursuit of legitimacy led the Hungarian and Polish elites to try to achieve a rapprochement with society in a form of consensus policy. Similar endeavours can also be seen in Romania in the guise of nationalism as a substitute for liberalization.

Still, besides the metamorphosis of the party leaderships there was another even more important and unforeseeable effect. Liberalization attempts relaxed the strict control over citizens, and the people started to pursue more freely their own goals in the new space the regimes made available. All of a sudden their actions became more difficult to anticipate and control. The best example of this is Poland, where the Solidarity movement was a direct result of the more liberal atmosphere of the Gierek administration that was favourable to the development of civic action.

According to Kieran Williams, the liberalizing political elite's aim was 'of preserving and improving, not destroying, existing institutions, liberalisation should not be studied in the same terms as revolution'.[68] Well, if we observe the important changes in the power structure we must disagree with Williams' statement. The reforms, with the exception of Romania, were all deregulation projects that weakened the central power to some extent. The Czech case is naturally the ideal example of this process, but the same held true for Poland and

Hungary, where the revision of central planning and management resulted in the polarization and depoliticization of the economic sphere. Moreover, the relaxation of control over society was a sign of the gradual withdrawal of state power from the social sphere. Reforms reintroduced the idea of pluralist interests inside of the parties and in society at large. Consequently, these endeavours weakened the authoritarian state considerably instead of strengthening it, as Williams' definition suggests.

Khrushchev's significance for the gradual opening and diversification of the Eastern bloc is unquestionable even though the processes he launched reached their full maturity only later. We argue that he actually prepared the region for Gorbachev's *perestroika* two decades later – an era that probably would never have taken place without Khrushchev. Although the Brezhnev period represented a huge setback, the processes that started under Khrushchev could not be halted.

Notes

1 H. Gordon Skilling described this phenomenon as 'a group of man-made satellites which travel in pre-ordained orbits from the Moscow launching pad', in H. Gordon Skilling, *Communism National and International. Eastern Europe after Stalin*, Toronto: University Press, 1964, p. 131.

2 K. Waltz, *Theory of International Politics*, Reading, MA: Addison-Wesley, 1979; Fred Halliday, *Rethinking International Relations*, London: Macmillan, 1994, pp. 31–46, 94–102; S. Hanson, 'Sovietology, post-sovietology, and the study of postcommunist democratization', *Demokratizatsiya*, Winter 2003.

3 Q. Skinner, 'Language and political change', in T. Ball, J. Farr and R.L. Hanson (eds), *Political Innovation and Conceptual Change*, Cambridge: Cambridge University Press, 1989, pp. 6–23.

4 According to Margot Light, internationalism at that time followed explicitly the original thought of Karl Marx. M. Light, *The Soviet Theory of International Relations*, Brighton: Wheatsheaf Books, 1988, pp. 159–65.

5 B. Király, 'A magyar hadsereg szovjet ellenőrzés alatt' [The Hungarian Army under Soviet Control], in I. Romsics (ed.), *Magyarország és a nagyhatalmak a 20. században* [Hungary and the Great Powers in the Twentieth Century], Budapest: Teleki László Alapítvány, 1995, p. 236.

6 L. Holmes, *Politics in the Communist World*, Oxford: Clarendon Press, 1986, pp. 356–7, 363–4; M. McCauley (ed.), *Khrushchev and Khrushchevism*, London: Macmillan Press, 1987, pp. 156–7, 172–3.

7 The 'honeymoon' period between the Soviet Union and Yugoslavia did not last long; however, a kind of rapprochement could be detected from the Belgrade Declaration (1955) until the International Congress of the Communist and Workers Parties in Moscow (1960). The Yugoslavian representatives were obviously not invited to this event.

8 S. Clissold (ed.), *Yugoslavia and the Soviet Union 1939–1973*, London: Oxford University Press, 1975, pp. 254–7; J.R. Lampe, *Yugoslavia as History. Twice there was a Country*, Cambridge: Cambridge University Press, 2000, pp. 267–71.

9 Although the speech was delivered in the absence of foreign delegations, copies had already been sent to other communist leaderships in March 1956, to inform them carefully about the changes.

10 Z.A. Medvedev and R.A. Medvedev (eds), *The 'Secret' Speech Delivered to the Closed Session of the Twentieth Congress of the Communist Party of the Soviet Union by Nikita Khrushchev*, Nottingham: Spokesman Books, 1976, pp. 61–62.

11 'The Declaration of the Soviet Government at 30 October 1956', E.H. Gál et al. (eds), *A 'Jelcin-dosszié'. Szovjet dokumentumok 1956-ról* [The Jelcin Dossier. Soviet Documents from 1956], Budapest: Századvég – 1956-os Intézet, 1993, pp. 65–7.

12 The statement by the Government of the Soviet Union on 30 October 1956; *Pravda*, 31 October 1956.

13 The Soviet Union used armed force to restore order at the 'request' of the Hungarian government. See the official statement by the Government of the Soviet Union in G.T. Korányi (ed.), *Egy népfelkelés dokumentumaiból, 1956*. [Documents of an Uprising, 1956], Budapest: Tudósítások Kiadó, 1989, p. 86.

14 *Novaya Vremya*, November 1956.

15 The Soviet Union was ready to negotiate with the coalition government, which meant the temporary acceptance of having a multiparty system. See the official statement by the government of the Soviet Union. Korányi (ed.), *Egy ...*, p. 86; C. Horváth, *Magyarország 1944-töl napjainkig* [Hungary from 1944 up to the Present], Pécs: Prezident, 1993, p. 106.

16 V. Zubok and C. Pleshakov, *Inside the Kremlin's Cold War. From Stalin to Khrushchev*, Cambridge, MA and London, England: Harvard University Press, 1996, p. 184.

17 See speech by Imre Nagyin in: Korányi (ed.), *Egy ...*, p. 90.

18 *A kommunizmus építőinek kongresszusa* [The Congress of the Builders of Communism], Budapest: Kossuth, 1961, pp. 13–19; Robert Bideleux and Ian Jeffries, *A History of Eastern Europe: Crisis and Change*, London: Routledge, 2007, pp. 477–87.

19 P. Márer, 'A szovjet bloc mint integrációs model: gazdasági, politikai és katonai aspektusok', in D.A. Bán, et al. (eds), *Integrációs törekvések Közép- és Kelet-Európában a 19–20. században* [Integrative Tendencies in Central and Eastern Europe in the 19th and 20th Centuries], Budapest: Teleki László Alapítvány, 1997, pp. 239–44.

20 K. Kulcsár, *A modernizáció és a magyar társadalom* [Modernization and the Hungarian Society], Budapest: Magvetö, 1986, p. 37.

21 K. Nyíri, *Európa szélén* [On the Edge of Europe], Budapest: Kossuth, 1986.

22 K. Kulcsár, 'East Central Europe and the European Integration', M. Szabó (ed.), *The Challenge of Europeanization in the Region: East Central Europe. European Studies 2*, Budapest: Hungarian Political Science Association and Institute for Political Sciences of the Hungarian Academy of Sciences, 1996, pp. 11–13.

23 *Filozófiai Kislexikon*, Budapest: Kossuth, 1972, p. 150.

24 A. Gramsci, *Filozófiai írások* [Philosophical Writings], Budapest: Kossuth, 1970, pp. 349–80.

25 I. Harsányi, 'Modernizáció és modernitás. Az 1875 utáni spanyol Restauráció és korszerûsítés' [Modernization and Modernity. The Spanish Restoration and Reform after 1875], in J.L. Nagy (ed.), *A modernizáció határai. Tradíció és integráció Kelet-Európa (hazánk) és a Mediterránium történetében a 19–20. században* [The Limits of Modernization. Tradition and Integration in the History of Eastern Europe (Hungary) and the Mediterranean in the 19th and 20th Centuries], Szeged: Szegedi Tudományegyetem Kiadó, 1972, pp. 21–34.

26 G. Hunya, T. Réti, A.R. Süle and L. Tóth, *Románia 1944–1990. Gazdaság- és politikatörténet* [Romania from 1944 to 1990. Political and Economic History], Budapest: Atlantisz Kiadó, 1990, pp. 30–7.

27 Holmes, *Politics ...*, p. 357; I.T. Berend, *Central and Eastern Europe 1944–1993. Detour from Periphery to Periphery*, Cambridge: University Press, 1996, pp. 95, 104; Vladimir Tismaneanu, *Stalinism pentru eternitate. O istorie politica a comunismului romanesc*, Iasi: Politrom, 2005.

28 The plan was officially accepted at the III. Congress of the Romanian Worker Party. *A Román Munkáspárt III. kongresszusa* [The Congress of the Romanian Worker Party], Bukarest: Politikai Könyvkiadó, 1960. See also D. Deletant, *Communist Terror in*

Romania. Gheorghiu-Dej and the Police State 1948–1965, New York: St Martin's Press, 1999, pp. 282–4.

29 This Western orientation of Romania clearly paid off: during 1959–62, Western imports tripled (of which the import of industrial machinery increase five-fold) and this was vital for the industrial programme. The import of Western machinery was so significant that it made up two-thirds of the total machinery imports. This figure also shows the importance of Western contacts to Romania's development strategy. The Declaration of Independence made a big difference, especially in relation to the United States. The month after the announcement, Washington started bilateral trade talks with Bucharest; the first time it had done so with an Eastern bloc country. J. Harrington, 'American–Romanian Relations, 1953–1998', in A.R. DeLuca and P.D. Quinlan (eds), *Romania, Culture and Nationalism. A Tribute to Radu Florescu*, New York: Columbia University Press, 1998, pp. 111–12.

30 The Declaration was published in, for example, *Scinteia*, 18 April 1964 and on the front page of *Elôre*, 19 April 1964.

31 Deletant notes that, actually, Ceausescu's image of being brave enough to oppose the Soviets was a significant factor when it came to the choice for the successor of the previous party chief, Gheorghe Gheorghiu-Dej. D. Deletant, *Romania under Communist Rule*, Iasi, Oxford and Portland: Center for Romanian Studies, 1999, pp. 105–6.

32 *Scinteia*, 20 July 1965.

33 In January, Bucharest established diplomatic relations with the Federal Republic of Germany. How revolutionary this was can be understood if it is considered that the Soviet Union only followed Romania's example after three years, in 1970. A similarly bold step was taken later in June 1967. Romania did not join the other European socialist countries in breaking off diplomatic relations with Israel as a protest over the occupation of new territories in the 'Six Days' War'.

34 See Manescu's speech in *Scinteia*, 20 September 1967, and his other statements in *Scinteia*, 30 December 1967; and of Deputy Malita in *Lume*, 2, 1968.

35 The communists received 38 per cent of all votes. *Dejiny Ceskoslovenska v datech* [History of Czechoslovakia in Data], Praha: Svoboda, 1968, p. 468.

36 Debates were published in *Literárni noviny* in 1956–58. See, for example, 21 April 1956; 17 November 1956; 1 December 1956; 9 March 1957; 16 March 1957, 4 January 1958.

37 J. Satterwhile, 'Marxist Critique and Czechoslovak Reform', in Raymond Taras (ed.), *The Road to Disillusion. From Critical Marxism to Postcommunism in Eastern Europe*, New York: M.E. Sharpe Inc., 1992, pp. 115–34.

38 *Nová mysl*, 5/1968.

39 K. Williams, *The Prague Spring and its Aftermath. Czechoslovak Politics 1968–1970*, Cambridge University Press, 1997, pp. 14–20.

40 His model resembled the Hungarian plans that were developed a few years earlier. The 'New Economic Mechanism' started officially in January 1968, but the preparations began as early as December 1964 and the final version of the programme was accepted in May 1966 in a Central Committee meeting. *Magyar Szocialista Munkáspárt (MSZMP) határozatai és dokumentumai 1963–1966* [The Decisions and Documents of the Hungarian Socialist Worker Party (HSWP) from 1963 to 1966], Budapest: Kossuth, 1975, pp. 98, 265–6, 348–64. The reform programme was introduced for the first time in public in the daily, *Népszabadság*, 25 April 1965.

41 A. Dubcek, *Viimeisenä kuolee toivo. Poliittiset muistelmat* [Hope Dies Last. Political Memoirs], Helsinki: WSOY, 1993.

42 Especially fast-developing sectors were the milling industry, the canning industry and tractor production.

43 Miklós Szabó suggested that a socialist version of the Danish economy, that is the dominance of a modern, technically high-level agriculture, could have been accepted

easily and would have helped the adjustment of communism. M. Szabó, *Politikai kultúra Magyarországon 1896–1986* [Political Culture in Hungary from 1896 to 1986], Budapest: Atlantisz Program – Medvetánc, 1989, pp. 285–6.

44 Király, 'A Magyar …', pp. 232–3.
45 Centralization was huge. For example, 13 industrial branches (paper, glass, sugar, rubber, beer, etc.) had only one factory each. Another telling indicator was that, in 1965, half of all the employees worked in plants that employed over 1,000 people (corresponding numbers in the Soviet Union were 34 per cent of the workers). S. Szakács, 'A kádár-korszak gazdaságtörténetének fôbb jellemzôi' [The Main Characteristics of Economic History of the Kadar-Era], in *Society and Economy*, XVI, 1994/5, pp. 188–94.
46 *Társadalmi Szemle*, October 1965, article by vice prime minister Jenö Fock.
47 Hungarian National Archive OL-M-KS-288-15/115.öe., pp. 46–7. Report to the Committee of Economic Policy of the Politburo.
48 J. Kornai, *A szocialista rendszer. Kritikai politikai gazdaságtan* [The Socialist System. Critical Political Economy], Budapest: Heti Világgazdaság, 1993, pp. 496–534.
49 G. Földes, *Az eladósodás politikatörténete. 1957–1986* [The Political History of Indebtedness from 1957 to 1986], Budapest: Maecenas, 1995, p. 43.
50 Hungarian National Archive OL-M-KS-288-4/88.öe., pp. 38–40. Minutes of the meeting of HSWP CC 14.6.1967. The report of R. Nyers.
51 G. Kolankiewicz and G.P. Lewis, *Poland. Politics, Economics and Society*, London and New York: Pinter, 1988, p. 103.
52 Davies mentions licences purchased from the Fiat and Berliet companies for Polish car and bus production, from Massey Ferguson for the tractor factories, from Grundig for the electronics industry, from Leyland engine technology and from Jones to manufacture cranes. N. Davies, *God's Playground. A History of Poland, Vol. II From 1795 to the Present*, Oxford: Oxford University Press, 2005, p. 627.
53 T. Garton Ash, *In Europe's Name: Germany and the Divided Continent*, London: Vintage, 1994, pp. 219–20.
54 H. Sjursen, *The United States, Western Europe and the Polish Crisis, International Relations in the Second Cold War*, New York: Palgrave-Macmillan, 2003, pp. 110–16.
55 Davies, *God's Playground*, pp. 616–17.
56 Davies, *God's Playground*, p. 625.
57 Kolankiewicz and Lewis, *Poland*, pp. 66–8.
58 J.M. Ouimet, *The Rise and Fall of the Brezhnev Doctrine in Soviet Foreign Policy*, Chapel Hill and London: University of North Carolina Press, 2003, pp. 105–7, 109–11.
59 P. Machewicz, 'Social Protest and Political Crisis in 1956'; A. Kemp Welch (ed.), *Stalinism in Poland, 1944–1956*, London: Macmillan, 1999, pp. 99–118.
60 Published for the first time in *Pravda*, 13 November 1968.
61 E. Moreton, 'Foreign Policy Perspectives in Eastern Europe', in K. Dawisha and P. Hanson (eds), *Soviet–East European Dilemmas: Coercion, Competition and Consent*, London: Heinemann, Royal Institute of International Affairs, 1981, pp. 172–94.
62 The decision to stop the 'New Economic Mechanism' and to return to the old command system of economy with the re-centralization of the industrial plants was made at the session of the Central Committee (14–15 November 1972). The main reformers who were members of the Central Committee also had to go. György Aczél, Jenô Fock and Rezsô Nyers, the chief designers of the reform, were dismissed from their posts. This 're-groupment' of Kádár's trustees in March 1972 meant that they were transferred far from the top decision-making to lesser posts. Hungarian National Archive OL-M-KS-288-4/119; also *A MSZMP határozatai és dokumentumai 1971–1975* [The Decisions and Documents of the Hungarian Socialist Worker Party (HSWP) from 1971 to 1975], Budapest: Kossuth, 1978, p. 377.

63 This analytical model is based on K. Miklóssy, *Manoeuvres of National Interest*, Helsinki: Kikimora, 2003.

64 W. Guo and Z. Yunling, *China, US, Japan and Russia in a Changing World*, Beijing: Social Sciences Documentation Publishing, 2000, p. 8.

65 K. Palonen, 'Retorinen käänne poliittisen ajattelun tutkimuksessa. Quentin Skinner, retoriikka ja käsitehistoria' [The Rhetorical Change in the Study of Political Thought], in K. Palonen and H. Summa (eds), *Pelkkä retoriikka. Tutkimuksen ja politiikan retoriikat* [Bare Rhetorics. Rhetorics of Research and Politics], Tampere: Vastapaino, 1996, pp. 137–60.

66 C. Tilly, *European Revolutions 1492–1992*, Oxford: Blackwell, 1993; A.J. Motyl, *Revolutions, Nations, Empires. Conceptual Limits and Theoretical Possibilities*, New York: Columbia University Press, 1999.

67 G. Ekiert, *The State against Society. Political Crises and their Aftermath in East Central Europe*, Princeton, New Jersey: Princeton University Press, 1996.

68 Williams, *The Prague Spring*, p. 3.

11 The economy of illusions

The phenomenon of data-inflation in the Khrushchev era[1]

Oleg Khlevniuk

One of the best-known episodes of Soviet history in the Khrushchev period is the so-called 'Ryazan Case'. Towards the end of the 1950s the agricultural successes of the Ryazan region were presented by official propaganda as an example to be imitated. It was declared that in 1959 alone this region tripled its deliveries of meat to the state. A.N. Larionov, the first secretary of the Ryazan regional committee, was awarded the title of Hero of Socialist Labour, and awards also rained down on other Ryazan workers. Although indications that the Ryazan successes were based on deceit did reach Moscow, no attention was paid to them. But at the end of 1960, agricultural failures in Ryazan region and in the country as a whole compelled Khrushchev to repudiate the 'meat leap-forward'. In many regions, including Ryazan, checks began to be made that uncovered massive cheating.[2] Larionov could not cope with his unmasking and killed himself. In October 1964 the Ryazan scandal would be one of the accusations laid against Khrushchev as he was removed from power.[3]

The 'Ryazan Case' was an important reason for the wide anti-cheating struggle carried out in the late 1950s and early 1960s. To understand the place of this campaign in the history of the Khrushchev period it is important to emphasize its two distinguishing features. The first was that the campaign had a negative character. That is, it was carried out because there was little choice, even though to a large degree it discredited both the regime and Khrushchev himself. This implied that the cheating was on a really substantial scale and entailed social and economic damage that was more significant than the political damage. The second feature of the campaign was its close connection with two key stages of the Khrushchev regime. On the one hand, the campaign was a specific result of Khrushchev's initiatives in the second half of the 1950s. On the other hand, it gave a push to new reorganizations (especially to the restructuring of the *apparat* on the branch principle) that were an impulsive attempt to overcome internal defects of the system and that, in the final analysis, became fateful for Khrushchev. In this way the anti-cheating campaign reflected typical features and basic tendencies of the development of the Soviet model of socialism linked with Khrushchev's name. Simultaneously, and independent of the temporal and political context, the massive cheating of the 1950s and 1960s can be seen as an organic part of the Soviet economic system as a whole.

From these premises, this chapter will look at three questions. The first relates to the causes of cheating as an element of the Soviet command-planning system, including its Khrushchev variant. The second deals with the campaign against cheating as a reflection of the mechanisms of political leadership characteristic of the Khrushchev period. The third looks at cheating as an adaptation mechanism put into practice by the lower links in the Soviet economic system.

Plans, forward leaps and cheating

Various book-keeping falsifications and the concealment of the true state of affairs can be encountered in any social-economic system. However, in the Soviet Union there were additional incentives for falsification: the unlimited power of the state in the economic sphere, economic adventurism and the cruel (at some periods extremely criminalized) methods of managing the economy. Putting forward unattainable economic goals, the state gave the producer a choice: either suffer a harsh punishment for failure to fulfil the plan, or look for fraudulent ways out. The best-known and most extreme form of this model of adaptation to the system (or rather, survival inside it) is the so-called *tufta* (trash – production of false results) in the Gulag economy.

The intensity of cheating was directly connected with the phases of continually repeating cycles in the Soviet economy, where the leap forward (*skachok*) is standard.[4] Having taken form in the pre-war five-year plans, the mechanism of economic leaps was also used in the Khrushchev period. This mechanism of leaps, or shock campaigns and 'storming', had several elements. The first of these was the putting forward, from above, of ambitious targets, which as a rule were politically motivated economic adventurism. The second was the bringing forward of leading workers and initiators, who worked to those targets and received appropriate rewards. The third was the mobilization of the *apparat* on the principle: do as these advanced workers do. The shock campaigns, in this way, were a method of bringing forth reserves and overcoming the reality of sabotage by the producers, who in the command-planning system preferred to conceal the true possibilities of production.

The most notable of Khrushchev's shock campaigns occurred in agriculture, which reflected the particular anxiety of the Soviet authorities about the poor development of this branch of the economy. After a relative improvement of the grain balance as a result of opening up the virgin lands, the most urgent problem was the livestock crisis. On Khrushchev's initiative an attempt was made to solve this problem with another leap-forward campaign. A competition was announced in the country under the slogan 'Catch up and overtake the USA in per-capita production of milk and meat'. In his report at the plenum of the CC (Central Committee) of the CPSU (Communist Party of the Soviet Union) in December 1958, Khrushchev introduced the key figures for this competition. The USSR's annual meat production per head according to official figures was 38 kg and in the USA it was 94 kg. To catch up with the USA, the USSR would have to produce 20–21 million tonnes of meat. But the seven-year plan as adopted had laid down that meat production in 1958–65 would grow from 8 million to 16 million tonnes per annum. In this connection Khrushchev made it quite clear that this plan did not suit him and he required 20–21 million tonnes as the envisaged output. Khrushchev at the same time threatened that those leaders who did not secure 'a boost of the economy' would lose their jobs.[5]

Most likely both Khrushchev and his colleagues were aware that the impending campaign contained a certain element of economic adventurism; it was not by accident that the figure of 16 million tonnes had been included in the official plan. But the wager was that local officials, facing a difficult task, would be compelled to seek a way to increase output and move to more efficient technologies in cattle raising and in feed production. Evidently definite hopes were also placed on an increase in the number of cattle in state and collective farms on account of the individual plots. Repeated summons to reduce individual livestock raising by transferring animals to the collective farms had in fact been sabotaged both by the peasants and by the *kolkhoz* (collective farm) administrations. The meat campaign could be a stimulus for a compulsory reduction of individual-plot husbandry.

It was decided to entrust the leaders of Ryazan region with the initiation of the 'meat leap-forward'. This decision had a certain justification in that the first secretary of the Ryazan regional party committee, A.N. Larionov, was a senior party functionary, having occupied a high position in the CC *apparat* for several years under Stalin. In 1948 he had been transferred from Moscow to the lagging Ryazan region. The energetic Larionov not only succeeded in achieving definite successes in the new appointment but also maintained his connections in the CC *apparat* both under Stalin and Khrushchev. He enthusiastically supported all of Khrushchev's initiatives. At the end of 1958 a personal envoy of Khrushchev was sent to organize the new campaign in Ryazan. This was V.P. Mylarshchikov, who headed the agricultural department for the RSFSR of the CPSU CC. In the name of Khrushchev he suggested to the Ryazan leaders that they should accept a commitment to fulfil in 1959 three annual plans for meat production (150,000 tonnes) and come forward with a challenge to other regions to follow their example. What Mylarshchikov may have firmly promised Larionov is not known. But most likely it was a matter of all-round help from the centre, in accordance with the political importance of this campaign.[6]

Larionov enthusiastically accepted for fulfilment the directive from Moscow. Every effort was thrown into meat procurement. A special 'meat HQ' was set up in the party regional committee. But further events demonstrated that a miracle did not take place. As a subsequent inspection revealed, the *kolkhozy* and *sovkhozy* (state farms) of the region delivered, with difficulty, 60,000 tonnes. Another 40,000 tonnes of meat were purchased from people of the region and beyond. And even in that 100,000-tonne figure there was a certain amount of cheating that was simply difficult to bring to light. The final 50,000 tonnes (third plan) were entirely a matter of cheating. Moscow helped Larionov carry off this massive misrepresentation. Wishing to avert political opprobrium, the leadership of the bureau of the CC of the CPSU for the RSFSR and of the RSFSR council of ministers (doubtless by agreement with Khrushchev) agreed to so-called 'juggling'. With this, cattle were formally delivered to the state, but remained 'in the air' at sovkhozes and kolkhozes, allegedly in order to be fattened. This enabled the 'record breakers' to double-count one and the same animal as delivered to the state at the same time as still belonging to a kolkhoz. Other ways of cheating apart from 'juggling' were also being used.[7]

In essence, the shock campaigns offered two possible action models for producers. The first was the use of optimal reserves, and making up for insufficient plan-fulfilment resources with the help of cheating. In this case the producers conserved their basic assets and the possibility of developing future production, but they made themselves liable to the risk of being unmasked as cheats. The second model assumed the use of resources in excess of the optimal reserves at the expense of a greater exploitation of basic assets. In this case, in the short-term perspective, it was possible to avoid, or minimize, cheating, but at the same time undermine the prospects for developing production in the future.[8] It can be assumed that in most cases the producers, drawn into the shock campaigns, would strive to find the optimal balance between these two models. This entailed a balance of risks with the aim of minimizing both the loss of assets and the extent of cheating. However, if the campaigns acquired the same scale as in Ryazan, achieving such a balance was impossible. The trebling of output, when in reality it was scarcely possible to fulfil a third of the plan, led both to a high level of cheating and to an intensive destruction of basic assets.

Notwithstanding the energetic purchase of animals from the local population and from other regions, the cattle stock in Ryazan's kolkhozes and sovkhozes, according to official figures, diminished from 302,000 to 255,200 (by 15.5 per cent) from 1 January 1959 to 1 January 1960, and pigs fell from 248,800 to 114,500 (by 54 per cent). But neither did these figures correspond to reality, because a proportion of the cattle was misrepresented.[9] The mass purchase of cattle from the population (which mostly had an obligatory character) led to a reduction of livestock numbers on the private plots of the Ryazan region. The number of cattle belonging to the population diminished by 15.5 per cent from 1 January 1959 to 1 January 1960, and pigs by 10 per cent.[10] Already in 1960 the bled-white Ryazan economy was having difficulty delivering meat to the state. Although the accepted obligation to deliver meat amounted to 200,000 tonnes, by 1 November 1960 only 32,800 tonnes had been delivered.[11] Calculations showed that, even if the region in 1960 delivered all the cattle and poultry belonging to its kolkhozes, sovkhozes and individuals, the obligatory 200,000-tonne delivery would have been fulfilled by only 90 per cent.[12]

Since a large part of the meat was purchased by Ryazan in 1959 at market prices and delivered to the state at the low procurement prices, enormous expenditure was involved. The money was drawn out of the kolkhozes or received as credits from the state. As a result, the kolkhozes and sovkhozes of Ryazan went into bankruptcy. On 1 October 1960, whereas kolkhoz bank accounts amounted to 2.6 million roubles, kolkhoz indebtedness came to 350.2 million. According to figures produced in November 1960, the monetary income of kolkhozes in 1960 was expected to amount to 491 million roubles, whereas expenses, including the redemption of debt, were 1,040 million roubles. Since the kolkhozes could not pay for the work of their peasants, the latter in many farms stopped going to work.[13]

Although the situation in Ryazan was extraordinary, a serious crisis also affected a lot of other regions that had energetically involved themselves in the campaign. For example, in Kirov region, a raised pledge that entailed a 1960 delivery of 180,000 tonnes of meat was likewise accepted. As in Ryazan, these obligations were fulfilled at the expense of a mass slaughter of cattle in the kolkhozes and sovkhozes, the purchase of cattle from the population, and by cheating. Despite the straining of all forces, the region officially delivered 93,300 tonnes of meat in 1960. At the same time the number of all types of domestic animals declined sharply in all farms of the region. Even horses were delivered as meat, and their number declined by 23 per cent from 1 October 1959 to 1 October 1960. A great number of cattle were bought on the side. The Kirov kolkhozes owed the state enormous sums. Having no funds, they paid no wages for three to five months.[14]

In his report at the plenum of the CPSU CC in January 1961, Khrushchev publicly acknowledged that there were negative tendencies in livestock farming. For example, the growth of cow numbers, which was (in all types of farm) 2.4 million in 1957 and 1.9 million in 1958, fell sharply to 0.6 million in 1959 and 0.9 million in 1960.[15] From the first months of 1961 there was a noticeable reduction in meat deliveries: on March 20 they were recorded as 24 per cent less than the corresponding period in the previous year, and on 20 May 15 per cent.[16] At the beginning of May the government conducted an enquiry from which it emerged that, even if government reserves were freed, by mid-1961 there would not be enough meat to supply current requirements.[17] The leadership of the Procurement Committee indicated to the government quite frankly in May 1961 that, with a sharp fall in the growth of the cattle stock, the meeting of even the 1961 planning tasks 'can create serious difficulties for livestock growth in the current year and the increase of meat deliveries in the following years'.[18]

The anti-cheating campaign

The meat shock-campaign of 1959–60 on the one hand created the conditions for a wide extension of cheating and other abuses, while on the other hand it did not permit a struggle against them. One of the main causes for this was the mutual understanding between the central and regional *apparats*, united in the urge to bring the campaign to an end, in line with the principle that 'the winner is not judged'. The special solidity of this mutual pact is underlined by the fact that even Khrushchev was drawn into it. The exit from the leap-forward campaign, for these reasons, was accompanied by considerable political difficulties and required the operation of a new campaign, this time directed against cheating and adventurism. To the extent that the initiator of the 1959 leap-forward was Ryazan, that region was the first whose machinations were uncovered by this counter-campaign.

Initially Khrushchev, by all accounts not realizing the real scale of cheating, tried to quieten the scandal. Despite the death of Larionov on 22 September

1960,[19] and the uncovering of a host of facts about the true state of affairs in Ryazan region, the original targets for checking for cheating were not Ryazan, but other regions of the USSR. On 11 October 1960 the Presidium of the CPSU CC in the absence of Khrushchev examined the draft of a resolution of the CC and Council of Ministers titled 'On crude violations in the carrying out of cattle deliveries which have been allowed in a number of RSFSR and Kazakh SSR regions'. Among the Russian regions that were subjected to criticism, above all for 'juggling' schemes and double registration of cattle in the plans, were Kurgan, Voronezh, Sverdlovsk regions and Krasnodar krai. There was nothing about Ryazan. Meanwhile the course of the Presidium session, recorded in the short protocol notes, allows us to conclude that the supreme leadership of the country took a decision on toughening the struggle against cheating. Those members of the Presidium who came forward at the session criticized the proffered draft for its 'liberalism' and sent it back for reworking.[20] No doubt the Presidium members were aligning their actions in accordance with Khrushchev's intentions.

This was confirmed by subsequent actions of Khrushchev himself. On 29 October he sent a note to the CC Presidium titled 'Against complacency, self-tranquilization, and conceit arising from the first successes in agricultural development'. This was published. Khrushchev's note was devoted to various agricultural problems and contained only a few paragraphs condemning cheating. On Ryazan, Khrushchev was rather indefinite: 'There is talk throwing doubts on the Ryazan people's successes ... I think there should be a check so that the CC can know the true state of affairs – does this talk have some solid soil or whether it is a libel against the Ryazan people.'[21] Khrushchev, who knew of the true state of affairs both in Ryazan and other regions, was compelled to lie. Declaring the need for supplementary checks on how the high obligations were fulfilled, he suggested that the central leadership itself was the victim of deceit.

In November 1960 a check on the situation in Ryazan gave a new push to the development of the scandal. It was carried out by members of the CPSU CC *apparat* together with the new first secretary of the Ryazan regional committee K.N. Grishin, transferred to Ryazan from his post as first secretary of the Vladimir regional committee. The facts about cheating brought to light by the commission were studied on 2 and 3 December in sessions of the CPSU CC RSFSR bureau,[22] and then on 23 December at a session of the CPSU CC Presidium.[23] At the insistence of the Presidium, on 4 January 1961 the RSFSR bureau of the CC affirmed the resolution titled 'On the results of the check on the fulfillment by Ryazan region of the obligations for meat deliveries to the state in 1959'. This document, which was intended for distribution in all the regional and krai party committees, was the result of a definite political compromise. The resolution actually affirmed that mass cheating and various machinations were used by the Ryazan leadership only with the third-plan delivery; that is, for the final 50,000 tonnes. The purpose of this lie was to try to stop the discrediting of the very idea of the 'meat leap-forward', initiated by Khrushchev. With the help of this approach there was a division in the resolution between the 'self-sacrificing labour' of the Ryazan kolkhozniks who had fulfilled two meat delivery plans, and the 'criminal schemes' of the regional leadership, which had rushed after its self-interested careerist interests under the cover of the third plan. The main culprits in the 'anti-state activities and the organized deceit of the party' were acknowledged to be two regional party secretaries (in addition to Larionov) and the regional sovnarkhoz's head of administration of the meat and dairy industry. They were expelled from the party. The chairman of the regional executive committee received a severe reprimand and lost his job. Other regional officials received various penalties.[24]

The results of this beginning stage of the anti-cheating campaign were brought up in the CPSU CC plenum in January 1961. In his speech at the plenum, N.S. Khrushchev again acknowledged the existence of cheating, but this time in Ryazan region as well.[25] In connection with the decisions of the plenum some measures were taken for strengthening central control over accountability in agriculture. The Council of Minister's State Procurement Committee was created on 25 February 1961.[26] The very same day a joint resolution by the CPSU CC and the Council of Ministers brought into action a new arrangement for agricultural procurements. The aim was to closely control each stage of the delivery procedure and not allow kolkhozes and sovkhozes to lower their plans. Deliveries had to be carried out on the basis of contractual agreements with the government. With this, agreements were subject to constant correction according to the study of the situation of kolkhozes and sovkhozes and the content of their production plans. To execute this detailed control, a branched system of district state delivery inspectorates was created.[27] These measures testify to the fact that Khrushchev, disappointed in the mass enthusiasm organized from above, had again moved the accent on to the tested levers of centralized administrative control.

In the beginning stage of its development, the anti-cheating campaign had three politically sensitive points. The first was the explanation of the role of the centre and of Khrushchev in particular who, thanks to many public activities, appeared to be deeply involved in the meat adventure. The second was the attitude towards rank-and-file participants in cheating and above all to leading kolkhozniks, who had been generously rewarded for exaggerated (but by no means entirely fictional) achievements. The third was the limits of punishment for the vast army of officials, from the highly placed functionaries of the centre and regions down to the kolkhoz chairmen, who had emerged as the main motive force of the 'leap forward'. A real punishment of all the guilty would mean a wide-ranging purge of the *apparat*. Such a purge was not only politically undesirable, but also organizationally impossible.

The anti-cheating campaign was structured in accordance with these circumstances. Above all, the central leadership was sheltered from the blow. On the one hand the centre held in its own hands the initiative for unmasking the cheating. On the other hand those fragments of Khrushchev's public discourses in which he spoke of caution in accepting raised obligations were propagandized in all kinds of ways.[28] In this way the old differentiation between the leader, drawn into mistakes, and the 'regenerated bureaucrats', was once more put to use.

From the very beginning there was an almost total ban on the punishment of rank-and-file participants in the 'leap forward', especially of leading kolkhozniks.

In public speeches criticizing careerist administrators involved in cheating, Khrushchev continued to praise the work achievements of ordinary leading workers. As a result of this arrangement, as already mentioned, only a small group of administrators in Ryazan were punished, mainly at the regional level. This happened in spite of the fact that on 8 January alone 3,481 people in Ryazan region received awards (32 of them received Hero of Socialist Labour titles). Most of them were ordinary kolkhozniks, but there were also quite a few officials: 89 secretaries of district party committees, 32 secretaries of district Komsomol committees, 31 chairmen of district and town executive committees, and so on.[29] A similar policy was carried out in other regions.

On the whole, a quite small group of bureaucrats was punished directly for cheating. Several highly placed leaders suffered in Moscow. For example, straight after the January plenum of 1961, A.B. Aristov, Khrushchev's deputy in the RSFSR bureau of the CPSU CC, was sacked.[30] In this way the people, and the *apparat*, were shown who was personally guilty for agricultural failures at the very top. Despite the universal extent of the corruption, only a few regional leaders were specifically punished for cheating. In February 1961 the first secretary of the Kirov regional party A.P. Pchelyakov, whose work Khrushchev had sharply criticized in his report at the January CC plenum, was removed in disgrace from his post. That same February in Kazakhstan the first secretaries of the Pavlodar[31] and Kzyl-Ordinsk[32] regional committees were removed from their jobs and excluded from the party. As a result of a major scandal over vast cheating in cotton deliveries the first CC secretary and the chairman of the Tadzhik council of ministers were removed.[33] At the end of April the first secretary of the Tyumen regional committee lost his job because of cheating.[34] Several secretaries, clearly mixed up in cheating, were retired without fuss on formal grounds.

In fact, between Moscow and the regional leaderships there was an unspoken agreement by which the regional, krai and republic secretaries received the right to pursue the anti-cheating campaign at the expense of officials at lower levels, thereby diverting the blow from themselves. In many regions and republics, cases connected with cheating were organized on a district or individual kolkhoz scale.[35] The exemplary punishment of selected scapegoats also took place at that level. In these instances, leaders at republican or regional level, fully acquainted with the real mechanisms of cheating and aware of their own degree of guilt, tried not to go further than the campaign's narrow framework that sufficed for accounting with Moscow.

Indicative in this respect was the situation that developed in Kirov region. Under pressure from the centre, the Kirov regional committee's bureau in March 1961 punished the leaders of several districts and kolkhozes for cheating. P.E. Shelest, at that time the first secretary of the Kiev regional committee, later commented about the pressurized decision: 'They themselves pushed them into copying the Ryazan people and buying animals for meat. And now they blame them and make them responsible. What's the use of that?'[36]

Evidently feeling a certain resistance, the top Soviet leadership decided to give the campaign a fresh impulse. On 12 May 1961 the Presidium of the

CPSU CC affirmed the latter's letter: 'On the tasks of party organizations in the decisive eradication of the deception of government, eyewash, and cheating in the fulfilment of national economic plans and socialist obligations'. The letter was intended for distribution to all party organizations right down to the primary.[37] The letter enumerated a few of the best-known scandals involving cheating (Ryazan, Kirov, cotton cheating in Tadzhikistan, and others) and put forward to party and state organizations a demand to intensify the struggle against such phenomena. This was a political signal that made known the intention of the top leadership to prolong the campaign against cheating. Discussion of this CC letter in the localities produced the revelation of a certain number of supplementary facts about cheating. But this campaign did not gain the same wide extent as that of the beginning of 1961.

The preparation of new government anti-cheating decisions indicated the transformation of the campaign into the routine phase of 'universal strengthening of control'. Study of this question and the working out of appropriate documentation was entrusted to the Central Statistical Administration (TsSU). On 8 April 1961 the material presented by the TsSU was examined at a session of the Presidium of the USSR Council of Ministers. In the adopted decision, a commission under the chairmanship of the deputy chairman of the USSR Council of Ministers, A.N. Kosygin, was entrusted with the preparation of an appropriate draft resolution. The result of this commission's work was the resolution (19 May 1961) of the CPSU CC and USSR Council of Ministers 'On measures for averting factual deception of the state and for strengthening control over the trustworthiness of figures relating to the fulfillment of plans and obligations'. Along with formal summons to the struggle against cheating, the resolution noted certain measures for strengthening administrative control over the trustworthiness of accounts. The responsibility of the chief accountants for the accuracy of accounts, along with the leaders of an enterprise, was established. A routine was introduced for the withdrawal of rewards given out as a result of cheating. Provision was made for the withholding of publication of information in the press about plan fulfilment until a provisional assessment was made by the appropriate statistical organs. The posts of head of krai, regional and autonomous region statistical administrations became part of the CPSU CC *nomenklatura*, which substantially elevated their status and gave greater independence from the influence of local power. Transfers of these workers would henceforth be carried out only with a joint presentation by local party organizations and the TsSU.[38]

On 24 May 1961 the decree of the USSR Supreme Soviet Presidium 'On the responsibility for cheating and other distortions in accounts of plan-fulfillment' was adopted. The decree required that cheating in governmental accounting and other wilful distortions of numerical data about plan fulfilment should be regarded as 'anti-state activity, bringing harm to the national economy'. Deprivation of liberty for up to three years was the sanction envisaged.[39]

By degrees the campaign progressed from the stage of big cases and political declarations to the revelation of individual instances of cheating and the bringing of individual lower-level leaders to party and legal responsibility. In total for

1961, the prosecuting organs of the RSFSR alone saw 1,113 papers on cheating and 724 criminal cases were brought, of which 91 cases were under the 24 May 1961 Supreme Soviet decree; 350 cases involving 603 people (including 31 cases against 55 persons under that decree) went to court; 255 cases were terminated at the investigation stage in connection with the guilty parties being brought to party or administrative responsibility, or for other reasons.[40] Leaders of economic organizations made up the bulk of the accused. In the first half of 1961, cases brought to court by RSFSR prosecuting organs involved 35 kolkhoz chairmen, 15 sovkhoz directors, 28 responsible persons of the delivery organizations, 23 leaders of industrial enterprises, 18 leaders of consumer organizations, 13 heads of construction and work superintendents. A few party workers also went to court. Thus the former first secretary of the Irtysh district committee of the party in Kazakhstan was condemned to two years' deprivation of liberty for compelling directors of sovkhozes to cheat.[41]

An obvious leaning towards soft punishments, and predominance of measures of party and administrative responsibility, was reflected in the overall totals of condemnations for cheating. According to data for the USSR Supreme Court more than 1,200 people were condemned in 1961–63 for cheating and other accounting distortions. Of these, 254 were in the peak period of the campaign (May – December 1961). The maximum number of condemned was in 1963 (670), after which followed a decline: in nine months of 1964, 289 were condemned.[42] Most likely, the quoted figures involved those condemned to deprivation of liberty. The number of those receiving reprimands without loss of liberty was higher. Thus in 1961 in 36 regions of the RSFSR 58 people were condemned to deprivation of liberty for cheating, and 248 to correctional labour or suspended punishment.[43]

Taking into account the scale of cheating and the number of leaders and ordinary workers drawn into various contrivances, these data on those made to answer for them seem rather modest. Those in power, from the very start of the campaign, applied a selective principle of punishment. The essence of this was explained by the USSR public prosecutor R.A. Rudenko in a report note at the CPSU CC of 24 July 1961 in the following way:

> Prosecutors are advised not to allow the raising of criminal cases and imputation of responsibility against those workers who found themselves involved in anti-state activities involuntarily, by the fault of careerists and self-seekers, and to carefully sort things out and distinguish premeditated conscious deception from mistakes allowed by accident.[44]

Despite the relatively moderate number of punishments directly for cheating, the anti-cheating campaign had more significant consequences for the mutual relationship of central and regional leaders. Compelled to give up plans for a leap-forward-type boost of agricultural production, Khrushchev in actual fact placed the entire responsibility for the failure on the *apparat* and, above all, its regional structures. At the end of 1960 and in 1961 there was a massive rotation of regional committee secretaries. This marked the beginning of an offensive against the *apparat*, which ended with the division of party organizations into industrial and agricultural branches at the end of 1962.[45]

The functions and consequences of cheating

On 5 August 1966, almost two years after the overthrow of Khrushchev and more than five years after the beginning of the anti-cheating campaign, the USSR Public Prosecutor, R.A. Rudenko, sent to the USSR Council of Ministers a routine note on this question, in which it was said: 'The facts and the checks testify to the extensiveness of abuses of the 19 May 1961 resolution of the CPSU CC and USSR Council of Ministers.' The concrete examples of cheating enumerated in this report were analogous to those that figured in the documents of the 1961 campaign period. The reaction to Rudenko's note was feeble and formal. It was sent to the councils of ministers of the union republics, to ministries and departments 'for taking measures'.[46] All this once more bears witness to the fact that cheating was a phenomenon not eradicated, and inherent in the Soviet economy.

Although each branch of the economy had its own most characteristic and concrete forms of accounting falsification, in terms of functional criteria all cheating could be divided into three groups. The major part of the cheating (at least by variety of methods) was directed at the distortion of quantitative numerical data. This could be direct cheating applied to figures of production which, in fact or in general was not produced, or did not possess the required qualities. Frequent cases of cheating that involved unproduced production used double-counting of output already delivered or the enumeration, in the accounting data of a producer, of output which had already been reported by other producers. In agriculture, the already-mentioned purchase of meat and butter in shops and organizations followed by their second delivery to the state belonged to this class of cheating. A type of quantitative cheating, characteristic of industry, was the inclusion in accounting data of incomplete or substandard products. In construction, this type of cheating was widely employed in the form of fictitious deliveries of entities that were either entirely unfinished, or had numerous defects and unfinished parts that needed to be put right.

The second category of cheating was distortion of qualitative indicators. For example, the most widespread method of improving yield-per-hectare figures (for grain and especially cotton)[47] was to remove from the accounts a certain part of the sown area. In the 1958–60 shock campaign to raise the milk yield, the usual device was similar, involving the concealment from the figures of some of the cows: reclassification as heifers of cows actually giving milk, and so on. In industry and building there was a widespread practice of reducing the number on paper of workers, thereby allowing enhanced indicators of productivity and cost reduction.

The third group involved misrepresentation of allocation indicators and above all the concealment of part of the working capital and existing resources (metal, building materials, etc). This allowed requests for additional funds to be overstated. Raising the reported number of workers in industry and building was similar in character, and allowed additional funding of the wage bill.

This categorization of cheating leads us right into the question of causes and, hence, the functions of such a phenomenon. In Soviet economic literature, limited as it was by its well-known ideological framework, the existence of cheating ('distorted economic information') was attributed to two basic premises. The first was the 'striving of individual members of socialist society to elevate the results of their labour' with the aim of receiving undeserved material benefits (including the benefits of theft) and also career advancement. The second was 'defects in the economic mechanism, mistakes in the planning and administration of the economy, in particular insufficiently realistic planning tasks'.[48] Despite the obligatory caution of such formulations, they reflected important features of the phenomenon, and can serve as a starting point for investigating the role of cheating in the Soviet system.

It is evident that cheating actually in many cases facilitated various criminal acts and above all theft. Examples of this can often be found in the documents. Thus, in the Klintsy district of Bryansk region a criminal group carried out the following scheme of theft in 1959–60. Officials in charge of meat procurements, having received from kolkhozes money for purchases of cattle on the market and delivery of them to the local meat combine, in reality bought no cattle. The money received from the kolkhozes was divided by the officials between themselves, the meat combine workers and certain district bosses. A fictitious receipt included the undelivered animals in the figures for meat deliveries by the district's kolkhozes. To make up for the meat shortfall, the meat combine workers lowered the average weight that they recorded of the animals actually delivered by the kolkhozes. That is, in effect they robbed the latter twice over. With the money they received, the plunderers bought houses, private cars (even at that time very scarce in the USSR), and so on.[49] This case is a good example of how the widespread practice of mass purchase of cattle on the side in order to fulfil unattainable plans could be used for the purpose of robbery.

Despite such examples, it would be a mistake to treat cheating as a primarily criminal activity. By its nature cheating was linked with the cardinal sins of the Soviet economic system. For the producers, cheating was a means of adapting to this system, of observing a definite balance between the demands of economic expediency and centralized administration. Like many other illegal and semi-legal (by the standards of Soviet legality) elements of Soviet reality, cheating activities stand out mainly as 'quasi-market' regulators, assuring the survivability of the system. Without such regulators the system, restricted by the narrow framework of rigid centralization, simply could not exist.[50] An example of this could be Khrushchev's demand for the universal cultivation of maize. Threatened by an absence of cattle feed, the economic leaders of those regions where maize could not grow because of climatic conditions in many cases continued to sow other feed crops, naming them as maize in their accounts.[51]

With the help of cheating, the leaders in economic fields overcame such negative phenomena of the planning system as the weak links between producer and consumer that led to storming rhythms of production. The constant breaks in deliveries made it impossible for enterprises and building sites to fulfil plans on time. This led to various administrative and material sanctions on the part of the state, which could only make things worse for an enterprise, for example by causing an exit of qualified workers. To stop this, in many cases with the help of book-keeping distortions, the indicators of the actual fulfilment of plan were redistributed. For example, output produced in more favourable times was held back and included in the figures for a period of interrupted deliveries and unfulfilled plans.[52] This allowed hold-ups, repair of sub-standard products, overtime work, and so on, to be retrieved.[53]

Book-keeping distortion was a common way for producers to correct the extremely inefficient system of the centralized distribution of resources with funding fixed from above. This set-up, as a rule, caused a massive shortage of resources and also production failure. To insure against this, reserves were established by enterprises.[54] Since the formation of reserves led to a sterilizing of considerable resources, the state tried to fight against this phenomenon by demanding the production of accounts for the presence of unused materials, raw or otherwise. In their turn the enterprises, being unable to function without reserves, struggled against the state with their sole possible method: they concealed their stocks with the help of false accounting. Thus the check by the TsSU in January–March 1964 (selective, no doubt) revealed that enterprises and building sites had concealed from accounts about 40,000 tonnes of ferrous metal.[55]

Although the fact itself of widespread cheating in the USSR is not open to any doubt, the question remains of its scale and the degree of its effect on official economic statistics. Above all, one must consider those facts that could serve as arguments for a wary approach to the influence of cheating. The first of these is the difference between branches. The literature has noted that in different branches there were different opportunities for cheating. Depending on the branch, the level of state control over accounting could differ. It was easiest to carry out cheating in agriculture and building. A different world was that of the priority branches (for example, the defence industry) where tight control made cheating difficult.[56]

Former workers in the statistical organs who had emigrated to the USA were questioned in 1984 and did not consider cheating to be too big a problem for the Soviet economy. In their opinion, the existing control mechanisms in accounting were sufficient to prevent significant cheating.[57] Carrying this further, one might note that the Soviet planning system itself did not permit enterprise leaders to cross a certain line with distorted book-keeping. For example, the exaggeration of figures for output could bring noticeable dividends in the way of rewards, but this led inevitably to an increased plan target for the following year.

On the strength of a whole complex of causes one can thus propose that the most important type of book-keeping distortion was short-term cheating of a compensatory nature; for example, end-of-month cheating to create the appearance of plan-fulfilment, which would be compensated for in the following month. Documents confirm the wide extent of this type of distortion. In aggregate, such cheating could not substantially distort the general indicators in economic statistics.

Along with that group of factors that were an obstacle to the spread of cheating, there are also contrary arguments. Above all, it should be noted that powerful stimuli for spreading distorted book-keeping were embedded in Soviet economic reality. For example, the chronic shortage of most products and services deserves particular attention. As is well known, such shortages compelled consumers to accept from producers any standard of quality. In this connection, types of cheating – such as the inclusion in output of the production of substandard or incomplete units – were widespread.[58] The housing shortage allowed builders to easily give consumers rubbishy apartments, while the transport shortage obliged users to pay cheat-money when their freight was carried,[59] and so on.

The predetermination of cheating by the Soviet economic system itself, and likewise its essentially ineradicable character ('in the interests of the matter in hand'), do much to explain the more than tolerant attitude to cheating that became an unwritten rule of Soviet economic relationships. This meant that the very mechanisms of accounting control were not as rigid or perfect as it might seem at first glance.

As frequently noted in the documents, there was mutual understanding between producers and their administrators at the branch and territorial level. None of the *sovnarkhozy*, the ministries or the regional party committees were interested in the manifestations of cheating. Thus in the TsSU's note about the coming into play of production capacity in 1963, there was mention that, in many cases, assets that were seriously incomplete or otherwise defective were accepted for use by state commissions appointed by republic councils of ministers and sovnarkhozes.[60] In such cases, when cheating nevertheless became manifest, the regional powers and sovnarkhozes did everything possible to drag the guilty people away from harm and put brakes on examination of the case.[61] The periodic anti-cheating campaigns, as we have seen, dragged in only a small number of scapegoats.

The facts outlined above allow us to suggest that in reality only a certain proportion of cheating was revealed. So far, no archival material has been found to demonstrate a serious study of the cheating problem or attempts to evaluate its overall size. As a rule, documents give individual examples of cheating, which are difficult to put together to produce an overall picture. Despite this, these scattered facts and data in many cases demonstrate the significant volume of cheating. Above all, cheating concerns agriculture. As already mentioned, the check carried out in 1960 in Ryazan brought to light cheating in 1959 meat deliveries to the tune of 50,000 tonnes. Unfortunately, everything points to the fact that such detailed checks did not take place in other regions. Ryazan cheating amounted in 1959 to about 0.7 per cent of the total volume of meat deliveries (which was 7.5 million tonnes by the official figures).[62] But it has to be kept in mind that Ryazan region and its agriculture were not the country's biggest. Cheating over cotton in the Central Asian republics was significant. Thus, just in Tajikistan, the revealed cheating amounted to 28,700 tonnes in 1958, 37,100 the following year, and 55,600 in 1960.[63] The overall delivery of cotton according to the official figures was 4.34 million tonnes in 1958 and 4.65 million in 1959.[64] Thus the Tajik cheating alone represented about 12 per cent of the overall growth of cotton production in 1959 over 1958. In just three districts of Latvia 5,000 tonnes of recorded potato deliveries were the product of cheating, which was 4 per cent of the delivery plan of the whole republic.[65]

There are examples of quite a few cases of cheating in other branches of the economy, too. A selective check of the USSR Stroibank revealed that, in 1961, 3.2 per cent of recorded building work was covered by cheating in Kirgizia, 2.6 per cent in Kazakhstan, 3.4 per cent in Tajikistan and 2.2 per cent in the RSFSR. In individual regions these amounts were even higher, amounting, for example, to 6.3 per cent in Dagestan and 4.8 per cent in Voronezh regions;[66] 1.5 per cent of the traffic recorded by the enterprises of Latvia's motor transport ministry was falsified.[67]

The revelation and systematization of such data will perhaps in future allow a more complete general picture to be presented, as well as an evaluation of the degree of distortion in Soviet economic statistics that resulted from cheating.

* * *

The policy of forward leaps in agriculture during 1959 and 1960, and also the anti-cheating campaign that followed, demonstrated many important features of the Soviet system as a whole and also of its variant, which was established in the period of Khrushchev's rule. Despite their rejection of many of the most odious manifestations of Stalinist tyranny, the new Soviet leaders in intellectual and political matters remained to a considerable degree heirs of Stalin. Having been trained in the storming period of the Stalinist five-year plans, they had completely assimilated the programmatic ways and methods of the leadership that were born in this epoch. The characteristic feature of Khrushchev's politico-economic mind was an absolute belief in the all-powerfulness of the party state. Rejecting such a key element of the Stalinist system as mass repression, the new Soviet leaders tried to compensate with methods of administrative pressure and manipulated enthusiasm.

The course and the mechanisms of the meat campaign provide extra arguments for the theory of close links between economic leaps and the phenomenon of the mass cheating which was basically inherent in the Soviet economy. Indeed, in the periods of leaps-forward the form of cheating that was most obvious involved the containment of the destructive consequences of the centre's economic adventurism.

In essence, the anti-cheating campaign that began at the end of 1960 was a mechanism for exiting the leap-forward stage in agriculture. Despite a loud propaganda accompaniment and the organization of numerous showcases, on the whole the campaign even in its initial and most active stage was not very harsh either in scale or in the kind of sanctions imposed. This course of the campaign, its operation according to the principle of punishing a relatively small number of scapegoats, was the result of a definite compromise between the centre, the regional party and economic leaders. The relative softness of the penalties is also explicable in that the anti-cheating struggle came up against a certain resistance both from the local powers and from the prosecuting and judicial organs.

Together with this, one important result of the campaign against cheating was the accumulation of significant materials about the realities of how the Soviet economy functioned. These documents allow us to examine cheating as one of the mechanisms by which producers adapted themselves to the realities of the Soviet planning system, a means of overcoming its contradictions and bottlenecks as a whole, and also in part as a means of moderating the destructive consequences of leap-forward policies.

As the documents show, there were many barriers on the road towards uncovering cheating, mainly due to the mutual understanding the producers had with the highest party-state and economic structures. All the same, the documents bear witness to the large scale of cheating, especially in agriculture. The aim of this chapter has not been the systematic revelation of cases of cheating and the creation on that foundation of a base for generalized evaluations of the scale of cheating. Such a work would need the efforts of a collective of researchers. However, even individually discovered facts allow us to say that cheating cannot be ignored as a factor influencing the trustworthiness of Soviet statistics.

The scandalous failure of the leap-forward in agriculture played an important role in determining the later policy of Khrushchev. Although the massive extension of cheating was a natural reaction of regional leaders to the economic adventurism of Khrushchev, he himself no doubt regarded cheating as sabotage on the part of the *apparat*. Soviet agriculture, despite the positive results of the virgin lands programme and the weakening of the taxation grip on the peasantry, nevertheless could not attain the necessary dynamic of development in face of the internal weaknesses of the *kolkhoz/sovkhoz* structure. Numerous experiments in the countryside initiated by Khrushchev, the most negative of which was the systematic offensive against peasants' individual plots, made things worse. The shortage of food, in its turn, was a catalyst in the tense social situation in the country.

Encountering such challenges Khrushchev, however, could not respond with Stalinist methods. The Khrushchev response was pressure on officials, mass transfers of leading cadres in the centre and regions, the limitation of time-in-office for the leading positions of responsibility, and division of the party and Soviet apparatus according to the branch principle – industrial and agricultural. The growing conflict between Khrushchev and the *apparat* led to his removal in October 1964.

Notes

1 Thanks to R.W. Davies, Melanie Ilic, Mark Harrison and Yoram Gorlizki for help and substantial advice. In preparing this chapter use was made of a range of documents that made their appearance within the project 'Networks and Hierarchies in the Soviet Provinces: The Role and Function of Regional Party Secretaries from Stalin to Brezhnev (1945–1970)', directed by Yoram Gorlizki with the support of

the UK Economic and Social Research Council (grant 000230880). This chapter was translated by John Westwood.

2 In administrative documents and propaganda great use was made of terms such as 'deceiving the state', 'eyewash', *pripiski*. They described a wide range of practices that basically were the same, as will be examined in more detail in the last part of this chapter. For the bulk of this chapter *pripiski* will be the term used. (Translator's note: *pripiski*, statistical misrepresentation, will usually be rendered as 'cheating' in this chapter.)

3 Nikita Khrushchev, 1964, *Stenogrammy plenuma TsK KPSS i drugie materialy*, Moscow: Materik, 2007.

4 S. Shenfield, 'Pripiski: False Statistical Reporting in Soviet-type Economies', in M. Clarke (ed.), *Corruption. Causes, Consequences and Control*, London: Frances Pinter, 1983, pp. 253–4.

5 N.S. Khrushchev, *Stroitel'stvo kommunizma v SSSR i razvitie sel'skogo khozyaistva*, vol. 3, Moscow: Gospolitizdat, 1962, pp. 412–15.

6 A. Agaryev, *Tragicheskaya avantyura*, Ryazan: Russkoe slovo, 2005, pp. 48–9.

7 RGANI, f. 13, op. 1, d. 793, ll. 2–8, 25–30, 61–120.

8 I am grateful to M. Harrison for bringing this problem to my attention.

9 RGANI, f. 13, op. 1, d. 793, l. 15.

10 RGANI, f. 13, op. 1, d. 793, l. 32.

11 RGANI, f. 13, op. 1, d. 793, l. 31.

12 RGANI, f. 13, op. 1, d. 793, l. 28.

13 RGANI, f. 13, op. 1, d. 793, l.12–13, 28.

14 RGANI, f. 13, op. 1, d. 805, ll. 4–5; RGASPI, f. 17, op. 91, d. 1787, ll. 72–4.

15 N.S. Khrushchev, *Stroitel'stvo kommunizma v SSSR*, vol. 4, Moscow: Gospolitizdat, 1963, p. 204.

16 GARF, f. R-5446, op. 95, d. 808, l. 15; d. 774, l. 9.

17 GARF, f. R-5446, op. 95, d. 808, l. 52.

18 GARF, f. R-5446, op. 95, d. 774, l. 11.

19 Taking everything into consideration, Larionov may be said to have committed suicide. Suffering from alcoholism in the last part of his life, he died from a sedative overdose.

20 *Prezidium TsK KPSS, 1954–1964*, vol. 1, Moscow: Rosspen, 2003, pp. 443–4, 1076.

21 N.S. Khrushchev, *Stroitel'stvo kommunizma v SSSR*, vol. 4, p. 167.

22 RGANI, f. 13, op. 1, d. 790, ll. 61–120.

23 RGANI, f. 13, op. 1, d. 829, ll. 4, 84.

24 RGANI, f. 13, op. 1, d. 802, ll. 1–4.

25 N.S. Khrushchev, *Stroitel'stvo kommunizma*, vol. 4, pp. 362–7.

26 *Postanovleniya Soveta Ministrov SSSR, fevral' 1961*, Moscow: Upravlenie Delami Soveta Ministrov SSSR, 1961, pp. 214–15.

27 *Postanovleniya ... fevral' 1961*, pp. 216–22.

28 Especially frequent were quotations of a sentence from Khrushchev's speech at a Kremlin reception of a Ryazan *kolkhoznik* delegation on 16 October 1959, in which he said 'If you are now calculating and see that you will not reach fulfillment of the third plan, then you must say so honestly, and the Party's Central Committee will not blame you', *Pravda*, 17 October 1959.

29 A. Agaryev, *Tragicheskaya avantyura*, p. 16.

30 *Prezidium TsK KPSS*, 1954–1964, vol. 1, p. 479.

31 RGANI, f. 5, op. 31, d. 169, ll. 10–18.

32 RGANI, f. 5, op. 31, d. 169, ll. 94–6.

33 Details of the case were widely discussed at the plenum of the Tajikistan Communist Party Central Committee (RGASPI, f. 17, op. 91, d. 597, ll. 1–152).

34 RGANI, f. 13, op. 1, d. 842, ll. 68–70.

35   See, for example, cases of cheating in the Yegorlyk district of Rostov region (RGASPI, f. 556, op. 14, d. 185, ll. 19–22), in the Gus'-Khrustal'nyi district of Vladimir region (RGASPI, f. 556, op. 14, d. 183, ll. 118–22), in the Alarskii district of Irkutsk region (RGASPI, f. 556, op. 14, d. 187, ll. 2–16), in the Galyachskii district of Poltava region (RGASPI, f.17, op. 91, d. 1074, ll. 115, 126–51) and in the districts and kolkhozes of Azerbaidzhan (RGANI, f.5, op. 31, d. 172, ll. 3–12), and so on.

36   P. Shelest, *Da ne sudimy budete*, Moscow: Edition Q, 1995, p. 146.

37   RGANI, f. 3, op. 12, d. 907, ll. 18–33.

38   *Postanovleniya Soveta Ministrov SSSR, mai 1961*, Moscow: Upravlenie Delami Soveta Ministrov SSSR, 1961, pp. 102–6.

39   *Spravochnik partiinogo rabotnika*, vol. 4, Moscow: Gospolitizdat, 1963, p. 525.

40   RGASPI, f. 556, op. 23, d. 147, l. 4.

41   RGANI, f. 5, op. 31, d. 167, ll. 110–13.

42   GARF, f. R-5446, op. 96, d. 348, l. 48; op. 98, d. 421, l. 52; op. 98, d. 429, l. 30.

43   RGASPI, f. 556, op. 23, d. 147, l. 4.

44   RGANI, f. 5, op. 31, d. 167, l. 103.

45   Y. Gorlizki, 'Political Reform and Local Party Interventions under Khrushchev', in P.H. Solomon (ed.), *Reforming Justice in Russia, 1864–1996*, New York, London: M.E. Sharpe, 1997, pp. 270–1.

46   GARF, f. R-5446, op. 100, d. 220, ll. 146–50.

47   According to incomplete data of the TsSU, in Uzbekistan alone the kolkhozes and sovkhozes concealed 30,000 hectares of cotton cultivation in 1961, 38,000 in 1962 and 42,000 in 1963 (GARF, f. R-5446, op. 98, d. 429, l. 24).

48   G.I. Khanin, *Dinamika ekonomcheskogo razvitiya SSSR*, Novosibirsk: Nauka, 1991, p.13. Khanin refers to I.P. Suslov, *Osnovy teorii dostovernosti statisticheskikh pokazatelei*, Novosibirsk: Nauka, 1979, and G.G. Edel'gauz, *Dostovernost' statisticheskikh pokazatelei*, Moscow: Statistika, 1977. Khanin himself, whose book appeared as Gorbachev's perestroika was in decline, could include in his work a proposition such as 'the most substantial causes of distortion of economic information lurks in the command system of administration. While being economically extremely inefficient, it strives to hide this inefficiency with the help of "felicitous" economic information. Thus as long as the administrative-command system survives, so will large-scale distortion of economic information remain unavoidable' (p. 14).

49   RGASPI, f.17, op. 91, d. 1447, ll. 70–1.

50   This problem is well known in Western literature (see J.S. Berliner, *Factory and Manager in the USSR*, Cambridge, MA: Harvard University Press, 1957). 'Quasi-market' regulators in the pre-war Stalinist economic system are analysed in detail in the works of R.W. Davies (see for example R.W. Davies, M. Harrison and S.G. Wheatcroft, *The Economic Transformation of the Soviet Union, 1913–1945*, Cambridge University Press, 1994, pp. 18–20).

51   S. Shenfield, *The Functioning of the Soviet System of State Statistics*, CREES Special Report SR-86-1, Birmingham: Centre for Russian and East European Studies, University of Birmingham, 1986, p. 45.

52   GARF, f. R-5446, op. 95, d. 330, l. 18.

53   For more on storming and supply problems, and violations of wages and norm-setting regulations, see D. Filtzer, *Soviet Workers and Destalinization. The Consolidation of the Modern System of Soviet Production Relations, 1953–1964*, Cambridge: Cambridge University Press, 1992, pp. 19–22, 110–11.

54   J.S. Berliner, *Factory and Manager*, Harvard MA, Harvard University Press, pp. 12–13, (1957).

55   GARF, f. R-5446, op. 95, d. 429, l. 29.

56   S. Shenfield, 'Pripiski: False Statistical Reporting …', pp. 256–7. Commenting on my chapter, M. Harrison made the important observation that in industry, compared

to agriculture, there were more opportunities for enhancing reported figures. For example, by raising prices.

57 S. Shenfield, *The Functioning*, pp. 44–7.
58 GARF, f. R-5446, op. 95, d. 330, l.17; op. 98, d. 429, l. 25. For the mass output of such products see D. Filtzer, *Soviet Workers and Destalinization*, pp. 162–7.
59 In 1965, Leningrad region motor transport added about 2 million cheated tonne-kilometres of non-transported freight and received from clients 108,000 excess roubles (GARF, f. R-5446, op. 100, d. 220, l. 148).
60 GARF, f. R-5446, op. 95, d. 328, ll. 102–3; op. 98, d. 421, ll. 1, 19–22, 28–30.
61 GARF, f. R-5446, op. 98, d. 429, ll. 31–2.
62 N.S. Khrushchev, *Stroitel'stvo kommunizma v SSSR*, vol. 4, p. 108.
63 RGANI, f. 3, op. 12, d. 907, l. 23.
64 N.S. Khrushchev, *Stroitel'stvo kommunizma v SSSR*, vol. 4, p. 57.
65 RGANI, f. 4, op. 16, d. 1062, l. 154.
66 GARF, f. R-5446, op. 96, d. 348, ll. 81–3.
67 RGANI, f. 4, op. 16, d. 1062, l. 154.

12 The modernization of Soviet railways traction in comparative perspective

John Westwood

A large part of social and economic development under Khrushchev can be categorized as the adoption of new trends and technologies, the fairly routine assimilation of progress that was already part of world history and that was taking place all over the contemporary world; trends like prefabricated building construction, the assimilation of jet propulsion, replacement of wood and metal by plastics. Because such forward steps had a natural-course-of-events appearance they might excite little comment, and escape evaluation. They were not adventurous, and their outcome was unsurprising, but in their way they could be achievements.

An important part of evaluating such achievements must surely be a recognition, not so much of whether a given transformation was carried out, but of the elegance with which it was obtained. In particular, it would be instructive to reflect on the mistakes and misjudgements that were *not* made. Whenever moves are made in a new direction there are false steps to be taken, and success in a given transformation can be assessed not only in terms of what was achieved but also in the pitfalls that were avoided. The untrodden banana skin, as it were, becomes a success-indicator.

Researching things that did not happen has obvious problems but one way of doing this is to compare a given process in one country with that in others. This chapter looks at one sector of the railway modernization on which most of the post-war world embarked: the replacement of steam traction by more advanced forms of power. For international comparisons, the British experience has been given some weight. Like Soviet railways, the British railways were state-owned and under a high degree of state control, while the respective modernization plans were almost concurrent. (Additionally, transport historians have long treasured the late-twentieth-century British railways as a source of case studies, for not only did successive British governments step resolutely on existing banana skins but they also succeeded in creating entirely new ones.)

The onset of modernization

From the end of the nineteenth century, world railways had begun to adopt electric traction for a limited number of lines with specific characteristics, in particular for routes of exceptional topographic problems or lines where heavy traffic intensity could demand and support the infrastructural costs of electrification.

That the Soviet railways would introduce new forms of traction had in effect been decided in the 1920s, with the railways' association with the Leninist plan for nationwide electrification. Railway electrification had featured in all the five-year plans. The targets set were never achieved, but experience had been gained, a modest electric locomotive production established and real data collected. The lines that were electrified had been chosen for physical rather than financial reasons – suburban lines in the two capitals where enormous traffic growth and short routes would have strained steam traction, a difficult mountain line in the Caucasus, part of the Murmansk Railway where electrification helped solve winter problems and also enabled the postponement of double-tracking, a couple of congested sections in the Urals and Siberia. Cost saving was not really a factor, more a promised fringe benefit.

With the exception, perhaps, of Switzerland with its cheap hydro-electricity, nations did not envisage an end to steam traction, but the picture changed in the 1940s when diesel locomotives became an attractive proposition. Railways then began to look forward to the total replacement of steam traction by a mix of electric and diesel locomotives; the question henceforth was how, and how soon.

In the Soviet Union, while Stalin-era electrification schemes instilled confidence in further electrification, diesel traction did not have quite the same foundation. In the 1920s and 1930s the USSR had some claim to be the country most advanced in the development of diesel traction. But the decision in the 1930s to allocate the small diesel fleet to Central Asia, where watering steam locomotives was a problem, meant that teething troubles of the new locomotives had little chance of resolution.

Lazar Kaganovich was peoples' commissar for railways from 1935 to 1944, apart from two short breaks (he was the last railways commissar to have a place in the Politburo; his successors were professional railwaymen, technical rather than political operatives). He decided to end diesel locomotive construction, placing his faith in condensing steam locomotives, an innovation that had all the glamour, and subsequent disappointments, of the quick fix. If, as was later believed, Kaganovich had a soft spot for steam locomotives, this could be seen as evidence; the protagonists of diesel traction could only have seen him as a wrecker at that time, although they kept their mouths shut. In retrospect, his action cannot really be described as entirely negative (among other things, the rigid-wheelbase configuration of Soviet mainline diesels had no future, as later became evident).

During the war, about a hundred US-built diesel locomotives were imported. These proved the effectiveness of their concept and, according to reliable anecdotal evidence, attracted the approval of Stalin. En route to the Potsdam Conference, his steam locomotive slipped to a standstill near Mozhaisk; one of the new diesels was attached instead, arousing his benign curiosity, and the US diesel took him all the way to Potsdam. Two months later he approved the building of diesel locomotives based on the US design. Allusions to this chain of events appeared occasionally in popular publications, and achieved confirmation in Rakov's definitive work on Russian locomotives.[1]

By the 1950s a new breed of diesel locomotives, based on US models, was in charge of a few selected sections and proving to be a viable form of traction. Here again, experience and data were being accumulated.

Railway modernization in Russia neither began nor ended with the Khrushchev period. In 1955, 14 per cent of gross tonne/kms were already hauled by non-steam locomotives (8 per cent by electric and 6 per cent by diesel traction).[2] The 1956 plan is crucial, however, because it marked the change from piecemeal introduction of electric and diesel traction to a clear decision to completely eliminate steam traction; modernization moved from a retail to a wholesale operation

Putting the economic case

Whereas most railways examined traction substitutions on a line-by-line basis, both Britain and the USSR at this time had moved to nationwide appreciations. This was one basic similarity, and another was that both displayed methodological shortcomings in their cost analyses, although whether these shortcomings had much importance, or indeed whether the financial appreciations themselves had much ultimate significance, is open to question, especially in the Soviet case.

At that time there were no sophisticated techniques like discounted cash flow appraisals, and cost-benefit analysis was little developed. Rate of return was the financial marker, for better or for worse.[3] In comparing the costs of the different types of traction, assumptions could determine the conclusions of a study. For example, the figure chosen for locomotive longevity could be crucial, and the widespread use of a 20-year life for all locomotives rather favoured the diesel locomotive in cost comparisons (due to early obsolescence, 20 years was rather optimistic for the diesel but unfair for the steamer, whose life could approach 40 years). This assumption determined depreciation costs. That other 'hidden' cost, interest on capital invested, tended to be bypassed in Soviet cost studies.

In the post-war decade there was an increasing flow of articles and books recommending the virtues of electric and diesel traction for the USSR. They were all of a certain pattern. There is no single study or book that can be cited as a key influence; each had the basic premise that new forms of traction were the future, and each brought in new evidence for the same argument, typically introducing fresh data from actual experience.[4] Comparative cost studies for steam, electric and diesel traction, supported by quite detailed tables, were the norm; however, the key argument was not financial but operational: the enhanced traffic capacity that modernization would bring. For purposes of making choices, the rate of return (as expressed in the time taken for a given investment to be fully recovered by the resulting cost reduction) was the yardstick. Or, as one appraisal concluded, 'Consequently, savings in operating costs on the preceding electrifications will cover fully the capital invested in the new electrification.'[5]

Unlike the Soviet modernization strategy, the British scheme had as its absolute top priority the elimination, or at least reduction, of the losses threatening the Treasury. Modernization as a means of enhancing capacity did not come into the equation, except insofar as the government welcomed any relief from press complaints about quality of service. A major weakness of the British plan, surprisingly, was its inability, or refusal, to clearly and accurately forecast the return on capital that railway investments would bring. It could hardly be blamed for its failure to correctly predict how traffic would develop, but it could be blamed for what might have been subterfuges (but might also have been incompetence) in some of its claims. In effect, despite the financial imperatives, the advocates of traction modernization seem to have based themselves on technical rather than commercial or financial possibilities. An astonishing (at least, it should have been astonishing) feature of the British plan was provision of considerable extra trackage to increase route capacity, which in fact would not be needed with the heavier and faster trains promised by traction modernization.

It is interesting that, while Western critics were pointing out how by ignoring interest on the capital invested Soviet planners could enhance the apparent financial benefit of projects, the British Transport Commission (BTC) planners were themselves not without sin when they decided to exclude from their calculations interest on that part of the capital raised from British Railways' own resources. Moreover, future depreciation costs were not taken into account when the rate of return was calculated. Even more remarkable is that the British added into the financial betterment expected from capital invested in modernization those losses that would be eliminated by the proposed closure of certain lines. And there is the general observation applicable to many such projections, that the cost reductions of a given scheme might well bring a seductive return on capital, but the object of the investment could still remain a loss-maker.

In the case of Soviet Railways the accuracy or inaccuracy of cost comparisons is interesting but, unlike with other railways, of secondary relevance. Traction modernization in the USSR was not aimed at reducing operating costs; it was part of that policy, dating from at least the First Five-Year Plan (FYP) (1928–32), of minimizing investment in new infrastructure and thereby increasing traffic density on existing lines to unprecedented (and, some would say, unacceptable) volumes. By the mid-1950s it was becoming clear that with increasing traffic volumes the policy was coming unstuck; congestion, not just on the main lines but also on the single-track sand-ballasted routes, was posing a danger of repeated hold-ups that could threaten the economy as a whole. With the existing policy, the only solution for handling rapidly expanding traffic was greater line capacity, which could be achieved by running heavier and faster trains. Some extra-powerful steam prototypes had been built, using the articulation concept as favoured by US railroads, but such locomotives needed reconstructed depots with longer turntables. Moreover, as US experience seemed to be indicating, in the longer term steam traction had little room left for further development, whereas modern forms of traction, where there was theoretically no limit to the number of power units that could be attached to a train, were still at the threshold of their development. This is why the succession of Soviet modernization studies, unlike those in other countries, while laying out the usual cost comparisons, emphasized calculations about how much extra capacity traction modernization could produce. Even if electric and diesel traction had not reduced operating costs, the case for their adoption might still have been overwhelming in Soviet circumstances.

The modernization plans

The decisions to totally eliminate steam traction, in both the USSR and the UK, were wrapped up inside plans of much wider scope. The wide-ranging BTC Modernisation Plan was introduced in January 1955 and was therefore only about a year in advance of the Soviet sixth Five-Year Plan (1956–60). Like the latter, it was acclaimed at first but then abandoned when it became clear that it was some way distant from reality. In its rolling stock pages the BTC plan acknowledged that technical progress with diesel traction meant that the steam locomotive could be eliminated, and proposed that steam passenger locomotive construction should cease after the completion of the 1956 programme and that all steam construction should end within a few years.

The Soviet decision to end steam traction was embodied in the transport and transport engineering sectors of the sixth Five-Year Plan.[6] The railway articles specified the numbers of diesel and electric locomotives to be acquired, without mentioning steam, while the transport engineering articles said it out straight: '… to develop the production of electric and diesel locomotives and terminate the output of mainline steam locomotives …'.[7] Mainline steam construction ended promptly, in 1956. Imports from Poland ended in 1957.

There was a degree of hesitancy still, evidenced by the exclusion of steam yard locomotives from the bar against new construction. In the event, this exclusion fell by the wayside, with steam yard locomotive construction ceasing a year after that of mainline units. British railways had dabbled with diesel yard locomotives since the 1930s because they gave a very great economy of operating costs. However, they did nothing for the Soviet preoccupation, line capacity.

The directives of the sixth Plan had been preceded by considerable discussion designed to produce a general plan for railway electrification. In November 1955 a joint submission by the Railways Ministry (MPS) and Gosplan had been presented to the Central Committee. The latter decided that a faster rate of electrification was required, and Gosplan and the MPS submitted their amended plan, this time in conjunction with the Ministry of Power Stations, in late December. This was discussed by the Presidium in early January 1956 (but only as the final item on a well-filled agenda, which also included hydrogen bomb testing and wage increases). According to notes on the meeting, Khrushchev found the proposals acceptable and complained that this was a matter that had been put off for far too long, adding platitudinously that the scale of the work would depend on 'possibilities', and demanding to know precisely what material shortages were holding things back. The Presidium thereupon instructed the Gosplan, MPS and power representatives to amend the plan as indicated in the discussion and to clarify the obstacles to a faster rate of electrification. In late January these three duly submitted their report. The main obstacles to a rapid electrification were seen to be a shortage of equipment for substations, and a general lack of electric power in certain regions (as well as run-of-the-mill problems like a shortage of high-quality wire). The MPS was also anxious about the production rate of electric locomotives and was not happy with Gosplan's recommendation that a proposed new locomotive works at Novosibirsk be dropped; there was to be an approach to the trade ministry with a view to importing electric locomotives from the peoples' democracies.[8] (As things turned out, electric passenger locomotive production would be shifted to Czechoslovakia and d c freight locomotive production to a new works created out of the Tbilisi repair shops.)

The sixth Five-Year Plan came to a premature end, being followed by an interregnum and then a new seven-year plan. These apparent upheavals made virtually no difference to the progress of railway traction modernization. Khrushchev's platitude about the scale of work depending on possibilities was only too true: railway reconstruction had been proceeding as fast as possible, and would continue to do so irrespective of targets.

The British modernization plan lasted little longer than the sixth FYP, but the consequences of its piecemeal abandonment were far more serious. It was falling apart by 1957 and its end was hastened in 1959 when the government published the White Paper 'Reappraisal of the Plan for the Modernisation and Re-equipment of British Railways'. The plan was itself faulty, as should have been realized at the time. Seen by the government as a means to reduce financial losses, it was compiled by numerous subcommittees of railwaymen who saw it as a means of exploiting the latest technical possibilities, and paid little more than lip-service to commercial and financial considerations. The Treasury could, and did, impose cuts on planned annual investment (and in 1961 would demand a cut of no less than 30 per cent). There was also, as will be seen, government interference in the placing of orders and then a decision to go for a breakneck dieselization. In 1960 came the governmental change of course from which the 'Beeching Plan' emerged, entailing the wholesale closure of lines. To decide on the size of the railway system after, rather than before, the plan for modernization, delighted all critics of central planning and was a sad case of governmental inadequacy. The spectacle of new locomotives designed for low-traffic lines being delivered after those lines were closed was only the most visible of the resulting absurdities. Meanwhile, both the trades unions and the private supply trade associations struck hard bargains that ultimately were borne by the rail user.[9]

The Beeching Plan and the government-imposed acceleration of the dieselization programme were the two major factors enabling flags to be waved and whistles to be blown in 1968 when the last British Rail steam train was run, just eight years after the last steam locomotive had been built. It would be about two decades later that Soviet Railways finally ended steam traction, without whistle-blowing and flag-waving, but as the end of a process that was measured, consistent and relatively efficient in the use of resources.

Questions of timing

It was easy enough for Khrushchev to remark that the change had been delayed too long, but in fact the best time to embark on such a transformation and how long the process should take were questions almost impossible to answer at the time and very difficult to answer even in retrospect. It is, however, possible to broadly distinguish those railways that more or less got it right and those that did not.

'Getting it right', in this context, means avoiding mistakes, and what those mistakes might have been can best be seen by looking at the experience of railways outside the Soviet sphere. For some, the new traction was a quick fix, a chromium-plating of dysfunctional basics and a way to avoid new thinking. Several US railroads fell into this trap and were weakened rather than strengthened by the new investment before finally being swallowed up by their stronger brethren. Other railways, particularly state-owned systems in the Third World, eagerly bought diesels without providing money for back-up in the form of maintenance facilities, trained personnel and, crucially, spare parts. The result can still be seen in lines of rusting diesel locomotives and, sometimes, the return of steam locomotives to fill the gaps.

In North America the large number of private railroads meant a variety of approaches. On the one hand, rather like airlines going for wholesale adoption of jet propulsion, some railroads chose a prompt total dieselization for emotional or publicity reasons, despite the financial burdens this imposed. At the other extreme was the Norfolk & Western RR, most of whose traffic was coal and most of whose clients were coal companies. This company postponed dieselization for a decade, before finally accepting that diesels would bring reduced costs. By then, diesel locomotives were better value for money. Partly thanks to this careful husbanding of resources, the N&W went on to purchase less successful companies and survived to become one of the current 'big five' US railroads.

Most of the continental European countries are reckoned to have managed the transition wisely. In France the government's policy was to electrify steadily and entrust the diminishing non-electrified lines not so much to diesel traction as to the remaining modern steam locomotives. In Germany there was more dieselization, but here again the more modern steam locomotives were retained for some years, cascaded down to the less important routes as modernization squeezed them off the more intensive services.

So what are the decisions that railway boards and governments make in order to bring about the change? First comes the decision to actually introduce the new traction; this is no light step, because the new locomotives need specialized facilities (and considerable infrastructural investment in the case of electric locomotives). The motivation here is the urge to modernity (rarely mentioned, possibly because it is taken for granted) and, most emphasized and supportable by numerical data, reduction of operating costs. Even outside Russia the increase of traffic capacity claimed for the new motive power can be a motivation, although this argument, if deployed at all, usually has force only when substantial traffic growth is expected.

Having compared this cost reduction with the investment required, the decision is taken to go forward. Then comes the question of timescale; almost always an immediate start is accepted, but the speed of transformation is controversial, and can be linked with the choice between total and partial modernization, the latter choice implying a continued though limited role for the steam locomotive. Finally, when electrification is chosen, there is the question of which lines to electrify and which to leave to diesel (maybe steam) traction, the salient point here being that electric locomotives cost about half the price of diesels, but require massive investment in electricity supply and distribution. In countries where some lines are already electrified, a decision is needed on whether to adopt the new high-voltage alternating current systems (more efficient, and much less demanding of substations and copper) even though they are incompatible with the old.

All these thought processes were travelled by Soviet railway managers, economists inside and outside the railway industry and, finally, by the party and government decision-makers in the 1950s. By that time their choice was easy because they no longer had much choice: either the railways went through their traction revolution or economic growth would choke. The choice facing British and most European railways was less clear-cut – there was at least a choice between modernization or increased subsidy, and a wrong choice would not be catastrophic.

Neither the British nor the Soviet modernization plans actually specified a deadline for the elimination of steam operation, as opposed to steam locomotive construction. Although the final steam run and the subsequent complete closure of the steam establishment brings a definite one-off financial benefit, the premature write-off of young steam locomotives is not necessarily balanced by the reduced running costs of the new traction. What seems to have been ignored (or, perhaps, deliberately obscured) is that the cost benefits of the new traction over the old diminish and may even reverse as the transformation proceeds, simply because the first phase of modernization sees the new locomotives allocated to the most intensive duties, but when it comes to replacing traction on low-volume lines the enhanced availability and productivity of the expensive new units is devalued. A good case can be made for keeping some steam units in service until such time as their maintenance costs become prohibitive. Most railways, not always for the right reasons, did just this. Some US railroads, the British and some Third World systems did not, and misinvestment comes at a price, even though the precise price may be hard to pin down. In the USSR the slower, and usually wiser, path was taken; in Britain the steam locomotive lasted for only a dozen years after the decision was taken to eliminate it, whereas in Russia it lasted more than double that time.

These considerations might just possibly give some validity to the view that Kaganovich was propounding in the mid-1950s. During a long discourse to the railwaymen's conference in 1954 he included a passage welcoming new forms of traction but pointing out that there was some life yet in the steam locomotive: 'I am against those who imagine we shall have no steamers.'[10] In 1956, according to several sources, he was the only member of the Presidium to oppose the railway electrification plan. Allusions to this crop up from time to time in articles. A 2006 example is in an interview with P.M. Shilkin, who was in charge of electrification works at the time: 'the well-known opponent of diesel and electric locomotives Lazar Kaganovich called this electrification plan wrecking'.[11] Again, in 2007, a former journalist of the railway newspaper *Gudok* reminisced:

> We were in the middle of the twentieth century, and the great man was laying down that the main locomotive of Soviet Railways was and remained the steam locomotive. Maybe here and there electric and diesel locomotives might have their attractions but we soviets had our own general line and technical policy for transport: the steam locomotive.[12]

Khrushchev later utilized this indication of conservatism in his political campaign against Kaganovich, but how far he was justified in doing so would depend on what Kaganovich actually said; until that is determined it might be equally arguable (just!) that Kaganovich's was a lonely voice of reason.

Technology transfer

Foreign technology had much to offer, but the responses of the respective governments were widely different. The USSR had long seen the benefit of purchasing foreign models for study, and in the 1930s had bought steam locomotives from both Britain and the USA with the aim of assimilating best practice. The 1940s diesel acquisitions had been made less to meet a wartime need than as a way of making acquaintance with a promising new form of traction. Britain had always been less interested in foreign practice, even though the study of US design and the acquisition of two French locomotives had enabled its Great Western Railway to revolutionize British locomotive technology before the First World War. In the 1950s the Soviet railways were able to acquire several advanced Western locomotives, but the British railways were denied a similar opportunity.

At that time the USA was a clear leader in diesel locomotive design, manufacture and operation. To some extent the Soviet railways acquired part of this know-how from their American imports of the 1940s (although it has to be said that those imports were far from being the last word in diesel technology). British engineers, meanwhile, were conscious that General Motors had a massive record of success and experience, but, in effect, they were not permitted to import US diesels either as samples or as part of the general dieselization programme: import of US locomotives could be prevented simply by government refusal to make available the foreign currency. Several reasons for the non-acquisition of GM locomotives were advanced at the time, including the slowness with which GM responded to suggestions of a licensing agreement, the different dimensional restrictions of US and British railways, and the alleged fuel inefficiency of the two-stroke engines that GM was currently using. None of these reasons seems particularly convincing. Misplaced nationalism and lobbying by the British locomotive industry seem more likely factors. It would be several decades before GM locomotives were imported, not by British Rail but by new private companies.

The most noteworthy Soviet search for foreign technology was for electric locomotive data, as this was a time when rapid advances were being made, particularly in the use of high-voltage ac and in the accompanying design of rectifiers. In 1957, 40 ac units were ordered from Alsthom in France. The quantity signified that this was not a purchase made for purposes of dissection, but that it was intended to hand over a section of line to this class to study the problems that could arise in daily operation (they went to the Krasnoyarsk Railway). Two years later 20 locomotives were ordered from Siemens, equipped with the latest silicon rectifiers; the Alsthom units with their ignitron rectifiers had become outdated in just two years. With dc locomotives there was evidently less to learn, although a couple of such units were ordered from the German Democratic Republic. (The hundreds of Skoda-built units, both ac and dc, that were ordered at this time for passenger services, were not intended for study but for filling a deep production need.).

Both the British and Soviet railways dabbled in German-inspired hydraulic drives for diesel locomotives. In Leningrad the Proletariat Works was reorientated to produce mainline diesel-hydraulics, and one of the constituent regions of British Railways adopted such locomotives wholesale. Both eventually came to the same conclusion: the concept was viable, but not superior. Both in due course abandoned diesel-hydraulics, and the Leningrad works reverted to its previous tasks. In this field the USA did better: two railroads ordered prototypes and after trials decided not to place further orders.

Locomotive procurement

The usual Soviet practice when introducing new designs was to build a handful of prototypes, which were then subjected to months if not years of monitored running. This practice continued in the years of modernization. The mainline diesel locomotive chosen for mass production was already through its early phases when modernization began, and for the first few years it was the sole design; almost 6,800 units were built eventually, with the works at Kolomna, Kharkov and Lugansk cooperating in their construction. Electric locomotive construction could not be similarly concentrated on a single model because there were two systems of electrification and in any case technological advance was very fast. The perceived need for a greater output was met on the one hand by entrusting Czechoslovakia with all electric passenger locomotive construction (although Soviet passenger locomotive designs had been prepared, the Ministry of the Electro-technical Industry had declared itself unable to find the capacity to build them),[13] and by the establishment of the new electric locomotive works at Tbilisi.

The British experience was very different. Traditionally, most locomotives had been built in the workshops of the railway companies, which were often more efficient than those of private locomotive builders, which concentrated on the export market. Because the railway workshops had little experience with diesel traction, and thanks to pressure from the Locomotive Manufacturers' Association, a large proportion of the initial new orders was channelled to the private builders. For political reasons the famed North British Locomotive Works was given big orders even though it had little experience of non-steam work; the excess weight of the resulting products was said to be about 50 per cent.

The initial order was for 174 units of various types divided among 22 works, which it was intended to try out for three years before choosing which designs and which builders would get the future orders. However, a worsening national financial situation soon induced the government to change the plans. Future electrification was curtailed and there was to be a drive for complete dieselization of the other lines. The result was that large numbers of diesel locomotive types were ordered before they had been properly evaluated. Many of these turned out to be mediocre performers.

The result was that BR had about 40 different diesel locomotive designs on its books, when three or four were all that were necessary. Apart from the mediocrity of many designs, the variety meant not only problems with repairs and spare parts but also necessitated extended training of locomotive crews so that they could cope with all the designs they were likely to encounter.

Some conclusions

In the last half of the twentieth century railways all over the world were making the transition from steam traction to electric and diesel locomotives. This was a fundamental transformation, affecting the lives of railway workers, requiring substantial capital investment, securing the future of railways against competition, and making it possible to provide a better service for railway users. Different railway systems handled this change each in their own way. Some, it can now be seen, did it well and others less so. This is not to say that any failed; it required no great talent to buy new locomotives and scrap old ones. Nevertheless it was possible to make a mess of things, and it was also possible to win every trick. From the evidence presented above it seems that, in the USSR, the process was managed with a more steady and effective hand than in some other countries – notably Britain, where the changeover took place in an ambience of dithering and flapping.

As to the basis for decisions, the Soviet situation benefited from the general acknowledgment that technical modernization had one priority: prevention of a transport breakdown. The British, on the other hand, never quite settled on what the priority was. For the Treasury, it was to reduce or avoid financial losses. To others it was to apply available new technology and see what happened. At other times it was to enable the railways to win back traffic. And overall there was the annual Treasury review, which meant that sums planned for investment could be spirited away. Although in Britain in this crucial period the same party remained in government, the composition and the concepts of that government could fluctuate, and one transport minister could succeed another at short intervals; the tenure of British transport ministers could sometimes be measured in months.

In contrast, B.P. Beshchev, a professional railwayman, headed the Soviet railway ministry from 1948 to 1977. Such longevity could, of course, engender conservatism. On the other hand it ensured that transport ministers had the self-confidence imparted by deep knowledge of their subject. In this limited sector at least, the Soviet structure of decision-making and execution seems to have worked better than the British. This may be a provisional judgement because, whereas on the British side there is ample published evidence about how decisions were arrived at, there is less material on the Soviet side. However, on the face of it, the Soviet practice of letting the involved ministries and Gosplan argue out the details, reserving the participation of the Presidium to the fundamentals, worked out far better than the British structure with the transport minister in the cabinet, where quite detailed questions were decided. And re-decided.

Notes

1 V.A. Rakov, *Lokomotivy otechestvennykh zheleznykh dorog 1845–1955*, Moscow: Transport, 1995, p.374.
2 *Narodnoe khozyaistvo SSSR v 1958g*. Moscow: Gosstatizdat, 1959, p. 553.
3 J. Majumdar, *The Economics of Railway Traction*, Aldershot: Gower, 1985, is a very full discussion of how railways evaluate the merits of different forms of traction. It is based on another big state-owned system, Indian Railways, but has a few sections devoted to the Russian experience.
4 A good example, replete with numerical data for the Omsk, Tashkent and other railways, is R.Ya. Shvarts, 'Opyt ekspluatatsii elektrovozov i teplovozov', in E.A. Ashkenazi (ed.), *Metody ekspluatatsii tyagovykh sredstv na transporte*, Moscow, 1957, pp. 63–75. Among books, two of the most relevant are A. P. Mikheev, *Effektivnost' elektricheskoi i teplovoznoi tyagi na zheleznodorozhnom transporte*, Moscow, 1960, and S.S. Ushakov, *Poysheniye effektivnosti novykh vidov tyagi*, Moscow: Transzheldorizdat, 1959.
5 I.S. Sal'nikov, 'Elektrifikatsiya zheleznykh dorog', in S.S. Minsker and L.I. Krishtal' (eds), *Voprosy razvitiya zheleznodorozhnogo transporta*, Moscow: Gos. transportnoe zheleznodorozhnoe izd-vo, 1957, p. 21.
6 *Direktivy XX s"ezda KPSS po shestomu pyatiletnemu planu razvitiya narodnogo khozyaistva SSSR na 1956–1960 gody*, Moscow: Pravda, 1956, pp. 21, 41.
7 *Direktivy*, p. 41.
8 Sources for this paragraph are A.A. Fursenko (ed.), *Prezidium TsK KPSS 1954–1964, T.2*, Moscow: Rosspen, 2006, pp. 174, 175–7, and A.A. Fursenko (ed.), *Prezidium TsK KPSS 1954, T.1*, Moscow: Rosspen, 2004, pp. 84–5, *Protokol* of 5 January 1956.
9 A blow-by-blow account of this period is to be found in T.R. Gourvish, *British Railways 1948–1973*, Cambridge: Cambridge University Press, 1986, pp. 256–304.
10 The full text is in L.M. Kaganovich, *Uluchshit' rabotu i organizovat' novyi podem zh.d.transporta*, Moscow: Gos. izd-vo politicheskoi literatury, 1954, p. 70, and the essential extract can be found in J.N. Westwood, *A History of Russian Railways*, London: Allen and Unwin, 1964, p. 275.
11 'Tekhnicheskaya revolyutsiya', *Gudok*, 3 Feb 2006, p. 4.
12 L. Vichkanova, 'Sochinenie slavnykh pochinov', *Gudok*, 4 July 2007, p. 7.
13 V.A. Rakov, *Lokomotivy otechestvennykh zheleznykh dorog 1956–1975*, Moscow: Transport, 1999, p. 48.

13 From Khrushchev (1935–1936) to Khrushchev (1956–1964)

Construction policy compared

R.W. Davies and Melanie Ilic

Anyone who has walked around the inner suburbs of Russia's major cities and towns will have noticed what are known locally as *khrushchevki* (five-storey blocks of flats). The mass housing programme and the provision of the single-family flat were central to Khrushchev's policy making in the late 1950s and early 1960s. Khrushchev's interest in construction and urban development, however, stretched back to the 1930s, when he was secretary of the Moscow Communist Party. A well-documented meeting on construction was convened in Moscow in December 1935, which discussed the current difficulties being experienced in the building industry, and its future. This chapter examines the economics of construction from the 1930s and how changes made in the building industry impacted on the principles and practices that underpinned both house-building projects and the construction industry in the Khrushchev period. Drawing parallels between the 1930s and the years from 1956 to 1964, when Khrushchev was in office, it investigates three specific areas of the economics of construction: the stabilization of the labour force, rationalization and organization of the building sector, and oversight of construction costs.

The background: rapid growth, 1928–1936

During the early 1930s the Soviet building industry expanded extremely rapidly.[1] The number of people employed increased from about one million in the economic year 1927/28 to over three million in 1932.[2] This enabled the doubling of the capacity of the capital goods industries during these five years.

The labour force in the industry had always been mainly seasonal. The new building workers consisted very largely of peasants without industrial skills migrating from the countryside. This was a labour force in flux. The annual turnover reached 306 per cent in 1932; the average worker remained in the same post for only four months.[3] Many of the new workers were engaged in purely manual jobs such as earth moving. They would move to other building sites in search of better conditions, return to their village in the winter months or transfer to jobs in industry, where conditions were better. A leading building official stated at the end of 1935:

Until the present day building has added to its trained workers from seasonal workers and collective farmers. A huge section of these cadres came to the site without any qualification, and were trained up as fitters, workers with concrete, or as carpenters, but were then lost to construction, moving over to become permanent workers in new factories.[4]

The authorities made valiant efforts to regularize recruitment and stabilize the labour force by signing contracts with collective farms to acquire workers from them for a season, or one or two years. Most peasants, however, moved into building spontaneously and independently. According to Kaganovich: 'Rural customs and habits (*derevenshchina*) are strong. Putting it bluntly, some of the individual peasants and even some kulak elements have found in railway building "a place for isolating themselves from collectivization" [*Laughter*].'[5] In the large towns many builders undertook additional work in the evenings for high pay (so-called *shabashki*) and looked on their main work as a 'sad necessity', undertaken in order to obtain ration cards and accommodation. They left their main work early, borrowing tools from the site, to do what they regarded as their 'properly paid' work.[6]

Very little building machinery was available. One prominent official recalled:

I remember the construction of the Nizhnii Novgorod vehicle works, when it was a great event to receive a dozen pieces of building machinery from abroad ...

Only two years ago [i.e. in 1933] it was a uniquely difficult problem to obtain 15 excavators from the Moscow-Volga canal. The excavators, mostly foreign, and varying by type and capacity, had to be brought together from various sites.[7]

Building was carried out by a large number of organizations under different auspices. The economic commissariats established firms for carrying out major projects, working directly for the branch of the economy concerned. They were usually temporary. When a project was completed, efforts were made to transfer staff and building equipment to a similar site, with limited success. According to the head of Gosplan, 'With the old direct-labour arrangements, we were unable to secure a permanent building staff. The project was completed, the personnel dispersed, the accumulated experience was lost. Only the leading personnel were retained, and then not always.'[8]

In *Narkomtyazhprom* (the People's Commissariat for Heavy Industry) a Chief Administration for the Building Industry (*Glavstroiprom*) was responsible for the general oversight of construction in heavy industry. It managed some building organizations directly; it was responsible, for example, for the construction of aircraft factories.[9] Much major building, however, was managed by the chief administration of the industry concerned and was devolved in turn to a variety of organizations. The head of *Glavenergoprom*, responsible for the electric power industry, reported in 1935:

This year the overwhelming majority of our building work (90 to 95 per cent) has been carried out by direct labour – by district administrations, energy combines, individual power stations, and in some cases by building organizations directly subordinate to us. Our building experience shows that with such direct labour it is very difficult to establish a strong organization, Work is poorly mechanized, the achievements of the best sites are not generalized, overheads are high, and financial discipline is weak.[10]

The arrangements for the production and supply of building materials were equally complex. Cement was produced at a relatively small number of factories managed by *Narkomtyazhprom* and allocated centrally. Most materials, however, were produced partly in factories attached to the main building sites and organizations, partly by the republican People's Commissariats for Local Industry, and partly by small factories attached to the local soviets, and by artisan cooperatives. In 1935, 1,700 million of the total production of 5,959 million bricks were manufactured by the People's Commissariat for Local Industry of the Russian Socialist Federative Soviet Republic (RSFSR).[11] The building sites acquired their materials partly from their own factories, and partly by purchasing them from elsewhere.

As seen by the authorities, the crucial problem was the high cost of construction and of its three major components: labour, materials and capital equipment.[12] In the early 1930s construction costs rose considerably. The huge increase in the number of unskilled workers resulted in a substantial fall in labour productivity, greater than in the rest of industry. Despite this decline, money wages rose inexorably: by 14.9 per cent in 1931, 21.4 per cent in 1932, and 8.7 per cent in 1933, an increase of 51.7 per cent over the three years. Throughout these years, in spite of the decline in average skills, building wages were approximately the same as in industry as a whole.[13] Food, accommodation and other facilities were far poorer than in other industries. One major building site employed 18,000 workers but had only 300 places in the canteen. 'High earnings', according to one building official, 'have been until now the only attractive force for a building worker.'[14]

The slowdown in the rise in wages in 1933 was accompanied by a drastic cut of 25 per cent in the number of persons employed in the industry, from 3.1 million in 1932 to 2.4 million in 1933.[15] In 1933, labour productivity increased from its very low level, and it increased further in the following two years.

Building materials, the second major element in construction costs, also increased in cost in the early 1930s, whether purchased on central allocations, or produced in-house by the building industry itself, or purchased from local or artisan industry. By 1935 the costs of cement and other centrally allocated materials had been brought under control. According to Mezhlauk, cement cost 47 rubles a ton to produce, less than the planned 48 rubles 70 kopeks.[16] The prices of locally produced materials varied considerably, and the building sites frequently complained that they were higher than the cost of the same materials produced by their own organization. The head of the administration of the Red

Army responsible for building accommodation complained that local industries 'supply us with building components which are 20 to 25 per cent dearer than if we make them in artisan fashion on our own sites'.[17]

In 1935 the cost of local materials continued to rise. This is not surprising, as the abolition of rationing meant that food prices rose substantially for the workers in the industry, leading to an increase in wages. According to Mezhlauk, 'Voronezh region is the only region in the USSR which achieved a reduction in the cost of bricks'.[18] It was estimated that the cost of building materials as a whole increased by about 3 per cent in 1935.[19]

The cost of capital equipment, the third main element in construction costs, is particularly difficult to measure both because of its heterogeneous and changing nature and because a large proportion of it was imported in the early 1930s, and charged to the building industry at varying prices. In general, the cost of internally produced equipment, like the cost of all machinery, increased relatively slowly, but there were important exceptions. According to the head of the Azov iron and steel project, the cost of a ton of rolling-mill equipment had risen since 1930 from 800 to 3,000 rubles.[20]

In 1935, for the first time since 1930, the rise in construction costs was halted and perhaps reversed. Official estimates of the decline in the cost of pure building (that is, excluding the cost of equipment) varied from 1 to 4.2 per cent.[21] However, this reduction was much smaller than the planned 15 per cent!

The effort to modernize and reform the industry, 1935–1936

By the mid-1930s the authorities were making a desperate effort to achieve three competing objectives: to increase consumption, to switch resources to defence, and to continue to build up the capital stock of both heavy and light industries, and of the railways.[22] The Stakhanovite movement was directed primarily at the more efficient use of the existing capital stock; the campaign to reduce construction costs sought to obtain more resources per unit of investment.

A well-publicized discussion of building costs formed a major part of the general discussion on the building industry, which was widely publicized throughout the mid-1930s. The climax was the *Conference on Questions of Construction held in the Central Committee of the All-Union Communist Party (Bolsheviks)*. This took place between 10 and 14 December 1935, a few days before the plenum of the central committee which discussed Stakhanovism. It was attended by 350 prominent managers of large-scale building projects and from the building materials industries. Nine members of the Politburo took part, including Stalin, who was present on the last day. It heard reports from Mezhlauk, head of Gosplan, and Ginzburg, head of Glavstroiprom, the Chief Administration of the Building Industry of Narkomtyazhprom. Molotov summed up the proceedings, and the conference was addressed by 45 speakers, including Ordzhonikidze, Kaganovich, Mikoyan and Khrushchev.[23]

The conference was held only eight months before the Zinoviev–Kamenev trial, the first major public trial of the 'Great Purge', but it was almost free from attacks on the former oppositionists. Molotov criticized '1928 views on industrialization', but without mentioning Bukharin by name, and Khrushchev made a brief conventional attack on 'the Trotsky–Zinoviev opposition and the right-wing opportunists', but did not castigate them as class enemies.

Khrushchev, rapidly rising in the party hierarchy, was at this time secretary of the Moscow regional and Moscow city committee of the party. His was a major speech, longer than all the others except those by the two rapporteurs and Molotov.[24] It was the liveliest speech at a lively conference. The audience interrupted with applause and comments on no fewer than 27 occasions; those interrupting included Molotov, Lobov, Ordzhonikidze, Kaganovich, Ginzburg and Mezhlauk. Khrushchev concentrated on Moscow and its grandiose housing plan.[25] Yet, as Kaganovich called out from the platform, 'If an issue is raised for Moscow, it is also raised for the whole Soviet Union [*Applause*].'

Housing was certainly not completely neglected in the 1930s. The comprehensive provision of mass housing was always strongly emphasized in party doctrine. On pragmatic grounds, the commissariats, in order to attract workers to their factories – particularly skilled workers – had to offer them a place to live. Yet the housing programme was consistently under-fulfilled. The industrial and other commissariats gave lower priority to housing than to their major economic tasks; local government lacked resources; and the cost of house building rose inexorably. As a result, the urban population grew more rapidly than the stock of housing. Against this background, Khrushchev called for the expansion of housing. At the heart of his speech was a call for 'new advanced methods of construction and of the production of building materials'.

In Moscow this required that the large number of small building trusts working by direct labour must be replaced by specialized trusts working by contract. Most Moscow housing was constructed by the various commissariats for their own workers, and only 25 per cent by the Moscow soviet. Khrushchev was not so unwise as to advocate the transfer of house building from the commissariats to the soviet. Instead, he called on each commissariat to establish specialized housing trusts working for them under contract. He also proposed that all Moscow housing should be supervised by the soviet (Moscow did not possess this right at present) and form part of the state plan, receiving specific allocations of materials and equipment.

Khrushchev insisted that building materials and components – including bricks, doors, flooring and baths – should in turn be produced by specialized factories, attached either to the commissariats or to their building trusts, or to a specialized commissariat responsible for this production. *Narkomles*, the People's Commissariat of the Timber Industry, for example, should produce all components manufactured from timber.

The watchword for all these developments was mechanization. In the building process itself, excavators and lorries should replace manual labour, and the production of building materials and components should also be carried out by machinery: 'The only way out is the *mechanization* and *industrialization* of our construction.' Ultimately, 'the construction of a block of flats should be increasingly transformed into the assembly of building components'.

Khrushchev also called for the thorough revision of the system for preparing building projects and estimates. Good standards must be approved, with the participation of the architects, and applied throughout the industry. At present, 'putting it bluntly, architects often failed to consider the cost of building ... did not take care with the Soviet *kopek*, and their projects are completely detached from questions of economy'. Estimates must be compiled so that no supplementary resources are required (London Olympics 2012 please note): 'It is time to put an end to the criminal contempt for the approved estimates.'

Khrushchev's assertions and proposals were in accordance with two decrees promulgated on 11 February 1936. The major decree of *Sovnarkom* and the party central committee, which was already being drafted at the time of his speech, 'On the Improvement of Construction Activities and the Reduction in Construction Costs', was simultaneously approved by the Politburo by poll.[26] On the same day, the Council of Labour and Defence adopted a supplementary decree, 'On the Reduction of the Cost of Production of Building Materials and Components'.[27]

Khrushchev concentrated on the problems of providing accommodation for Moscow residents. The two decrees were primarily concerned with the much larger problem of industrial building. They resolved that building organizations should be established for the coal, hydro-power, thermal electric power, iron and steel, and oil industries, and that in each industry these should be supplemented by specialist building organizations for heating, sewage and water. Building materials should also be manufactured by specialized trusts, though these would often work for the major industrial building organizations. In all these activities, contracts between the client and the builder would replace direct labour. On each site a 'general contractor' would be responsible for the work as a whole, itself signing contracts with the specialized trusts.

Both decrees strongly emphasized the importance of mechanization. Some progress had already been made with the replacement of manual labour by machines. The total number of excavators increased from 700 to 1,000 in 1935, and the number of cranes from 310 to 330.[28] The *Sovnarkom* and central committee decree required that 'up to 60 per cent' of earth work should be mechanized in 1936, and set similar targets for quarries and for the transport and production of their output.

The decrees paid much attention to the need for economy. Under the influence of Stakhanovism, labour productivity in construction was planned to increase by 'at least 30 per cent' in 1936. To encourage this, following the example of industrial production, the output norms for workers in the industry should be substantially raised, so that the wages received per unit of output would be reduced. The provision of finance would be tightened up. While finance arrangements would be more flexible as a result of providing each building organization with its own working capital, the role of banks in controlling expenditure would sharply increase. According to the decree: 'Payment should be made in accordance with invoices based on acceptance certificates (*akty priemki*) for the work carried out, approved by the client.' The invoices should be prepared on the basis of the prices fixed in the cost estimate attached to the technical project, and reduced by the planned reduction in building costs.

With these reforms, the authorities hoped to secure the long-anticipated reduction in the cost of investment. In August 1935 the government agreed to reduce costs in 1936 'by at least 8 per cent in comparison with the estimate costs of 1935'.[29] The decree of 11 February 1936 proposed that pure building costs should be reduced by at least 14.5 per cent as compared with the estimate costs of 1935 and that all investment costs (including the cost of capital equipment) should be reduced by at least 11 per cent.

War preparations hinder progress, 1936–1940

The grandiose plans of 1935–36 were not achieved until after the economy emerged from the Second World War. The switch to defence construction and armaments production, together with the disarray resulting from the sweeping arrests and frequent executions of major and minor industrial officials, prevented the expansion of the building industry. In 1936 construction as a whole increased rapidly. In 1937 and 1938, however, total 'pure construction', measured by the Powell index, declined, and after a slight recovery in 1939 and 1940 was still 14 per cent less than the 1936 peak on the eve of the war (see Table 13.1). From this smaller total, a substantially larger proportion was allocated to construction work for the armed forces, to the construction of armaments factories, and to the provision of additional facilities for the production of armaments in existing factories.[30]

While 'pure construction' in the whole economy declined in the period from 1936 to 1940, investment in capital equipment increased. While accurate figures have not been available, the increase is indicated by the growth of the output of the machine-building and metal-working industries by 75 per cent in these years.[31] Very large increases took place in the production of specialized machine tools, mainly intended for new industries.[32]

By 1940 labour productivity (output per person employed), measured by dividing the Powell index by an index based on the number employed, had increased only slightly as compared with 1928 (see Table 13.2). During the first five-year plan, labour productivity declined greatly. The labour force increased much more rapidly than output. After 1932, however, the number employed in building declined while output greatly increased, so on the eve of the war productivity returned to approximately the 1928 level.

Despite the optimistic policies of the mid-1930s, housing remained a relatively neglected sector. During the second five-year plan (1933–37), only 26.8 million square metres of living space were completed, as compared with the plan of 64 million square metres. The housing completed per year declined throughout the period (see Table 13.3).

Table 13.1   The Powell Index of Soviet construction, 1928–1958
(1927/28=100; measured in 1937 prices)

1927/28	100
1929	124
1930	161
1931	174
1932	173
1933	156
1934	188
1935	232
1936	313
1937	273
1938	269
1939	275
1940	275
1945	117
1946	175
1947	216
1948	267
1949	368
1950	422
1951	480
1952	555
1953	591
1954	651
1955	719
1956	758
1957	826
1958	902

Source: calculated from data in R.P. Powell, *A Materials-input Index of Soviet Construction, Revised and Extended*, Santa Monica, CA, 1959, RAND Corporation Memorandum, RM-2454

Note
Following the practice of Simon Kuznetsk and others, Powell prepared this index of materials inputs into construction as a proxy for the output of the construction industry. For details, and discussion of why this index is preferable to other indices of the growth of construction, see R.P. Powell, *A Materials-input Index of Soviet Construction*, Santa Monica, CA, 1957, RAND Corporation Memoranda, RM-1872 and 1973.

Table 13.2 Comparison of the growth of the number employed in construction with the Powell Index, 1928–1958

	Personnel employed in construction (thousands)	Personnel employed in construction (1928=100)	Powell index (1928=100)
1928	984[a]	100	100
1932	3150	320	173
1940	2567	261	275
1945	2343	238	117
1950	4087	415	422
1956	5212	530	758
1958	5933	603	902

Source: labour employed derived from *Trud v SSSR*, Moscow: Statistika, 1968, p. 121; Powell index, see Table 13.1

Note
a Includes self-employed, which are not available for later years, but are believed to be small in number.

Table 13.3 Million square metres living space constructed 1933–1937

1933	1934	1935	1936	1937
7.2	6.0	4.6	5.5	3.5

Source: RGAE, 1562/1/1039, 79 (1939?)

War losses and growth, 1941–mid-1950s

During the Second World War, the German occupation of a large part of Soviet territory and widespread destruction elsewhere resulted in a vast decline in capital construction. 'Pure construction' had declined in 1945 to a mere one-seventh of the 1936 level. Recovery, however, was extremely rapid. In 1949, only four years after the war, 'pure construction' was already 17 per cent greater than in 1936, and at the time of Stalin's death in 1953 it was already 89 per cent higher than in 1936. Between 1950 and 1956, the Powell index increased by 79 per cent while the labour force increased by only 28 per cent. By 1956 the Powell index had already reached nearly 2.5 times the 1936 level, and was 7.5 times as large as in 1928, the year in which Soviet industrialization began. By the mid-1950s, following a large increase in the production of building machinery in the first decade after the war (see Table 13.15), despite the fact that the industry still exhibited many of the features characteristic of the pre-war period, it had made substantial progress towards modernization. While much manual labour was still involved, many major building processes were now mechanized.[33] Substantial resources had been devoted to training the labour force. Over two million trainees completed building trade school between 1940 and 1955.[34] Between 1941 and 1955, the percentage of the building labour force with higher or secondary specialized education in construction, engineering or project making increased from 4.5 to 6.6 per cent.[35] The foundations had been laid for Khrushchev to resume the ambitious plans for the industry set out at the Building Conference 20 years earlier.

In the years before Stalin's death, substantial efforts had already been made to repair the immense war-time damage to the stock of housing. Thus, in the four years from 1946 to 1949, 36.6 million square metres of living space were brought into operation, exceeding the living space completed in the five years of the second five-year plan.[36]

In 1949 Stalin spoke at the Politburo on the plan for the reconstruction of Moscow: 'Without a fine capital there can be no state. We need a beautiful capital, which everyone stands in awe … a capital which is a centre of science, culture and art.'

In this context, he called for blocks of flats which were eight to ten rather than four or five storeys high, and for 20 to 25 per cent of blocks of flats to have 12 or 14 floors. The approaches to the city should be lined with new eight- or ten-storey blocks of flats, 'which delight the eye'. When this mass building was complete, Stalin continued, it was essential to move on to the completion of the House of Soviets (postponed by the war).[37]

From the 1954 builders' conference to the 1957 housing decree

A major All-Union Conference of Builders, Architects and Workers in the Building-Materials Industry, in the Construction Machinery and Road Machinery Industries and in Design and Research Organizations was held in Moscow from 31 November to 7 December 1954. This was the first builders' conference of national significance to be convened in many years and was one of several held in the decade following Stalin's death. The Third All-Union Meeting on Construction was convened from 10 to 16 April 1958 and an All-Union Meeting on Urban Construction met from 7 to 9 June 1960.[38]

1954 Conference

On 3 July 1954, Khrushchev set out his thinking on the current state of the architectural profession and the construction industry in a memorandum sent to the presidium of the Central Committee and to the deputy prime ministers of the USSR Council of Ministers.[39] At the builders' conference later in the year, Khrushchev took the opportunity to set out publicly his own agenda for the future development of the construction industry and housing policy.[40] His policy, stressing the need for rapid development of construction without frills, was in implicit contrast to Stalin's. He concentrated on the need to industrialize the construction sector, especially by increasing the use of new building materials and techniques, and particularly those employing prefabricated reinforced concrete. This would, he claimed, bring about savings in the manufacturing and assembly processes, reduce overall costs and speed up the rates of project completion.

Khrushchev also called for the restructuring of the organization and financing of the construction sector, a process begun after the 1935 builders' conference, but that had undergone little change since the late 1930s, despite significant development in construction techniques since then. The Ministries responsible for major industrial projects (such as iron and steel, electric power, nuclear bombs and the space programme, for example) each undertook a large part of housing development. As a result, there were a large number of separate house-building organizations in one single location. Reform, with mixed results, sought to bring all of these under local management. When it was established in April 1954, *Glavmosstroi* (the Moscow Construction Board) 'amalgamated under one body 53 building trusts, 225 general and specialist building contractors and over 600 productive and auxiliary enterprises, previously controlled by 44 different ministries and departments'.[41]

The sector required greater levels of specialization to increase productivity and improve quality. Khrushchev called on the construction sector to become more flexible, mobile and efficient in its work. He argued that the very design of new buildings should be standardized, and that this should take place alongside a reduction in the use of costly and unnecessary adornments. Particular attention should also be paid to other ways of reducing costs, including the finishing of basic construction components in the factory rather than on site. Building projects should be more closely coordinated and construction plans should not exceed available capacity. Greater attention was to be paid to the training of a skilled workforce and reduction of labour turnover, which would result in increasing wage levels within the industry.

1957 Decree

Once he was in office, Khrushchev's proposals for the renewal of the Soviet Union's housing stock were set out in the Central Committee and Council of Ministers decree 'On the Development of Housing Construction in the USSR' issued on 31 July 1957.[42] The decree acknowledged two particular processes that now led Khrushchev to place housing high on his economic and social policy agenda: first, the devastating loss of residential accommodation suffered by the Soviet Union during the years of the Second World War, which had only partially been addressed in the period of post-war reconstruction; and, second, the ongoing process of urbanization, which had seen the number of people living in towns and cities triple in the space of the past 30 years. The decree further acknowledged that the existing housing stock was inadequate and in a poor state of repair, particularly in rural areas. In addition, if the industrial economy was to expand, then suitable housing needed to be provided for the new labour recruits. The stated intention of the decree was to overcome the Soviet Union's housing shortage in the course of the next 10 to 12 years. Speaking at the Central Committee plenum in May 1958, however, Khrushchev declared that the housing shortage should be overcome in an even shorter timeframe than that initially envisaged by the 1957 decree.[43]

How was this to be achieved? Demonstrating an awareness of the problems experienced in the building sector under Stalin, the 1957 decree called on those responsible for construction to use their initiative in the exploitation of raw materials and industrial output available locally in order to keep production and transportation costs to a minimum. The actual design and manufacture of housing was to move to industrial methods of production, despite the low level of technology currently available in the construction sector. Housing construction under Khrushchev was characterized by new building methods, including the industrial manufacture of standardized large-scale blocks and panels, which could then be transported to the construction site for erection and fitting on a time-efficient and cost-effective basis. New building materials were to be exploited, including reinforced concrete, cement, breeze blocks and asbestos. Responsibility for the management of construction projects was devolved to local government bodies, collective farms, building cooperatives and individuals.

The 1957 decree also recognized that further industrial growth would be needed to support the housing projects. The new residences would each need to be fitted with a bathroom – with bath, toilet and washbasin, all requiring plumbing materials. The kitchen would need to be fitted with cupboards and cooking equipment. Radiators would be needed for heating, glass for the windows and slate for the roofs. Different materials would be required for flooring, including, wherever possible, locally sourced wood for parquet designs, and also linoleum. Wood would also be needed for window frames and doors (and in timber-rich regions, the houses themselves could be built from wood). Existing communications networks and electricity supplies would need to be extended to the developing residential areas. Once built, the new apartment blocks would require furniture and other fittings for their interior design.

While these economies led to the appearance throughout the USSR of similar-looking and aesthetically uninspiring blocks of flats, the most important factor in the housing campaign was the use of labour power and materials on an unprecedented scale. The total stock of urban housing increased by 77 per cent between 1955 and 1963 (see Table 13.4), which was more than double the rate in the eight years from 1928 to 1936.[44]

Despite the great progress in both housing and industrial construction under Khrushchev, there were many complaints about the quality of both construction work and the fittings in new buildings.[45] One contemporary report noted that, because of delays in forwarding plans, construction work was often begun before estimates were complete. The proposed volume of construction did not take account of available resources and, as a result, shortages of supply held up the building projects. Many buildings were poorly constructed, with badly fitting window frames and doors, low-quality walls, floors that buckled and plaster falling from the ceiling. Kitchens were poorly supplied with cupboards and there was no attempt to landscape the exterior of the buildings.[46] Insufficient attention was paid in some regions to water supply for the new construction projects, especially in areas where natural sources of water were located at a long distance from the blocks of flats and would require lengthy piping.[47] The supply of other

Table 13.4 Growth in urban housing stock, 1926–1964

	1926	1940	1950	1955	1960	1961	1962	1963	1964
Urban housing (million m²)	216	421	513	640	958	1017	1074	1130	1182
Average living space per urban resident (m²)	8.2	6.5	7.0	7.3	8.8	9.1	9.3	9.5	9.7

Source: G. Andrusz, *Housing and Urban Development in the USSR*, Basingstoke: Macmillan, 1984, p. 22; *Narodnoe khozyaistvo SSSR v 1967g.*, Moscow: Statistika, 1968, p. 124; TsSU, *SSSR v tsifrakh v 1967 godu*, Moscow: Statistika, 1968, pp. 27, 138

local services and facilities – including shops, schools and hospitals – was often considered inadequate and led to increasing burdens being placed on housewives and working mothers.[48] In the late 1950s, some of the blame for poor-quality work was directed towards young workers with low levels of qualification for work in the building industry.[49]

The labour force in construction

Crucial to the success of Khrushchev's housing policy were the expansion, stabilization and increased specialization of the workforce in the building sector. The labour force employed directly in construction on building sites expanded rapidly, from 2.9 million at the time of Stalin's death in 1952, to just over four million in 1957 and to almost 5.4 million in 1964 (an increase of 2.5 million workers in just over a decade). In addition, a further 1.5 million workers were employed in the construction sector in subsidiary occupations, transportation and housing services (see Table 13.5).[50] In the industrial sector, by 1960, a little fewer than 1.5 million workers were employed in the building materials industry (recorded separately from the construction sector in the labour statistics data). This marked an increase of almost half a million workers over the numbers employed in 1955[51] (see Table 13.6). The number of manual workers employed directly in construction increased by 74 per cent between 1953 and 1964, and the number of engineering and technical workers (ITR) increased by 163 per cent. The numbers of ITR in the building materials industry more than doubled in the ten years from 1955 to 1965. During the years that Khrushchev was in office, among the manual labour force on building sites, workers came to be increasingly employed in specialized areas of construction, such as concrete layers, welders, painters, carpenters, plasterers and electricians, rather than as general navvies, joiners or road and repair workers (see Table 13.7).

Table 13.5 Labour in construction, 1955–1964 (000s; index)

	B		E												O		Q	
			Building organizations															
			Building workers															
	All		All		Blue collar workers		ITR		White collar workers		Other				Auxiliary workers		Trans + zhilkom	
	000s	Index	000s	Index	000s	index	000s	index	000s	index	000s	index			000s	Index	000s	Index
1955	4953	100	3210	100	2814	100	221	100	104	100	71	100			714	100	1029	100
1956	5212	105	3567	111	3137	111	247	112	110	106	73	103			584	82	1061	103
1957	5513	111	4017	125	3527	125	280	127	122	117	88	124			494	69	1002	97
1958	5933	120	4442	138	3921	139	311	141	128	123	82	115			469	66	1022	99
1959	6226	126	4819	150	4256	151	355	161	136	131	72	101			421	59	986	96
1960	6555	132	5143	160	4554	162	385	174	140	135	64	90			437	61	975	95
1961	6642	134	5270	164	4638	165	416	188	148	142	68	96			432	61	940	91
1962	6596	133	5172	161	4502	160	443	200	153	147	74	104			435	61	989	96
1963	6723	136	5237	163	4544	161	461	209	157	151	75	106			444	62	1042	101
1964	6896	139	5370	167	4640	165	492	223	163	157	75	106			449	63	1077	105

Source: Trud v SSSR, Moscow: Statistika, 1968, p. 121

Note: Col B = Cols E+O+Q

Table 13.6 Labour in the building materials industry (000s)

	1955	index	1960	index	1965	index
Total	1000.2	100	1493.4	149	1630.1	163
Inc:						
Workers	876.9	100	1309.9	149	1392.0	159
ITR	61.9	100	104.9	169	146.2	236

Source: *Trud v SSSR*, Moscow: Statistika, 1968, pp. 86–7

Table 13.7 Distribution of jobs in construction: 1959, 1962, 1965 (000s)

		1 Aug. 1959	1 Aug. 1962	2 Aug. 1965
armaturshchiki	Fitters	37	34	36
betonshchiki	Concrete layers	113	133	174
buril'shchiki/pomoshchniki	Drill operators/assistants	15	14	17
gazovarshchiki, elektrovarshchiki	Gas and electric welders	65	104	135
dorozhnye rabochie, mostovshchiki	Road and bridge builders	48	58	60
zemlekopy	Navvies	135	108	97
izolirovshchiki	Isolators	26	37	49
kamenshchki, pechniki, ogneuporshchiki, truboklady	Bricklayers, kiln operators, refactory workers, pipe layers	286	323	359
krovel'shchiki	Roofers	20	24	36
malyary	Painters	148	199	240
mashinisty, motoristy/ pomoshchniki	Drivers/assistants	218	336	479
montazhniki konstruktsii	Fitters	82	87	119
oblitsovshchiki	Tillers	16	18	24
plotniki	Carpenters	376	397	417
putevye rabochie, rabochie po remontu puti	Road layers, road repair workers	59	49	50
slesari	Fitters	281	315	376
stolyary	Joiners	48	45	44
transportnye (podsobnye) rabochie, gruzchiki, vozchiki, vagonetchiki, otkatchiki	Transport (subsidiary) workers, loaders, carters, wagoners, haulers	532	338	341
truboukladchiki	Pipe layers	26	36	46
shofery	Drivers	31	41	48
shtukatury	Plasterers	257	280	286
elektromontery	Electrician	119	132	199
elektroslesari	Electrical fitter	23	35	35
		2961	3143	3667

Source: *Trud v SSSR*, Moscow: Statistika, 1968, p. 228

Education and training

It is evident that the labour force was being better trained and becoming increasingly specialized under Khrushchev. Specialized training was an important part of Khrushchev's strategy for the expansion of the construction sector. Under Khrushchev, there was a growth in on-the-job training opportunities, and specialist courses for construction and building industry workers were delivered at technical vocational schools. It was widely believed that skilled workers were less likely to leave the construction site in search of better conditions and higher wages elsewhere. The skilling of the labour force was part of the aim to increase labour discipline and to reduce overall levels of labour turnover in industry. The ready availability of training also encouraged new workers to develop a specialist trade within the building industry. The proportion of specialist workers in construction, in both building and project organizations, with higher and secondary education increased from 6.6 per cent in July 1955 to 13.5 per cent in November 1964 (by which time Khrushchev had been removed from office) (see Table 13.8).[52]

Women workers

Although the numbers of women employed in construction rose steadily (from just under one million in 1955 to just over 1.5 million in 1963), female workers continued to constitute a stable proportion of the labour force at roughly 29 to 31 per cent (and they constituted between 45 and 48 per cent of the paid labour force as a whole in these years) (see Table 13.9). According to data for 1961, women held around one-fifth (22 per cent) of all professional and executive posts in construction. They were most widely represented among economists and book-keepers, rate setters, technicians and engineers. They were less likely to be found, not surprisingly, as shop floor supervisors and heads of enterprises (see Table 13.10). Like their male counterparts, increasing numbers of women employed in manual jobs in the construction sector and in the building materials industry had studied at technical vocational schools, where they were mainly trained as plasterers, masons, carpenters and joiners (see Tables 13.11 and 13.12). Although

Table 13.8 Specialists in the building industry

	1 January 1941	1 July 1955	15 November 1964
With higher education:			
in building organizations	16,900	50,700	143,700
in project organizations	22,400	58,000	178,900
With secondary education:			
in building organizations	23,700	67,600	302,900
in project organizations	10,600	37,000	100,600
Total:	73,600	213,300	726,100

Source: *Trud v SSSR*, Moscow: Statistika, 1968, pp. 264–5, 282–3

they were less likely to be found working directly on the construction site itself, in 1959 women constituted over half (54 per cent) of the workers employed in the building materials industry (see Table 13.13).

Labour discipline

Under Khrushchev, significant efforts were made to improve the levels of labour discipline in the construction sector. Rates of unauthorized absenteeism and poor

Table 13.9 Women in the national economy: construction and building assembly jobs

	National economy		Construction (construction and assembly work)	
	% women employed	% women in industry and construction	No.	%
1929	27	30	64,000	7.0
1930			156,000	9.6
1931			189,000	10.1
1932			380,000	12.8
1933	30		291,000	16.0
1934			454,000	18.7
1935			450,000	19.7
1936			402,000	19.1
1937			488,000	20.6
1940	38	40	359,000	23.0
1945	55		489,000	32.0
1950	47		845,000	33.0
1952			(948,000)	34.0
1955			(989,000)	31.0
1956			(1,064,000)	31.0
1958	47		1,335,000	30.0
1960			1,500,000	29.0
1961	48	39	1,544,000	29.0
1962			(1,494,000)	29.0
1963			(1,519,000)	29.0
1964			(1,035,000)	29.0

Source: *Women and Children in the USSR: Brief Statistical Returns*, Moscow: Foreign Languages Publishing House, 1963, pp. 98–102; 1958: *Women in the USSR: Brief Statistics*, Moscow: Foreign Languages Publishing House, 1960, pp. 33, 35, 37; see also: N.T. Dodge, *Women in the Soviet Economy: Their Role in Economic, Scientific, and Technical Development*, Baltimore: Johns Hopkins Press, 1966, pp. 178–9

Table 13.10 Women executives and professional workers: building organizations

	No. of women*	%	
1941	7,000	9	
1 December 1956	72,700	22	
1957	73,000	22	
1 December 1961		22	

	1 December 1956		1 December 1961
	No. of women	%	%
	72,700		
inc:			
Heads of enterprises	6,900		5
Engineers	13,200	31	39
Technicians	9,600	44	52
Foremen	13,500	19	13
Rate setters/dispatchers	300	30	62
Book-keepers	11,800	41	27
Economists	7,500	67	70
Other unspecified	9,900	19	

Source: *Women and Children in the USSR: Brief Statistical Returns*, Moscow: Foreign Languages Publishing House, 1963, pp. 120,122; *Women in the USSR: Brief Statistics*, Moscow: Foreign Languages Publishing House, 1960, pp. 47, 48, 50; see also: N.T. Dodge, *Women in the Soviet Economy: Their Role in Economic, Scientific, and Technical Development*, Baltimore: Johns Hopkins Press, 1966, p. 204

* 1941–57 equals 10 times increase

Table 13.11 Training of women at technical vocational schools, 1959–1961

			Graduates in employment:	
	Total	inc: Building schools	Building materials industry	Construction
1959	583,000	42,000		
1960	689,000	84,000		
1961	739,000	104,000	2,500	131,600

Source: *Women and Children in the USSR: Brief Statistical Returns*, Moscow: Foreign Languages Publishing House, 1963, pp. 156–7

work conduct were particularly high among young workers, who formed the bulk of new recruits to the industry, both on building sites and in the research institutes. Penalties were imposed and infringements of labour discipline were dealt with accordingly by the authorities, sometimes resulting in dismissals.[53] Attempts were also made to improve the living conditions of building workers.

Table 13.12 Training of women workers at technical vocational schools by trade, 1961

		%	%
Total no. graduates	429,270	100	
including:			
Building, wood-working and building materials industry	129,038	30.06	100
of whom:			
masons	28,320	6.60	21.95
masons in large-panel construction	891	0.21	0.69
firebrick liners	678	0.16	0.53
painters	11,639	2.71	9.02
plasterers	34,236	7.98	26.53
carpenters	23,567	5.49	18.26
joiners	15,696	3.66	12.16
wood-workers	609	0.14	0.47
concrete reinforcement assemblers	5,775	1.35	4.48
drivers, their assistants, and motorists of building machinery	3,157	0.74	2.45
building machinery fitters	923	0.22	0.72
[other not listed in source]	3,547	0.83	2.75

Source: Women and Children in the USSR: Brief Statistical Returns, Moscow: Foreign Languages Publishing House, 1963, pp. 158, 161

Note: % calculations are own calculations

Table 13.13 Number and percentage of women employed in construction involving physical labour, 1959

	Construction		*Building materials industry*	
	No.	*%*	*No.*	*%*
1959	905,400	18	290,200	54

Source: N.T. Dodge, *Women in the Soviet Economy: Their Role in Economic, Scientific, and Technical Development*, Baltimore: Johns Hopkins Press, 1966, pp. 178–9

Improvements in workplace facilities, such as the provision of washrooms and toilets for women workers in the building materials industry, contributed towards the stabilization of the labour force. Probably the most significant contributor to reducing the levels of labour turnover in construction, however, was the introduction of uniform wage rates across the building sector. This meant that it would no longer be profitable for a worker to leave the building site in search of higher wages elsewhere.

Building materials

The success of Khrushchev's construction policies was partly dependent on the development and expanded production of new types of building materials and the introduction of new techniques in their production.[54] The output of basic building materials was scheduled to take place, as much as possible, in an industrial setting and on a mass scale, with the aim of effecting significant reductions in the costs of production, improving the quality of output and speeding up the actual process of on-site construction. Some of the savings made in production costs and labour productivity were consequently passed on to workers in the form of wage increases, thereby broadening the appeal of the building materials industry and construction sector as areas of employment. In practice, the development of new building materials meant that concrete began to substitute for the use of timber and metals, and cement began to replace bricks as basic building materials. There was also some discussion of expanding the use of plastics to roofs, floors, windows and bathrooms.[55]

Concrete and cement

The broader use of reinforced concrete and cement had a number of benefits that aided the rapid expansion of housing construction under Khrushchev. By 1964, the Soviet Union had more than 2500 enterprises alone producing 55 million cubic metres of reinforced concrete. The output of cement rose from 15.9 million tons in 1953 to 64.9 million tons in 1964 (see Table 13.14).[56]

'Hard' (rather than 'plastic') concrete was considered to be more cost-effective, gave the building greater strength, could be produced more quickly and reduced

Table 13.14 Output of building materials, 1928–1965

	Cement (million tons)	Bricks (million)	Window glass (million m²)
1928	1.850	2,790	34.2
1932	3.478	4,900	29.5
1937	5.454	8,666	79.3
1940	5.675	7,455	44.7
1945	1.845	2,030	23.3
1950	10.194	10,240	76.9
1955	22.484	20,825	99.8
1957	28.896	24,671	120.9
1960	45.520	35,500	147.2
1963	61.018	35,600	169.1
1965	72.400	36,600	190.0

Source: *Promyshlennost' SSSR: statisticheskii sbornik*, Moscow: Statistika, 1964, pp. 318, 329, 343; TsSU, *SSSR v tsifrakh v 1967 godu*, Moscow: Statistika, 1968, pp. 48–9, 58

expenditure in the assembly process in comparison with its alternatives. The new building materials significantly reduced expenditure on basic products such as metals (including steel) that were more difficult to source and more costly to produce. Expenditure on the basic raw materials of construction, therefore, was cut by up to 50 per cent. Labour costs in production were also reduced by between 20 and 25 per cent. In addition, concrete and cement were resistant to corrosion and were considered to be more durable than their alternatives. These were also fire-resistant materials, and so not only did they strengthen the very fabric of the building, but it was also widely believed that they enhanced its overall safety.

Pre-fabrication

The use of pre-fabrication in housing construction, first publicized at the builders' conference in 1935, was developed in the Soviet Union under Stalin in the immediate post-war period. The Khrushchev era heralded its widespread use. Prefabricated panel design was used in only 3 per cent of house building in 1959, but by 1965 it accounted for 56 per cent of all state-sponsored housing construction.[57] The mass output of prefabricated panels meant that from 1958 to 1963 the cost of producing prefabricated materials was reduced by 18 per cent for each square metre.[58]

The use of girder and panel designs in housing construction meant that prefabricated floor, wall and ceiling panels could be mass produced in factories and then transported directly to the building site for erection. Two different types of prefabricated panel ('multi-hollow' and 'multi-ribbed') were in use. There were also discussions in some parts of the country about the possibilities of manufacturing entire one-storey buildings in the factory and delivering these complete and ready to live in to designated residential areas. Prefabricated panels again provided savings in raw materials (including metals), were relatively cheap to produce and could be made relatively quickly (unlike the on-site use of poured concrete, which would have to be given time to set). Basic painting and decorating could take place in the factory, thereby reducing the need for specialist labour on the building site itself. In addition, prefabricated panels were considered to provide good soundproofing.

Breeze blocks

The use of breeze blocks in construction was already evident in large cities in the Soviet Union by the 1940s.[59] Only 8 per cent of housing was built using breeze blocks in 1959, but this had increased to 12 per cent by 1965. Over the same time period, the use of bricks in construction declined from 52 to 12 per cent.[60]

By substituting the output of larger-sized breeze blocks for the more expensive and time-consuming production of bricks, overall costs of production were reduced by around 12 per cent. The extensive use of breeze blocks also brought about savings in the assembly process and helped to raise the levels of labour productivity on building sites. Unlike bricks, breeze blocks could be easily mass produced on site if necessary, thereby bringing about further savings in transportation costs. In addition, breeze blocks were significantly lighter in weight than bricks, which again aided the physical process of construction and reduced the overall weight of the building.

Interior and exterior design materials

Savings were also to be made in the materials used in the interior and exterior design of new buildings. Khrushchev extolled the virtues of the new floor covering, linoleum, in his speech to the 1954 builders' conference, claiming that is was 'not inferior to parquet'.[61] Parquet was time-consuming and costly to lay, whereas linoleum and vinyl could be mass produced cheaply and easily in a whole range of different designs.[62] The exteriors of the housing blocks were to be distinguished by the use of balconies and different designs on the facades.[63] Mass-produced ceramic tiles, which were 'durable, attractive and the colours do not fade', were widely used.[64] Wherever possible, tiles were to be affixed at the factory stage of production so that they were delivered to the building site already mounted on prefabricated panels.

Using the methods of industrial production to assist in the output of building materials as promoted by Khrushchev at the 1954 builders' conference and in the 1957 decree on construction, and with the aid of new building machinery to support on-site assembly, within a very short period of time a number of different areas of the country were able to provide favourable reports on the progress of house building at significantly reduced cost per square metre in their regions.[65] Such construction projects were considered to have the benefit not only of providing much needed accommodation in the localities, but they also strengthened the Soviet Union's regional industrial base.

Costs of production

Significant success was achieved in the aim to reduce the costs of production in construction through the rationalization of building design and sources of investment, the extensive introduction of mechanized means of production both on building sites and in factory-based manufacture combined with the reduction of transportation costs, raising overall levels of labour productivity and the introduction of a systematized wage scale for employees in the construction sector.

Rationalization of design

The rationalization of design, particularly for new housing, and the standardization of the basic components used in construction allowed for significant savings to be made in the overall costs of production. By the end of Khrushchev's period of office, 95 per cent of housing construction was based on standardized designs.[66] The government set standards for different areas of building design, such as establishing ceiling heights and the size of kitchens.[67] Building specifications divided the country according to four different climatic zones on the basis of a range of different temperature conditions, as well as the possibility of a region experiencing earthquakes.[68] The building norms and regulations, however, were not always considered to be adequately defined or to be suitable for the most extreme climatic conditions.[69] In Turkmenistan, for example, it was suggested that buildings should be restricted to three storeys, to withstand local seismic conditions, and that the increased use of panels would reduce internal temperatures, where the installation of windows could cause the temperature to rise by three to four degrees.[70]

By 1963, 62 per cent of all urban housing took the form of five-storey blocks with flats of between one and four rooms.[71] Some construction was also taking place of nine-, 12- and 16-storey blocks of flats and of lower-rise buildings. From the outside, these blocks were distinguishable mostly by the use of different-colour tiling on the facade. Their uniform and modular design allowed for little flexibility in the layout of the interiors.

The use of prefabricated materials in construction meant that the blocks were erected, it has been estimated, at an expenditure of between 35 and 40 per cent less than if bricks had been used.[72] The introduction of industrial methods of production and construction meant that a five-storey block containing between 60 and 80 flats could be completed in four to five months, during which time the actual construction itself could be completed in just 25 to 30 days.[73] One engineer based in Moscow claimed in 1958 that a five-storey building using large-panel design could be erected in only 93 days, and that future improvements in technology would reduce this timeframe even further.[74]

Capital investment

Under Khrushchev, around two-thirds of house building was funded directly by the government, with the remainder of investment coming from collective farms, house-building co-operatives and private individuals to whom the government offered credit.[75] Although the actual size of urban housing owned as personal property continued to expand under Khrushchev, it came to represent a declining proportion of overall construction in the years between 1960 (39.1 per cent) and 1964 (35.8 per cent).[76] The government also invested heavily in expanding the industrial base of construction. In six years (1959–64) of the Seven-Year Plan, 60 per cent of capital investment was directed towards the construction sector of the national economy, including 21 per cent allocated to house building. Some of the other money was spent on the expansion of public facilities such as schools, hospitals and shops.[77]

Mechanization of production

The most expensive elements in the costs of production in construction were machinery and transportation. In addition to the expansion of factory-based output of basic building components, machinery was also introduced for use directly on building sites. By 1963, most of the formerly manual, back-breaking work of ground preparation and the laying of foundations was now being carried out by mechanized means, made possible by the mass production of heavy building equipment (see Table 13.15).

Labour productivity

The increased mechanization of production and the automation of the construction process contributed to rising levels of labour productivity in the building industry. Increases in labour productivity were also the result of the more efficient use and effective organization of workers on building sites. Building site workers were encouraged to make the most economic use of available resources, as well as to reduce the levels of spoilage and accidental damage to building materials and equipment.[78] As indicated above, the labour force in construction was better trained and more highly educated under Khrushchev than it had been in previous decades and this in itself resulted in increases in labour productivity. Increases in productivity were supposed to stay in advance of increases in wages.

Wages

In 1955, largely because of the predominantly unskilled and non-specialized nature of the work on building sites, manual workers in the construction sector earned less on average per day than the national average for the economy as a whole and significantly less than workers employed in industry: manual workers in construction earned 70 rubles per day as compared with 71.5 rubles per day for the national average and 78.3 rubles per day for industrial workers (see Table 13.16). Five years later, however, by 1960, the wages of all workers in construction (including ITR and administrative personnel) outstripped that of both the national

Table 13.15 Growth in stock of mechanized building equipment, 1940–1965

	Excavators	Scrapers	Bulldozers	Cranes
1940	2,100	1,100	800	1,100
1950	5,900	3,000	3,000	5,600
1953	12,500	7,300	10,400	18,000
1960	36,800	12,200	40,500	55,000
1961	43,500	13,000	47,500	62,000
1963	56,500	15,800	56,000	71,500
1965	69,200	20,100	68,500	83,300

Sources: TsSU, *SSSR v tsifrakh v 1967 godu*, Moscow: Statistika, 1968, p. 110; N.V. Baranov (ed.), *Industrializatsiya zhilishchnogo stroitel'stva v SSSR*, Moscow: Gosstroi SSSR, 1965, p. 4; B. Ya. Ionas, *Ekonomika stroitel'stva*, Moscow: Gosudarstvennoe izdatel'stvo literatury po stroitel'stvu, arkhitekture i stroitel'nym materialam, 1963, p. 188

Table 13.16 Wages in construction (average per day)

| | Construction | | | | | | | | White collar | | National economy | | Industry | |
| | All | | Blue collar | | ITR | | | | | | | | | |
	Rs	Index	Rs	Index	Rs	Index	Rs	Index	Rs	Index	Rs	Index	Rs	Index
1955	74.2	100	70.0	100	136.9	100			79.6	100	71.5	100	78.3	100
1960	91.7	124	88.7	127	138.2	101			83.5	105	80.1	112	91.3	117
1961	96.8	130	93.4	133	144.2	105			92.0	116	83.4	117	94.5	121
1962	99.3	134	95.9	137	144.4	105			93.5	117	86.2	121	96.6	123
1963	101.6	137	98.3	140	146.6	107			93.0	117	87.6	123	98.4	126
1964	106.0	143	103.0	147	146.6	107			95.1	119	90.1	126	100.5	128

Source: Trud v SSSR, Moscow: Statistika, 1968, pp. 138–9, 145

average and the industrial sector: 91.7 rubles per day in construction; 80.1 rubles per day in the national economy as a whole; 91.3 rubles per day in industry.

During the years in which Khrushchev was in office, manual workers in the construction sector saw their wages rise by almost 50 per cent, from an average of 70 rubles per day in 1955 to 103 rubles per day in 1964. Their wages had overtaken the daily earnings of administrative personnel in construction by 1960. The earnings of ITR in construction, which were considerably higher than that of both blue- and white-collar workers, also continued to rise, though at a somewhat slower pace. Wages in the buildings materials industry were lower than on building sites, but they also saw a steady rise in the course of this decade.

Postscript

In the first decade of the twenty-first century, the remaining *khrushchevki* are receiving a mixed review in Russia. In some areas of the country – in Samara, for example, where some of the research for this chapter was conducted – *khrushchevki* are much sought after in some areas of the city, where they are being carefully renovated and turned into bijou apartments by rising young professionals. In Moscow, however, where land prices are sky-rocketing and the inner suburbs are highly prized sites for new construction projects, Khrushchev's experimental blocks of flats have sometimes been dubbed *khrushchoby* (playing on the Russian word for slums – *trushchoby*). The external fabric of the buildings is crumbling, with the danger that balconies will collapse into the street below, and concrete wall panels are in an evident state of decay. A mass demolition programme was due to be completed by 2010. The capital's Khrushchev-era architectural heritage will not be entirely lost, however – there are plans to turn one of the remaining *khrushchevki* into a museum!

Conclusion

Under Stalin, in the 1930s a huge expansion took place in capital investment, concentrated on the building up of heavy industry and the defence sector. Considerable efforts were made to increase the volume of urban housing, but these efforts failed to keep pace with the rapidly growing urban population. In 1935 the authorities sought to modernize the building industry in general and house building in particular, and they also planned to expand investment in housing. Khrushchev played a prominent role in these efforts. The priority given to defence from 1936 onwards greatly limited the effects of these plans. The plans for industrializing the building industry and expanding housing were resumed after the first stages of post-war recovery. The devastating impact wrought on living space by the war led priorities to begin to shift towards the housing sector after 1945. New and experimental methods of housing design and construction, already launched in the mid-1930s, were resumed in the late Stalin period but were not fully exploited.

When he came to office in the 1950s, Khrushchev made housing a central feature of his social, political and ideological agenda, and housing construction expanded extremely rapidly in the late 1950s and early 1960s. Khrushchev's housing policy placed great emphasis on achieving maximum living space at low cost, and was often criticized for the poor quality of construction that resulted.

Expansion of Soviet housing at relatively low cost was dependent on the rationalization of design, the widespread exploitation of new building materials and techniques, and the rapid expansion and stabilization of the labour force. Khrushchev's approach to housing, particularly in Moscow, strongly differed from Stalin's. He urged that housing should be built as economically as possible, and that it should consist mainly of four- or five-storey blocks (which would not require lifts). Unlike Stalin, Khrushchev had a 'no frills' approach to housing design, opting instead for more utilitarian and practical outcomes. His legacy in construction policy and housing design, with mixed public reception, was evident in the urban landscape until the end of the Soviet period and beyond.

Notes

1 The Soviet terms *kapital'noe stroitel'stvo* or *stroitel'stvo* usually cover all capital investment. Sometimes, however, they refer only to 'pure construction' (*chistoe stroitel'stvo*), building work excluding the cost of the equipment installed at the site. Here we translate *stroitel'stvo* both as 'building' and as 'construction'.

2 See Table 13.2 and R.W. Davies, *Crisis and Progress in the Soviet Economy, 1931–1933*, Basingstoke: Macmillan, 1996, pp. 441, 539. The 1927/28 figures include 684,000 employed in the industry and about 300,000 self-employed; the number of self-employed in 1932 seems to have been quite small, though the number of part-time *shabashki* (see below) is difficult to estimate.

3 See Davies, *Crisis and Progress*, p. 543.

4 *Soveshchanie po voprosam stroitel'stva v TsK VKP(b)*, Moscow: Partizdat, 1936, p. 49 (Ginzburg).

5 *Soveshchanie ... stroitel'stva*, p. 271 (Kaganovich).

6 *Za industrializatsiyu* (hereafter ZI), 5 January 1935 (D. Babitskii).

7 *Soveshchanie ... stroitel'stva*, p. 45 (Ginzburg).

8 *Soveshchanie ... stroitel'stva*, p. 28 (Mezhlauk).

9 ZI, 5 September 1935 (Rinberg, head of the building department of Glavaviaprom).

10 ZI, 5 September 1935 (K. Lovin).

11 *Soveshchanie...stroitel'stva*, p. 143; *Promyshlennost' SSSR: statisticheskii sbornik*, Moscow: TsSU, 1957, p. 291.

12 Much attention was also devoted to overheads, as in the case of all the aspects of the Soviet economy that the authorities were anxious to improve.

13 See Davies, *Crisis and Progress*, p. 544.

14 ZI, 5 January 1935 (D. Babitskii); for the canteen see the speech by Lyubimov, People's Commissar for Light Industry, in *Soveshchanie ... stroitel'stva* (1936), p. 255.

15 See Davies, *Crisis and Progress*, p. 542,

16 *Soveshchanie ... stroitel'stva*, p. 18.

17 *Soveshchanie ... stroitel'stva*, p. 178 (Levenson); see also p. 111 (Mamashvili, head of Makeevka construction). On the other hand, some locally produced materials were subsidized, which made it more advantageous for an All-Union building organization to buy them rather than manufacture them itself: p. 143 (Ukhanov).

18 *Soveshchanie ... stroitel'stva*, p. 18.
19 *Soveshchanie ... stroitel'stva*, p. 35 (Ginzburg).
20 *Soveshchanie ... stroitel'stva*, p. 111 (Gugel'); in response, Ordzhonikidze indignantly called out, 'None of you are trading, you pay what they tell you to.'
21 The lower figure was preliminary. *Soveshchanie ... stroitel'stva*, p. 11 (Mezhlauk); for the higher figure, see GARF, 1562/10/357, 2–4 (TsUNKhU estimate, n.d. [1937]).
22 For the investment situation, see R.W. Davies and O. Khlevnyuk, 'Stakhanovism and the Soviet Economy', *Europe-Asia Studies*, vol. 54, no. 6, 2002, pp. 867–78.
23 The conference was widely publicized in the daily press, and the verbatim report was published as *Soveshchanie po voprosam stroitel'stva v TsK VKP(b)* (1936).
24 *Soveshchanie ... stroitel'stva*, pp. 207–22.
25 See Simon et al., *Moscow in the Making*, London, New York and Toronto, 1937, on the Moscow ten-year plan, adopted in July 1935, and later expanded into a 20- to 25-year plan.
26 Published in *Sobranie zakonov*, 1936, art. 70.
27 Published in *Ekonomicheskaya zhizn'* (hereafter EZh), 14 February 1936.
28 *Soveshchanie ... stroitel'stva*, p. 20 (Mezhlauk).
29 EZh, 28 August 1935 (STO sitting of 23 August).
30 Capital investment by the armed forces, mainly 'pure construction', increased from 2,300 million rubles in 1936 to 6,000 million in 1940; see J. Barber and M. Harrison (eds), *The Soviet Defence Industry Complex from Lenin to Stalin*, Basingstoke and New York: Macmillan, 2000, p. 82 (Davies and Harrison), and M. Harrison, p. 254.
31 *Promyshlennost' SSSR:* (1957), p. 203.
32 *Promyshlennost' SSSR* (1957), p. 208.
33 For a summary of the state of the industry in the mid-1950s, see R.W. Davies, 'The Builders' Conference', *Soviet Studies*, vol. 6, 1955–6, pp. 443–5.
34 *Stroitel'naya gazeta*, 5 December 1954. An unknown number of these moved on to other activities.
35 Estimated from data in *Trud v SSSR*, Moscow: Statistika, 1968, pp. 264–5, 282–3.
36 The 1946–49 figure is taken from *Istochnik*, no. 1, 2001, p. 79, citing APRF, 3/31/17, 119–22. In 1949 alone 11.5 million square metres of living space were completed: see p. 88 of the same issue.
37 This account is from notes taken at the meeting of 17 June by G.M. Popov, president of the Moscow soviet: see *Istochnik*, no. 4, 2001, pp. 110–11. In his speech, Stalin insisted that the 20- to 25-year plan for Moscow, drawn up in 1935, should be replaced by a ten-year plan, because 'technology changes, conditions change, and even people's tastes change'.
38 See RGAE 339/3/576 and RGAE 339/3/1037 and 1038.
39 The memorandum was published for the first time in *Istochnik*, no. 1, 2001, pp. 89–102, citing APRF, 3/31/17, 141–69.
40 For Khrushchev's speech 'On the widescale introduction of industrial methods, improving the quality and reducing the cost of construction', see T.P. Whitney (ed.), *Khrushchev Speaks: Selected Speeches, Articles, and Press Conferences, 1949–1961*, Ann Arbor: University of Michigan Press, 1963, pp. 153–92. The speech was published in *Pravda*, 28 December 1954. See also Davies, 'Builders' Conference', pp. 443–57.
41 G.D. Andrusz, *Housing and Urban Development in the USSR*, Basingstoke: Macmillan, 1984, pp. 161–2.
42 'O razvitii zhilishchnogo stroitel'stva v SSSR', *Spravochnik partiinogo rabotnika*, Moscow: Gosudarstvennoe izdatel'stvo politicheskoi literatury, 1957, pp. 294–309. For an edited version of the decree, see also *Spravochnik profsoyuznogo rabotnika*, Moscow: Profizdat, 1959, pp. 538–42. On the implications of the decree for citizen's rights, see Mark B. Smith, 'Khrushchev's Promise to Eliminate the Urban Housing

Shortage: Rights, Rationality and the Communist Future', in M. Ilic and J.R. Smith (eds), *Soviet State and Society under Nikita Khrushchev*, London: Routledge, 2009.

43 D.L. Broner, *Sovremennye problemy zhilishchnogo khozyaistva*, Moscow, 1961, p. 95.

44 For the increase in 1928–36, see R. Moorsteen and R. Powell, *The Soviet Capital Stock, 1928–1962*, Homewood, IL: Yale University Press, 1966, p. 90.

45 For reports on aspirations and achievements in industrial construction, and plans to extend projects overseas, see, for example, RGANTD, f. R-7, op. 2–6, d. 213, ll. 7–10 (results for 1959); RGANDT, R-7/2-6/232, 6–15 (results for 1960); RGANDT, R-7/2-6/253, 7–18 (results for 1961); RGANDT, R-7/2-6/268, 18–19, 40 (results for 1962). For complaints about the quality of work and calls for greater regulation of the construction industry, see RGAE, 339/3/1038(3), 197–8 (speech by Ukhenkeli).

46 T. Sosnovy, 'The Soviet Housing Situation Today', *Soviet Studies*, vol. 11, no. 1, 1959, pp. 11–12.

47 See, for example, RGAE, 339/3/1038(3), 165 (speech by Dedunik).

48 RGAE, 339/3/1038(3), 209–13 (speech by G.A. Grudov) and 217–23 (speech by D.V. Popov).

49 For example, see GASO, 4370/1/23, 159 (1957), and RGAE, 339/3/1038(3), 165 (1960).

50 The data reveal an unexplained short-term dip in the number of manual workers employed in construction in 1962.

51 *Trud v SSSR*, pp. 86–7. Workers employed directly on building sites are listed under '*stroitel'stvo*'; those in the building materials industry are listed as a sub-category of industrial employment under '*promyshlennost' stroitel'nykh materialov*'.

52 Own calculations based on data from *Trud v SSSR*, pp. 24–5, 264–5, 282–3.

53 For examples, see RGANDT, R-7/2-6/220: *Rabota po ukrepleniyu trudovoi distsipliny* (1959); R-7/2-6/242: *Rabota po ukrepleniyu trudovoi distsipliny* (1960). See also GASO, 4370/1/23, 160–1.

54 On new techniques, including the vibration method in panel manufacturing, see RGAE, 339/3/1038(3), 154–6 (speech by V.V. Mikhailov).

55 RGANDT, R-149/5-4/27: *Primenenie plastmass v zhilishchnom stroitel'stve* (1958).

56 N. V. Baranov (ed.), *Industrializatsiya zhilishchnogo stroitel'stva v SSSR*, Moscow: Gosstroi SSSR, 1965, pp. 3–4.

57 Broner, *Sovremennye problemy*, p. 105 (Table 21).

58 Baranov, *Industrializatsiya zhilishchnogo stroitel'stva*, p. 27. See also the 4 April 1959 Council of Ministers decree 'On the development of large-panel housing construction'.

59 Baranov, *Industrializatsiya zhilishchnogo stroitel'stva*, p. 26.

60 Broner, *Sovremennye problemy*, p. 105 (Table 21).

61 Khrushchev, 1954 builders' conference, p. 178.

62 See also the arguments put forward by D.V. Popov in 1960: RGAE, 339/3/1038(4), 226–7.

63 The example of Novye Cheremushki was used at the 1960 meeting on urban housing development. See RGAE, 339/3/1088(1), 47–8 (speech by V.P. Lagutenko).

64 Khrushchev, 1954 builders' conference, p. 178.

65 See, for example, I. Panov, 'Narodnaya initsiativa v zhilishchnom stroitel'stve', *Partiinaya zhizn'* (PZh), no. 13, 1959, pp. 17–22 (Orenburg); 'Novyi metod organizatsii stroitel'stva zhilishch', *PZh*, no. 5, 1960, pp. 36–9 (Leningrad); M. Burka, 'Ukreplyaem material'no-tekhnicheskuyu bazu stroitel'stva', *PZh*, no. 21, 1960, pp. 22–7 (Ukraine).

66 Baranov, *Industrializatsiya zhilishchnogo stroitel'stva*, p. 18.

67 For discussion on specifications of design for new housing in Kramatorsk, see RGANDT, R-149/5-4/25, 3–45; for blueprints of the housing projects, see RGANDT, R-149/5-4/26, 4–12. For Khrushchev's justification of the cost savings involved in

lowering ceiling heights from 3.5 to between 2.7 and 2.5 metres (on the model used in England), see his speech to the V Congress of the International Union of Architects, 25 July 1958, published in *Istochnik*, no. 6, 2003, pp. 90–7 (p. 92), citing APRF, 52/1/545, 1–19.

68 Baranov, *Industrializatsiya zhilishchnogo stroitel'stva*, pp. 18–20.
69 For example, such concerns were raised at the 1958 All-Union Meeting on Construction: RGAE 339/3/576, 63–4 (speech by V.N. Glinka, from Turkmenistan), and 95–102 (speech by S.F. Agafonov, from Noril'sk).
70 RGAE, 339/3/1038(3), 161–3 (speech by O.V. Dedunik).
71 Baranov, *Industrializatsiya zhilishchnogo stroitel'stva*, p. 6.
72 RGAE, 339/3/1037(2), 77–8 (speech by A.V. Vlasov, 'Zastroika gorodov v usloviyakh dal'neishei industrializatsii stroitel'stva'). See also Baranov, p. 6.
73 Baranov, *Industrializatsiya zhilishchnogo stroitel'stva*, p. 28.
74 RGAE, 339/3/576, 87–8 (speech by Makrushin).
75 Baranov, *Industrializatsiya zhilishchnogo stroitel'stva*, p. 2.
76 Andrusz, *Housing and Urban Development*, New York: University of New York Press, 1985, p. 22.
77 Baranov, *Industrializatsiya zhilishchnogo stroitel'stva*, p. 22.
78 GASO, R-4370/1/22.

Select bibliography

Archives

Russia

Archive references are noted as f. (*fond*; fund), op. (*opis'*; list of files), d. (*delo*; file), l. (*list*; page).

GARF (Gosudarstvennyi arkhiv Rossiiskoi Federatsii): State Archive of the Russian Federation.

GASO (Gosudarstvennyi arkhiv Saratovskoi oblasti): State Archive of Saratov Region.

RGAE (Rossiiskii gosudarstvennyi arkhiv ekonomiki): Russian State Archive of the Economy.

RGANI (Rossiiskii gosudarstvennyi arkhiv noveishei istorii): Russian State Archive of Contemporary History.

RGANTD (Rossiiskii gosudarstvennyi arkhiv nauchno-tekhnicheskoy dokumentatsii): Russian State Archive of Scientific and Technical Documentation.

RGASPI (Rossiiskii gosudarstvennyi arkhiv sotsial'no-politicheskoi istorii): Russian State Archive of Social and Political History.

Ukraine

TsDAGO/TsGAOOU (Tsentral'nyi derzhavnyi arkhiv gromads'kikh ob'ednan' Ukraini/ Tsentral'nyi gosudarstvennyi arkhiv obshchestvennykh ob'edinenii Ukrainy): Central State Archives of Public Organizations of Ukraine.

TsDAVO/TsDAVOVU (Tsentral'nyi derzhavnyi arkhiv vishchikh organiv vladi ta upravliniya Ukraini): Central State Archives of Supreme Bodies of Power and Government of Ukraine.

Other

Archive of the Communist Party of Latvia
Hungarian National Archive

Journals, newspapers, serials

Ekonomicheskaya zhizn'

Gudok
Istochnik
Nová mysl
Partiinaya zhizn'
Pravda
Promyshlennost' SSSR
Scinteia
Skolotāju Avīze
Sobranie postanovlenii Pravitel'stva SSSR
Sobranie zakonov
Stroitel'naya gazeta
Társadalmi Szemle
Trud v SSSR
Za industrializatsiyu

Russian language sources

Agaryev, A., *Tragicheskaya avantyura*, Ryazan: Russkoe slovo, 2005.

Aksyutin, Yu.V. (ed.), *Nikita Sergeevich Khrushchev. Materialy k biografii*, Moscow: Izdatel'stvo politicheskoi literatury, 1989.

Andriyanov, V., *Kosygin*, Moscow: Molodaya gvardiya, 2004.

Baranov, N.V. (ed.), *Industrializatsiya zhilishchnogo stroitel'stva v SSSR*, Moscow: Gosstroi SSSR, 1965.

Biryukov, A., *Ya dumal, chto rodilsysa akrobatom...*, Moscow, 2004.

Bogdanov, Yu.N., *Strogo sekretno. 30 let v OGPU-NKVD-MVD*, Moscow: Veche, 2002.

Broner, D.L., *Sovremennye problemy zhilishchnogo khozyaistva*, Moscow: Vysshaia shkola, 1961.

Chebrikov, V. et al., *Istoriya sovetskikh organov gosudarstvennoi bezopasnosti: uchebnik*, Moscow: Vysshaya krasnoznamenskaya shkola komiteta gosudarstvennoi bezopasnosti pri sovete ministerov SSSR, 1977.

Danilov, A.A. and Pyzhikov, V.A., *Rozhdenie sverkhderzhavy: SSSR v pervye poslevoennyt gody*, Moscow: Rosspen, 2001.

Dashko, A.P. *Gruz pamyati: Trilogiya: Vospominaniya*, Kiev: Delovaya Ukraina Publishing House, 1997.

Direktivy XX s'ezda KPSS po shestomu planu razvitiya narodnogo khoziaistva SSSR na 1956–1960 gody Narodnoe khozyaistvo SSSR v 1958g., Moscow: Pravda, 1959.

Edel'gauz, G.G., *Dostovernost' statisticheskikh pokazatelei*, Moscow: Statistika, 1977.

Evenko, I., 'Proverka i ekonomicheskii analiz vypolneniya planov', *Planovoe khozyaistvo*, 1958, no. 10, pp. 3–15.

Fursenko, A.A. (ed.), *Prezidium TsK KPSS. 1954–1964. Chernovye protokol'nye zapisi zasedanii. Stenogrammy. Postanovleniya*, 3 vols, Moscow: Rosspen, 2003–2006.

Gregory, P., *Politicheskaya ekonomiya stalinizma*, Moscow: Rosspen, 2006.

Grishin, V.V., *Ot Khrushcheva do Gorbcheva*, Moscow: ASPOL, 1996.

Ismailov, E., *Azerbaidzhan: 1953–1956gg. Pervye gody ottepeli*. Baku: Adil'ogly, 2006.

Ivkin, V.I., *Gosudarstvennaya vlast' SSSR. Vysshie organy vlasti i upravleniya i ikh rukovoditeli. 1923–1991 gg. Istoriko-biograficheskii spravochnik*, Moscow: Rosspen, 1999.

Kaganovich, L.M., *Uluchshit' rabotu i organizovat' novyi podem zh.d.transporta*, Moscow: Gos. izd-vo politicheskoy literatury, 1954.

Khanin, G.I., *Dinamika ekonomcheskogo razvitiya SSSR*, Novosibirsk: Nauka, 1991.

Khlevnyuk, O.V., *Politburo: mekhanizmy politicheskoi vlasti v 1930–e gody*, Moscow: Rosspen, 1996.

Khrushchev 1964, Stenogrammy plenuma TsK KPSS i drugie materialy. Moscow: Materik, 2007.

Khrushchev, N.S. *Stroitel'stvo kommunizma v SSSR i razvitie sel'skogo khozyaistva*, vol.3, Moscow: Gospolitizdat, 1962.

Khrushchev, N.S., *Vremiya, Lyudi, Vlast'*. *Vospominaniya. kniga 1*, Moscow: TsK Moskovskie Novosti, 1999.

Korzhikhina, T.P., *Sovetskoe gosudarstvo i ego uchrezhdeniya. Noyabr' 1917–dekabr' 1991*, Moscow: RGGU, 1995.

Kozlov, V., *Kramola: Inakomyslie v SSSR pri Khrushcheve i Brezhneve 1953–1982*, Moscow: 'Materik', 2005.

Kozlov, V., *Neizvestnyi SSSR: Protivostoyanie naroda i vlasti 1953–1985*, Moscow: Olma-Press, 2006.

KPSS v rezolutsiiakh i resheniiakh s'ezdov, koferentsii i plenumov TsK, vols 9, 10, Moscow: Politizdat, 1986.

Kumanev, G., *Govoryat stalinskie narkomy*, Smolensk: Rusich, 2005.

Medvedev, R., *Nikita Khrushchev. Otets ili otchim sovetskoi 'ottepeli'?*, Moscow: EKSMO, 2006.

Mikoyan, N., *Svoimi glazami*, Moscow: 2003.

Mikoyan, S., *Vospominaniya voennogo letchika-ispytatelya*, Moscow: Tekhnika molodezhi, 2002.

Mitrokhin, N., *Russkaya partiya: Dvizhenie russkikh natsionalistov v v SSSR 1953–1985 gody*, Moscow: Novoe literaturnoe obozrenie, 2003.

Moiseenko, G., *Pamyat' plamennykh let*, Moscow: 2006.

Onikov, L., *KPSS: anatomiia raspada. Vzgliad iznutri apparata TsK*, Moscow: Respublika, 1996.

'Perestroika upravleniya promyshlennost'yu i stroitel'stvom i zadachi planovykh organov', *Planovoe khozyaistvo*, 1957, no. 5, May, pp. 3–11.

Petrov, N., *Pervyi predsedatel' KGB Ivan Serov*, Moscow: Materik, 2005.

Pikhoya, R.G., *Sovetskii Soiuz: Istoriia vlasti, 1945–1991*, Moscow: RAGS, 1998.

Politicheskoe rukovodstvo Ukraina, 1938–1989gg., Moscow: Rosspen, 2006.

Postanovleniya Sovieta Ministrov SSSR, fevral' 1961, Moscow: Upravlenie Delami Sovieta Ministrov SSSR, 1961.

Promyshlennost' SSSR: statisticheskii sbornik, Moscow: TsSU, 1957.

Pyzhikov, A. *Opyt modernizatsii sovetskogo obshchestva v 1953–1964 godax: obshchestvenno-politicheskii aspect*, Moscow: Izdatel'skii dom 'Gamma', 1998.

Pyzhikov, A.V., *Khruchshevskaia "ottepel'"*, *1953–1964*, Moscow: Olma Press, 2002.

Rakov, V.A., *Lokomotivy otechestvennykh zheleznykh dorog 1956–1975*, Moscow: Transport, 1999.

Rakov, V.A., *Lokomotivy otechestvennykh zheleznykh dorog 1845–1955*. Moscow: Transport, 1995.

Resheniya partii i pravitel'stva po khozyaistvennym voprosam, Moscow: Politizdat, 1968.

Sal'nikov, I.S., 'Elektrifikatsiya zheleznykh dorog' in S.S. Minsker and L.I. Krishtal' (eds) *Voprosy razvitiya zheleznodorozhnogo transporta*, Moscow: Gos. transportnoe zheleznodorozhnoe izd-vo, 1957.

Shelest, P.E., *Spravzhnii sud istorii shche poperedu. Spogadi, shchodenniki, dokumenti, material,*. Kiev: Geneza, 2003.

Shelest, P.E., *Da ne sudimy budete. Dnevnikovye zapisi, vospominaniya chlena Politburo TsK KPSS*, Moscow: Edition q, 1992.

Shevelev, V.N., *N.S. Khrushchev*, Rostov-on-Don: Feniks, 1999.

Shvarts, R.Ya, 'Opyt ekspluatatsii elektrovozov i teplovozov' in E.A. Ashkenazi (ed.), *Metody ekspluatatsii tyagovykh sredstv na transporte*, Moscow, 1957.

Smirnov, G.L., 'Malen'kie sekrety bol'shogo doma. Vospominaniia o rabote v apparate TsK KPSS', in V.A. Kozlov (ed.), *Neizvestnaia Rossiia. XX vek*, vol 3, Moscow: Istoricheskoe nasledie, 1993, pp. 361–82.

Soveshchanie po voprosam stroitel'stva v TsK VKP(b), Moscow: Partizdat, 1936.

Soveta Ministrov SSSR, mai 1961, Moscow: Upravlenie Delami Sovieta Ministrov SSSR, 1961.

Spravochnik partiinogo rabotnika, Moscow: Gosudarstvennoe izdatel'stvo politicheskoi literatury, 1957 ff.

Spravochnik profsoyuznogo rabotnika, Moscow: Profizdat, 1959.

Stukalin, B., *Gody, dorogi, litsa...*, Moscow: Fond imeni I.A. Sytina, 2002.

Suslov, I.P., *Osnovy teorii dostovernosti statisticheskikh pokazatelei*, Novosibirsk: Nauka, 1979.

Tregubenko, O.I., 'Sud'ba revolyutsionera', *Voprosy istorii*, no.8, 2007, pp.136–144.

Voslenskii, M., *Nomenklatura. Gospodstvuiushchii klass Sovetskogo Soiuza*, London: Overseas Publications Interchange Ltd, 1984.

Voznesenskii, L., *Istiny radi...*, Moscow: Respublika, 2004.

Yakovlev, A.N. et al. (eds), *Lavrentii Beria. 1953. Stenogramma iyul'skogo plenuma TsK KPSS i drugie dokumenty*, Moscow: MFD, 1999.

Yakovlev, A.N. et al. (eds), *Molotov, Malenkov, Kaganovich. 1957. Stenogramma iyun'skogo plenuma TsK KPSS i drugie dokumenty*, Moscow: MFD, 1998.

'Zadachi Gosplana SSSR v novykh usloviyakh upravleniya promyshlennost'yu i stroitel'stvom', *Planovoe khozyaistvo*, 1957, no. 7, July, pp. 3–11.

Zezina, M., *Sovetskaya khudozhestvennaya intelligentsia i vlast' v 1950–60 gody*, Moscow: Dialog MGU, 1999.

Zhirnov, E., 'Chisto chekisskaya chistka,' *Kommersant-vlast'*, 2006, no. 36, pp.64–9.

English and other non-Russian language sources

A MSZMP határozatai és dokumentumai 1971–5 (Decisions and Documents of the Hungarian Socialist Worker Party ((HSWP)) from 1971 till 1975, Budapest: Kossuth, 1978.

Ali, T. (ed.), *The Stalinist Legacy*, Harmondsworth: Pelican, 1984.

Andrusz, G.D., *Housing and Urban Development in the USSR*, Basingstoke: Macmillan, 1984.

Autio-Sarasmo, S. 'Soviet Economic Modernisation and Transferring Technologies from the West', in M.Kangaspuro and J.Smith (eds) *Modernisation in Russia since 1900*, Helsinki: SKS, 2006.

Barber, J. and Harrison, M. (eds), *The Soviet Defence Industry Complex from Lenin to Stalin*, Basingstoke and New York: Macmillan, 2000.

Barghoorn, F., *Détente and the Democratic Movement in the USSR*, London: Collier Macmillan, 1976.

Baron, S., *Bloody Saturday in the Soviet Union: Novocherkassk, 1962*, Stanford CA: Stanford University Press, 2001.

Berliner, J.S., *Soviet industry from Stalin to Gorbachev,* Cheltenham: Edward Elgar, 1985.

Berliner, J.S., *Factory and Manager in the USSR*, Cambridge, MA.: Harvard University Press, 1957.

Berman, H., 'The Struggle of Soviet Jurists Against A Return to Stalinist Terror', *Slavic Review*, June 1963, pp314–320.

Bilinsky, Y., 'The Soviet Education Laws of 1958–9 and Soviet Nationality Policy' *Soviet Studies*, vol.14, no.2, October 1962, pp. 138–157.

Blauvelt, T., 'Status Shift and Ethnic Mobilisation in the March 1956 Events in Georgia', *Europe-Asia Studies*, vol.61, no.4, June 2009, pp. 651–668.

Blitstein, P.A. 'Nation-Building or Russification? Obligatory Russian Instruction in the Soviet Non-Russian School, 1938–1953' in R. G. Suny and T. Martin (eds), *A State of Nations: Empire and Nation-Making in the Age of Lenin and Stalin,* Oxford: Oxford University Press, 2001.

Bloch, S. and Reddaway, P., *Russia's Political Hospitals*, London: First Futura Publications, 1978.

Boer, S. de, Dreissen, E. and Verhaar, H., *Biographical Dictionary of Dissidents in the Soviet Union, 1956 – 1975,* The Hague: Martinus Nijhoff Publishers, 1982.

Brinkley, G.A., 'Khrushchev Remembered: On the Theory of Soviet Statehood', *Soviet Studies*, vol. 24, no. 3, 1973, pp. 387–401.

Bukovsky, V., *To Build a Castle: My Life as a Dissenter*, London: Andre Deutsch, 1978.

Burford Jr., R., 'Getting the Bugs Out of Socialist Legality: The Case of Joseph Brodsky', *The American Journal of Comparative Law*, Summer 1974, pp.465–501.

Burlatsky, F., *Khrushchev and the First Russian Spring: The Era of Khrushchev Through the Eyes of his Adviser*: London: Weidenfeld and Nicolson, 1991.

Chamberlain, W., 'USSR: How Much Change Since Stalin', *Russian Review*, July 1963.

Chiesa, G., 'Perestroika: a revival of Khrushchevian reform or a new idea of socialist society?' in T. Taranovski (ed.), *Reform in modern Russian history. Progress or cycle?,* Cambridge: Cambridge University Press, 1995.

Clissold, S., (ed.), *Yugoslavia and the Soviet Union 1939—1973*, Oxford: Oxford University Press, 1975.

Connor, W., 'The Soviet Criminal Correction System: Change and Stability', *Law and Society Review*, February 1972.

Counts, G.S., *Khrushchev and the Central Committee Speak on Education,* Pittsburgh: University of Pittsburgh Press, 1959.

Crankshaw, E., *Khrushchev,* London: Collins, 1966.

Daniels, R.V., *A Documentary History of Communism volume 1*, London: I.B. Tauris, 1987.

Davies, N., *God's Playground. A History of Poland, Vol. II, from 1795 to the Present*, Oxford: Oxford University Press, 2005.

Davies, R. W., *Crisis and Progress in the Soviet Economy, 1931–1933*, Basingstoke: Macmillan, 1996.

Davies, R.W., 'The Builders' Conference', *Soviet Studies*, vol. 6, 1955–6.

Davies, R.W., *The Soviet Economy in Turmoil, 1929–1930*, Cambridge, MA.: Harvard University Press, 1989.

Davies, R.W., Harrison, M. and Wheatcroft, S.G., *The Economic Transformation of the Soviet Union, 1913–1945*, Cambridge: Cambridge University Press, 1994.

Davies, R. W. and Khlevnyuk, O., 'Stakhanovism and the Soviet Economy', *Europe-Asia Studies*, vol. 54, no. 6, 2002, pp. 867–78.

Dejiny Ceskoslovenska v datech (History of Czechoslovakia in Data), Prague: Svoboda, 1968.

Deletant, D., *Communist Terror in Romania. Gheorghiu-Dej and the Police State 1948–1965*. New York: St. Martin's Press, 1999.

Deletant, D., *Romania under Communist Rule*, Iasi, Oxford and Portland: Center for Romanian Studies, 1999.

Deutscher, I., *The Prophet Unarmed*, Oxford: Oxford University Press, 1959.

Deutscher, I., 'The Soviet Union Enters the Second Decade after Stalin' in I.Deutscher, *Russia, China and the West*, Oxford: Oxford University Press, 1970.

Dubcek, A., *Viimeisenä kuolee toivo. Poliittiset muistelmat (Hope Dies Last. Political Memoirs)*, Helsinki: WSOY, 1993.

Ducoli, J., 'The Georgian Purges 1951–53', *Caucasian Review*, vol.6, 1958, pp.54–61.

Ekiert, G., *The State against Society. Political Crises and their Aftermath in East Central Europe*, Princeton, New Jersey: Princeton University Press, 1996.

Face to Face with America. The story of N.S. Krushchev's visit to the U.S.A., Moscow: Foreign Languages Publishing House, 1960.

Fainsod, M., *How Russia is Ruled*, London: OUP, 1963.

Filtzer, D., *Soviet Workers and Destalinization. The Consolidation of the Modern System of Soviet Production Relations, 1953–1964*, Cambridge: Cambridge University Press, 1992.

Fireside, H., *Soviet Psychoprisons*, London: W.W. Norton & Company, 1979.

Fitzpatrick, S. *The Russian Revolution*. 2nd edition. Oxford: Oxford University Press, 1994.

Földes, G., *Az eladósodás politikatörténete. 1957–1986 (The Political History of Indebtedness from 1957 till 1986)*, Budapest: Maecenas, 1995.

Fursenko, A. and Naftali, T., *Khrushchev's Cold War: The Inside Story of an American Adversary*, New York: W.W.Norton, 2006.

Furst, J., 'Prisoners of the Soviet Self?: Political Youth Opposition in Late Stalinism', *Europe-Asia Studies*, Vol. 54, No. 3, May 2002, pp.353–375.

Gaddis, J.L., *The Cold War*, London: Allen Lane, 2005.

Gál, E.H., et al. (eds), *A 'Jelcin-dosszié'. Szovjet dokumentumok 1956–ról (The 'Jelcin Dossier'. Soviet Documents from 1956)* Budapest: Századvég – 1956–os Intézet, 1993.

Garton Ash, T., *In Europe's Name: Germany and the Divided Continent*, London: Vintage, 1994.

Gerovitch, S., *From Cyberspeak to Newspeak. A History of Soviet Cybernetics*, Cambridge, MA: MIT Press, 2001.

Gorbachev, M., *Memoirs,* London: Doubleday, 1996.

Gorlizki, Y., 'Anti-ministerialism and the USSR Ministry of Justice, 1953–56: A Study in Organisational Decline', *Europe-Asia Studies,* vol. 48, no. 8, 199, pp. 1279–318.

Gorlizki, Y., 'Party Revivalism and the Death of Stalin', *Slavic Review,* vol. 54, no. 1, 1995, pp. 1–22.

Gorlizki, Y., 'Political reform and local party interventions under Khrushchev', in P.H. Solomon (ed.) *Reforming Justice in Russia, 1864–1996*, New York, London: M.E. Sharpe, 1997.

Gorlizki, Y. and Khlevniuk, O., *Cold peace: Stalin and the Soviet ruling circle, 1945–1953*, Oxford: Oxford University Press, 2004.

Gourvish, T.R., *British Railways 1948–1973*. Cambridge: Cambridge University Press, 1986.

Graham, L., *Science in Russia and the Soviet Union,* Cambridge: Cambridge University Press, 1993.

Gregory, P.R. and Stuart, R.C. *Soviet and Post-Soviet Economic Structure and Performance*, Fifth Edition, London: HarperCollins, 1994.

Grigorenko, P., *Memoirs*, New York: W.W. Norton and Company, 1982.

Grix, J., *The Role of the Masses in the Collapse of the GDR*, Basingstoke: Macmillan Press, 2000.

Grushin, O., *The Dream Life of Sukhanov*, New York: Viking Press, 2006.

Guo, W. and Yunling, Z., *China, US, Japan and Russia in a Changing World*, Beijing: Social Sciences Documentation Publishing, 2000.

Häikiö, M., *Sturm und Drang. Suurkaupoilla eurooppalaiseksi elektroniikkyritykseksi 1983–1991. Nokia Oyj:n historia (A Major European Electronics Company. A History of Nokia, 1983–1991)*, part 2. Helsinki: Edita 2001.

Halliday, F., *Rethinking International Relations*, London: Macmillan, 1994.

Hanson, P., *The Rise and Fall of the Soviet Economy*, London: Longman, 2003.

Hanson, P., *Trade and technology in Soviet-Western Relations*, London: Macmillan, 1981.

Hanson, S., 'Sovietology, post-sovietology, and the study of postcommunist democratization', *Demokratizatsiya*, Winter 2003.

Harrington, J., 'American-Romanian Relations, 1953–1998' in A.R. DeLuca, and P.D. Quinlan (eds.), *Romania, Culture and Nationalism. A Tribute to Radu Florescu*, New York: Columbia University Press, 1998.

Harris, J., 'The Origins of the Conflict between Malenkov and Zhdanov: 1939–1941', *Slavic Review,* vol. 35, no. 2, 1976, pp. 287–303.

Harsányi, I., 'Modernizáció és modernitás. Az 1875 utáni spanyol Restauráció és korszerûsítés' ('Modernization and Modernity. The Spanish Restoration and Reform after 1875') in J.L. Nagy, (ed.), *A modernizáció határai. Tradíció és integráció Kelet-Európa (hazánk) és a Mediterránium történetében a 19–20, században (The Limits of Modernization. Tradition and Integration in the History of Eastern Europe (Hungary) and the Mediterranean in the 19th and 20th Centuries)*, Szeged: Szegedi Tudományegyetem Kiadó, 1972.

Hodnett, G. (ed.), *Resolutions and Decisions of the Communist Party of the Soviet Union. Volume 4. The Khrushchev Years 1953–1964*, Toronto: University of Toronto Press, 1974.

Hoffman, E.P. and Laird, R.F., *'The Scientific-Technological Revolution' and Soviet Foreign Policy*. Oxford: Pergamon Press, 1982.

Holliday, G.D., *Technology Transfer to the USSR 1928–1937 and 1966–1975*, Boulder, CO: Westview Press, 1979.

Holmes, L., *Politics in the Communist World*. Oxford: Clarendon Press, 1986.

Horváth, C., *Magyarország 1944-től napjainkig (Hungary from 1944 up to the Present)*, Pécs: Prezident, 1993.

Hosking, G., *Rulers and Victims. The Russians in the Soviet Union*, Cambridge, MA, London: Belknap Press, 2006.

Hunya, G., Réti, T., Süle, A.R. and Tóth, L., *Románia 1944–1990. Gazdaság- és politikatörténet (Romania from 1944 till 1990. Political and Economic History)*, Budapest: Atlantisz Kiadó, 1990.

Ilic, M. and Smith, J. (eds), *Soviet State and Society under Nikita Khrushchev*, London: Routledge, 2009.

Jackson, I., *The Economic Cold War: America, Britain and East-West Trade, 1948–63*, London: Palgrave Macmillan, 2001.

Jones, P. (ed.), *The Dilemmas of De-Stalinization: Negotiating Social Change in the Khrushchev Era*, London: Routledge, 2006.

Kagarlitsky, B., *The Thinking Reed: Intellectuals and the Soviet State from 1917 to the Present*, London: Verso, 1989.

Kähönen, A., *The Soviet Union, Finland and the Cold War. The Finnish Card in Soviet Foreign Policy*, Helsinki: SKS, 2006.

Kaiser, R.J., *The Geography of Nationalism in Russia and the USSR*, Princeton, NJ: Princeton University Press, 1994.

Kennedy-Pipe, C., *The Origins of the Cold War*, London: Palgrave Macmillan 2007.

Kharkhodin, O., *The Collective and the Individual in Russia: A Study of Practices*, Berkeley CA: University of California Press, 1999.

Khrushchev, S. (ed.), *Memoirs of Nikita Khrushchev. Volume 1. Commissar [1918–1945]*, University Park, PA: Pennsylvania State University Press, 2004.

Khrushchev, S. (ed.), *Memoirs of Nikita Khrushchev. Volume 2. Reformer [1945–1964]*, University Park, PA: Pennsylvania State University Press, 2006.

Khrushchev, S. (ed.), *Memoirs of Nikita Khrushchev. Volume 3. Statesman [1953–1964]*, University Park, PA: Pennsylvania State University Press, 2007.

Király, B., 'A magyar hadsereg szovjet ellenôrzés alatt.' (The Hungarian Army under Soviet Control) in I. Romsics (ed.), *Magyarország és a nagyhatalmak a 20. században* (*Hungary and the Great Powers in the Twentieth Century*), Budapest: Teleki László Alapítvány, 1995.

Knight, A., *Beria: Stalin's First Lieutenant*, Princeton NJ: Princeton University Press, 1993.

Kolankiewicz, G. and Lewis, G.P., *Poland. Politics, Economics and Society*, London and New York: Pinter, 1988.

Korányi, G.T., (ed.), *Egy népfelkelés dokumentumaiból. 1956 (Documents of an Uprising. 1956)*, Budapest: Tudósítások Kiadó, 1989.

Kornai, J., *A szocialista rendszer. Kritikai politikai gazdaságtan (The Socialist System. Crtical Political Economy)*, Budapest: Heti Világgazdaság, 1993.

Koropeckyj, I.S., 'Economic Prerogatives', in I.S. Koropeckyj (ed.), *The Ukraine Within the USSR: An Economic Balance Sheet*, New York: Praeger, 1977.

Kozlov, V.A., *Mass Uprisings in the USSR: Protest and Rebellion in the Post-Stalin Years*, Armonk, NY and London: M.E.Sharpe, 2002.

Kramer, M., 'The Soviet Union and the 1956 Crises in Hungary ad Poland: Reassessments and New Findings', *Journal of Contemporary History*, April 1998, pp. 164–206.

Kronvall, O., *Den bräckliga barriären: Finland i svensk säkerhetspolitik 1948–1962 (The Fragile Barrier: Finland in Swedish Security 1948–1962)*, Stockholm: Elander Gotab, 2003.

Kulavig, E., *Dissent in the Years of Khrushchev: Nine Stories About Disobedient Russians*, Basingstoke: Palgrave Macmillan, 2002.

Kulcsár, K., *A modernizáció és a magyar társadalom. (Modernization and Hungarian Society)*, Budapest: Magvetö, 1986.

Kulcsár, K., 'East Central Europe and the European Integration' in M. Szabó, (ed.), *The Challenge of Europeanization in the Region: East Central Europe. European Studies 2*, Budapest: Hungarian Political Science Association and Institute for Political Sciences of the Hungarian Academy of Sciences, 1996.

Kuusinen, O. (ed.), *Fundamentals of Marxism-Leninism*, London: Lawrence and Wishart, 1961.

Lampe, J.R., *Yugoslavia as History. Twice there was a country*. Cambridge: Cambridge University Press, 2000.

Lewis, R., *Science and Industrialisation in the USSR*. London: Macmillan, 1979.

Light, M., *The Soviet Theory of International Relations*, Brighton: Wheatsheaf Books, 1988.

Linden, C.A., *Khrushchev and the Soviet Leadership*, Baltimore, MD: Johns Hopkins University Press, 1990.

Machewicz, P., 'Social Protest and Political Crisis in 1956', in A. Kemp Welch (ed.), *Stalinism in Poland, 1944–1956*, London: MacMillan, 1999.

Maddrell, P., *Spying on Science. Western Intelligence in Divided Germany 1946–1961*, Oxford: Oxford University Press, 2006.

Magyar Szocialista Munkáspárt (MSZMP) határozatai és dokumentumai 1963–1966 (Decisions and Documents of the Hungarian Socialist Worker Party (HSWP) from 1963 to 1966), Budapest: Kossuth,1975.

Majumdar, J., *The Economics of Railway Traction*, Aldershot: Ashgate, 1985.

Marchenko, A., *My Testimony*, London: Pall Mall Press, 1969.

Martin, T., *The Affirmative Action Empire: Nations and Nationalism in the Soviet Union, 1923–1939*, Ithaca, NY and London: Cornell University Press, 2001.

McCauley, M. (ed.), *Khrushchev and Khrushchevism*, Basingstoke: Macmillan, 1987.

McCauley, M., *The Khrushchev Era, 1953–1964*, London: Longman, 1995.

Medvedev, R., *Khrushchev*, Oxford: Oxford University Press, 1982.

Medvedev, Z.A. and Medvedev, R.A., (eds), *The 'Secret' Speech Delivered to the closed session of the Twentieth Congress of the Communist Party of the Soviet Union by Nikita Khrushchev*, Nottingham: Spokesman Books, 1976.

Mickelsen K.-E. and Särkikoski, T., *Suomalainen ydinvoimalaitos (Finnish Nuclear Power)*, Helsinki: Edita 2005.

Miklóssy, K., *Manoeuvres of National Interest*. Helsinki: Kikimora, 2003.

Moorsteen, R. and Powell, R., *The Soviet Capital Stock, 1928–1962*, Homewood, IL: Yale University Press, 1966.

Moreton, E., 'Foreign Policy Perspectives in Eastern Europe', in K.Dawisha, and P. Hanson (eds), *Soviet – East-European Dilemmas: Coercion, Competition and Consent*, London: Heinemann – Royal Institute of International Affairs, 1981.

Motyl, A.J., *Revolutions, Nations, Empires. Conceptual Limits and Theoretical Possibilities*, New York: Columbia University Press, 1999.

Nironen, E., 'Transfer of Technology between Finland and the Soviet Union' in K. Möttölä, O.N. Bykov and I.S. Korolev (eds), *Finnish-Soviet Economic Relations*, London: Macmillan Press, 1983.

Nironen, E., 'Lännen embargopolitiikka murrosvaiheessa' ('The West's Embargo Polciy at a Crossroads'), *Ulkopolitiikka* 3/1990.

Norlander, D., 'Khrushchev's Image in the Light of Glasnost and Perestroika', *The Russian Review*, vol. 52, 1993, pp. 248–64.

Nove, A., *Stalinism and After*, London: HarperCollins, 1975.

Nyíri, K. *Európa szélén (On the Edge of Europe)*, Budapest: Kossuth, 1986.

Ofer, G., *Soviet Economic Growth: 1928–1985*, Los Angeles, CA: RAND/UCLA Center for the Study of Soviet International Behaviour, 1988.

Olesen, T.B. (ed.), *The Cold War – and the Nordic Countries*, Odense: University Press of Southern Denmark 2004.

Ouimet, J.M., *The Rise and Fall of the Brezhnev Doctrine in Soviet Foreign Policy*, Chapel Hill, NC and London: University of North Carolina Press, 2003.

Rindzeviciute, E., *Constructing Soviet Cultural Policy. Cybernetics and Governance in Lithuania after World War II*, Linköping: Linköping University Arts and Science no. 437, 2008.

Romanov, A.K., 'Suomen ja Neuvostoliiton välisen tieteellis-teknisen yhteistyön tuloksia' ('Results of Finnish-Soviet Scientific-Technical Co-operation') *Suomen ja Neuvostoliiton välinen tieteellis-tekninen yhteistoiminta 30 vuotta (30 Years of Finnish-Soviet Scientific-Technical Collaboration)*, Helsinki, 1985.

Satterwhile, J., 'Marxist Critique and Czechoslovak Reform' in R. Taras (ed.), *The Road to Disillusion. From Critical Marxism to Postcommunism in Eastern* Europe, New York: M.E.Sharpe, 1992.

Schapiro, L., 'The General Department of the CC of the CPSU', *Survey*, vol. 21, no. 3, 1975, pp. 53–65.

Schapiro, L., *The Communist Party of the Soviet Union*, 2 ed.; London: Methuen, 1970.

Scoblic, J.P., *US versus them: How a Half Century of Conservatism has Undermined America's Security*, New York: Viking, 2008.

Seppänen, J., *Tieteellis-tekninen informaatio Neuvostoliitossa (Scientific-technical information in the Soviet Union)*, Helsinki: Suomen ja Neuvostoliiton välisen tieteellis-teknisen yhteistoimintakomitean julkaisusarja 2, 1978.

Shelley, L., *Policing Soviet Society: The Evolution of State Control*, London: Routledge, 1996.

Shenfield, S., 'Pripiski: false statistical reporting in Soviet-type economies', in M. Clarke (ed.) *Corruption. Causes, Consequences and Control*, London: Pinter, 1983.

Shenfield, S., *The Functioning of the Soviet System of State Statistics*, CREES Special Report SR-86–1, Birmingham: Centre for Russian and East European Studies, University of Birmingham, 1986.

Shlapentokh, V., *Public and Private Life of the Soviet People: Changing Values in Post-Stalin Russia*, Oxford: Oxford University Press, 1989.

Shlapentokh, V., *Soviet Intellectuals and Political Power: The Post-Stalin Era*, Princeton, NJ: Princeton University Press, 1990.

Simon, E. *et al*, *Moscow in the Making*, London, New York and Toronto, 1937.

Simon, G., *Nationalism and Policy toward the Nationalities in the Soviet Union: from Totalitarian Dictatorship to Post-Stalinist Society*, transl. by Karen Forster and Oswald Forster, Boulder, CO: Westview Press, 1991.

Sjursen, H., *The United States, Western Europe and the Polish Crisis, International Relations in the Second Cold War*, New York: Palgrave Macmillan, 2003.

Skilling, H.G., *Communism National and International. Eastern Europe after Stalin.* Toronto: Toronto University Press, 1964.

Skinner, Q., 'Language and political change' in T.Ball, J.Farr and R.L.Hanson (eds), *Political Innovation and Conceptual Change*, Cambridge: Cambridge University Press, 1989, pp. 6–23.

Smith, J., 'Khrushchev and the Path to Modernisation through Education' in M. Kangaspuro and J. Smith (eds), *Modernisation in Russia since 1900*, Helsinki: SKS, 2006.

Smith, J., 'National Conflict in the USSR in the 1920s' *Ab Imperio* no.3, August 2001, pp. 221–265.

Smith, M.B., 'Khrushchev's Promise to Eliminate the Urban Housing Shortage: Rights, Rationality and the Communist Future', in M. Ilic and J. Smith (eds), *Soviet State and Society under Nikita Khrushchev*, London: Routledge, 2009.

Solzhenitsyn, A. *The Gulag Archipelago*, London: Fontana Books, 1978.

Sosin, G., *Sparks of Liberty: An Insider's Memoirs of Radio Liberty*, University Park, PA: Pennsylvania State University Press, 1999.

Sosnovy, T., 'The Soviet Housing Situation Today', *Soviet Studies*, vol. 11, no. 1, 1959.

Susiluoto, I. *Suuruuden laskuoppi. Venäläisen tietoyhteiskunnan synty ja kehitys (The History of the Russian information society),* Helsinki: WSOY, 2006.

Sutton, A., *Western Technology and Soviet Economic Development 1945 to 1965,* volume 3, Stanford, CA: Hoover Institution Press, 1973.

Szabó, M., *Politikai kultúra Magyarországon 1896–1986 (Political Culture in Hungary from 1896 to 1986,* Budapest: Atlantisz Program – Medvetánc, 1989.

Szakács, S., 'A kádár-korszak gazdaságtörténetének fôbb jellemzôi.' ('The Main Characteristics of the Economic History of the Kadar-Era'), *Society and Economy* Vol. XVI 1994/5, pp.188–194.

Taubman, W., Khrushchev, S., and Gleason, A., (eds), *Nikita Khrushchev,* New Haven, CA: Yale University Press, 2000.

Taubman, W., *Khrushchev: The Man and His Era,* London: Free Press, 2003.

The Road to Communism. Documents of the 22nd Congress of the Communist Party of the Soviet Union, Moscow: Foreign Languages Publishing House, 1961.

Tilly, C., *European revolutions 1492–1992,* Oxford: Blackwell, 1993.

Titov, A., 'The 1961 Party Programme and the fate of Khrushchev's reforms' in M. Ilic and J. Smith (eds), *Soviet State and Society Under Nikita Khrushchev,* London: Routledge, 2009.

Tompson, W., *Khrushchev. A Political Life* Basingstoke: Palgrave, 1995.

Volkogonov,D., *The Rise and Fall of the Soviet Empire,* London: HarperCollins, 1998.

Waltz, K., *Theory of International Politics,* Reading, MA.: Addison-Wesley, 1979.

Whitney (ed.), T.P., *Khrushchev Speaks: Selected Speeches, Articles, and Press Conferences, 1949–1961,* Ann Arbor, MI: University of Michigan Press, 1963.

Wilczynski, J., *Technology in COMECON. Acceleration of Technological Progress through Economic Planning and the Market,* London: Macmillan, 1974.

Williams, K., *The Prague Spring and its Aftermath. Czechoslovak Politics 1968–1970,* Cambridge: Cambridge University Press, 1997.

Zubok, V. and Pleshakov, C., *Inside the Kremlin's Cold War. From Stalin to Khrushchev,* Cambridge, MA.: Harvard University Press, 1996

Index